Partial Differential Equations: Advances in Spectral and High Order Methods

Partial Differential Equations: Advances in Spectral and High Order Methods

Editor: Patrick McCann

STATES
ACADEMIC PRESS
www.statesacademicpress.com

Published by States Academic Press
109 South 5th Street,
Brooklyn, NY 11249, USA
www.statesacademicpress.com

Partial Differential Equations: Advances in Spectral and High Order Methods
Edited by Patrick McCann

International Standard Book Number: 978-1-63989-404-8 (Hardback)

Cataloging-in-Publication Data

Partial differential equations : advances in spectral and high order methods / edited by Patrick McCann.
p. cm.
Includes bibliographical references and index.
ISBN 978-1-63989-404-8
1. Differential equations, Partial. 2. Differential equations, Partial--Numerical solutions. 3. Spectral geometry. 4. Differential equations--Methodology.
I. McCann, Patrick.

QA377 .P37 2021
515.353--dc23

Contents

Preface

This book was inspired by the evolution of our times; to answer the curiosity of inquisitive minds. Many developments have occurred across the globe in the recent past which has transformed the progress in the field.

Partial differential equations (PDEs) are mathematical equations that relate a function of several variables to its partial derivatives. A partial derivative of a function of many variables describes how fast the function changes when one variable is altered, keeping the others constant. The task of discovering a partial derivative can be applied to a function that is itself a partial derivative of another function to obtain a second-order partial derivative. The order and degree of partial differential equations are determined in the same way as ordinary differential equations. Different approaches, evaluations and methodologies on partial differential equations have been included in this book. It strives to provide a fair idea about this discipline and to help develop a better understanding of the latest advances within this field. Coherent flow of topics, student-friendly language and extensive use of examples make this book an invaluable source of knowledge.

This book was developed from a mere concept to drafts to chapters and finally compiled together as a complete text to benefit the readers across all nations. To ensure the quality of the content we instilled two significant steps in our procedure. The first was to appoint an editorial team that would verify the data and statistics provided in the book and also select the most appropriate and valuable contributions from the plentiful contributions we received from authors worldwide. The next step was to appoint an expert of the topic as the Editor-in-Chief, who would head the project and finally make the necessary amendments and modifications to make the text reader-friendly. I was then commissioned to examine all the material to present the topics in the most comprehensible and productive format.

I would like to take this opportunity to thank all the contributing authors who were supportive enough to contribute their time and knowledge to this project. I also wish to convey my regards to my family who have been extremely supportive during the entire project.

Editor

Part I
Invited Papers

1

An Adaptive Error Inhibiting Block One-Step Method for Ordinary Differential Equations

Jiaxi Gu and Jae-Hun Jung

Introduction

General linear methods have been extensively studied for solving ODEs. Among the large family of general linear methods the diagonally implicit multistage integration methods (DIMSIMs) in [1] are the special cases, which exhibit considerable potential for efficient implementation, providing the global error of the same order as the local truncation error. In [2], it was demonstrated that finite difference methods for PDEs can be constructed such that their convergence rates, or the order of their global errors, are higher than the order of the truncation errors. Following this idea, Ditkowski and Gottlieb devised the error inhibiting strategy in [3] by inhibiting the lowest order term in the truncation error from accumulating over time and thus showed that the global error of the scheme is one order higher than the local truncation error. The form of the error inhibiting scheme is inspired by the work of [7], where a block of s new step values is obtained at each step. The key idea of this method is to construct a coefficient matrix that has the null space where the local truncation error resides.

In this work, we further improved the original error inhibiting method by introducing an additional free parameter used in the radial basis function

J. Gu
Department of Mathematics, University at Buffalo, The State University of New York, Buffalo, NY, USA
e-mail: jiaxigu@buffalo.edu

J.-H. Jung (✉)
Department of Mathematics, University at Buffalo, The State University of New York, Buffalo, NY, USA

Department of Data Science, Ajou University, Suwon, South Korea
e-mail: jaehun@buffalo.edu

(RBF) approximations. The main idea of the proposed method is to adopt the free parameter in the reconstruction of the error inhibiting method and to control it for further possible error cancellations. This results in a higher order of convergence than the original method. One advantage is that the proposed method does not need any additional conditions, so it is efficient to implement.

The next section will review the explicit error inhibiting block one-step method. In other section, we will explain the RBF interpolation. We show how the new method can be derived followed by Section where numerical results are provided verifying that the convergence rate of the proposed method is increased by one order. A brief conclusion and an outline of our future research are presented in Sect.

Error Inhibiting Block One-Step Method

Consider the initial value problem for the first-order ODE below

$$\begin{aligned} u'(t) &= f(t, u(t)), \ t \geqslant a \\ u(a) &= u_a \end{aligned} \tag{1}$$

where we assume $f(t, u)$ is uniformly Lipschitz continuous in u and continuous in t. We choose a value h for the step size and set $t_n = a + nh$ a discrete sequence in the time domain. Denote the numerical approximation of the solution $u(t_n)$ by v_n.

Define the solution vector U_n by

$$U_n = \left[u_{n+\frac{s-1}{s}}, \cdots, u_{n+\frac{1}{s}}, u_n\right]^T,$$

where $u_{n+\frac{j}{s}} = u(t_{n+jh/s})$ is the exact solution at $t = t_n + \frac{jh}{s}$ for $j = 0, \cdots, s-1$. The corresponding approximation vector V_n is defined as

$$V_n = \left[v_{n+\frac{s-1}{s}}, \cdots, v_{n+\frac{1}{s}}, v_n\right]^T.$$

In [3], the scheme is formulated as

$$V_{n+1} = QV_n \tag{2}$$

where the operator Q is represented by the following

$$Q = A + hBf$$

and $A, B \in \mathbb{R}^{s\times s}$. There are 4 sufficient conditions imposed on the matrices A and B in order to be error inhibiting:

1. $\mathbf{rank}(A) = 1$.
2. The only non-zero eigenvalue of A is 1 and its corresponding eigenvector is
$$[1, \cdots, 1]^T.$$
3. A can be diagonalized.
4. The matrices A and B are constructed such that when the local truncation error is multiplied by the discrete solution operator, we have
$$||Q\tau_\nu|| \leqslant O(h) \cdot ||\tau_\nu||.$$
This is accomplished by requiring that the leading order term of the local truncation error is in the eigenspace of A associated with the zero eigenvalue.

We derive those matrices of A and B with symbolic computation. As an example of the derivation of the error inhibiting method, we consider the construction of the scheme with $s = 2$. The solution vector is then

$$U_n = [u_{n+1/2}, u_n]^T,$$

and the corresponding approximation vector is given by

$$V_n = [v_{n+1/2}, v_n]^T.$$

In order to satisfy those conditions listed above we first select

$$A = \begin{bmatrix} 1-\upsilon & \upsilon \\ 1-\upsilon & \upsilon \end{bmatrix}, \tag{3}$$

which can be diagonalized as

$$A = \begin{bmatrix} 1-\upsilon & \upsilon \\ 1-\upsilon & \upsilon \end{bmatrix} = \begin{bmatrix} \upsilon-1 & \upsilon \\ \upsilon-1 & \upsilon-1 \end{bmatrix} \begin{bmatrix} 1 & 0 \\ 0 & 0 \end{bmatrix} \begin{bmatrix} -1 & \frac{\upsilon}{\upsilon-1} \\ 1 & -1 \end{bmatrix}. \tag{4}$$

Then conditions 1, 2 and 3 are satisfied. Further suppose that

$$B = \begin{bmatrix} b_{11} & b_{12} \\ b_{21} & b_{22} \end{bmatrix}. \tag{5}$$

Then

$$V_{n+1} = \begin{bmatrix} 1-\upsilon & \upsilon \\ 1-\upsilon & \upsilon \end{bmatrix} V_n + h \begin{bmatrix} b_{11} & b_{12} \\ b_{21} & b_{22} \end{bmatrix} \begin{bmatrix} f_{n+1/2} \\ f_n \end{bmatrix} \tag{6}$$

where $f_{n+1/2} = f(t_{n+1/2}, v_{n+1/2})$ and $f_n = f(t_n, v_n)$. The components of V_{n+1} are

$$v_{n+3/2} = (1-\upsilon)v_{n+1/2} + \upsilon v_n + h(b_{11} f_{n+1/2} + b_{12} f_n),$$
$$v_{n+1} = (1-\upsilon)v_{n+1/2} + \upsilon v_n + h(b_{21} f_{n+1/2} + b_{22} f_n).$$

We write each difference equation in the form of error normalized by the step size and then insert the exact solutions to the ODE into the difference equation. Expanding $u_{n+3/2}$, u_{n+1} and $u_{n+1/2}$ around $t = t_n$ in Taylor series gives the local truncation error

$$\boldsymbol{\tau}_n = (\tau_{n+1/2}, \tau_n)^T,$$

where

$$\tau_{n+1/2} = \frac{1}{2}(2 - 2b_{11} - 2b_{12} + \upsilon)u_n' + \frac{1}{8}(8 - 4b_{11} + \upsilon)u_n'' h + \frac{1}{48}(26 - 6b_{11} + \upsilon)u_n^{(3)} h^2 + O(h^3), \tag{7}$$

$$\tau_n = \frac{1}{2}(1 - 2b_{21} - 2b_{22} + \upsilon)u_n' + \frac{1}{8}(3 - 4b_{21} + \upsilon)u_n'' h + \frac{1}{48}(7 - 6b_{21} + \upsilon)u_n^{(3)} h^2 + O(h^3). \tag{8}$$

Vanishing the coefficients of the constant term and the term h in (7) and (8), and equating the quotient of the coefficient of the terms h^2 in (7) and (8) to $\frac{\upsilon}{\upsilon-1}$, the condition 4 is satisfied.

Finally we have the desired scheme as in [3]

$$V_{n+1} = \frac{1}{6}\begin{bmatrix} -1 & 7 \\ -1 & 7 \end{bmatrix} V_n + \frac{h}{24}\begin{bmatrix} 55 & -17 \\ 25 & 1 \end{bmatrix} \begin{bmatrix} f_{n+1/2} \\ f_n \end{bmatrix}, \tag{9}$$

and correspondingly the local truncation error is 2nd order convergent as expected

$$\boldsymbol{\tau}_n = \frac{23}{576}\begin{bmatrix} 7 \\ 1 \end{bmatrix} u_n^{(3)} h^2 + O(h^3). \tag{10}$$

RBF Interpolation

Now we briefly explain the RBF interpolation in one dimension. Suppose that for a domain $\Omega \subset \mathbb{R}$, a data set $\{(x_i, u_i)\}_{i=0}^{N}$ is given where u_i is the value of the unknown function $u(x)$ at $x = x_i \in \Omega$. We use the RBFs $\phi : \Omega \to \mathbb{R}$ defined by $\phi_i(x) = \phi(|x - x_i|, \epsilon_i)$, where $|x - x_i|$ is the distance between x and x_i and ϵ_i is a free parameter. The reconstruction of a function, $u(x)$, is then made by a linear combination of RBFs

$$I_N^R\, u(x) = \sum_{i=0}^{N} \lambda_i \phi(|x - x_i|, \epsilon_i), \tag{11}$$

where λ_i are the expansion coefficients to be determined. Using the interpolation condition $I_N^R u(x_i) = u_i,\ i = 0, \cdots, N$, we could find the expansion coefficients λ_i by solving the linear system

$$\begin{bmatrix} \phi(|x_0 - x_0|, \epsilon_0) & \phi(|x_0 - x_1|, \epsilon_1) & \cdots & \phi(|x_0 - x_N|, \epsilon_N) \\ \phi(|x_1 - x_0|, \epsilon_0) & \phi(|x_1 - x_1|, \epsilon_1) & \cdots & \phi(|x_1 - x_N|, \epsilon_N) \\ \vdots & \vdots & & \vdots \\ \phi(|x_N - x_0|, \epsilon_0) & \phi(|x_N - x_1|, \epsilon_1) & \cdots & \phi(|x_N - x_N|, \epsilon_N) \end{bmatrix} \cdot \begin{bmatrix} \lambda_0 \\ \lambda_1 \\ \vdots \\ \lambda_N \end{bmatrix} = \begin{bmatrix} u_0 \\ u_1 \\ \vdots \\ u_N \end{bmatrix}. \tag{12}$$

If we choose the multiquadric RBF with all the free parameters equal, then the interpolation matrix, A, becomes a symmetric matrix with all diagonal entries 1,

$$A = \begin{bmatrix} 1 & \sqrt{1 + \epsilon^2 (x_0 - x_1)^2} & \cdots & \sqrt{1 + \epsilon^2 (x_0 - x_N)^2} \\ \sqrt{1 + \epsilon^2 (x_1 - x_0)^2} & 1 & \cdots & \sqrt{1 + \epsilon^2 (x_1 - x_N)^2} \\ \vdots & \vdots & & \vdots \\ \sqrt{1 + \epsilon^2 (x_N - x_0)^2} & \sqrt{1 + \epsilon^2 (x_N - x_1)^2} & \cdots & 1 \end{bmatrix}. \tag{13}$$

Consider the case of three equally spaced nodes x_0, x_1, x_2 with $x_0 < x_1 < x_2$. Let h be the grid spacing. Then the linear system becomes

$$\begin{bmatrix} 1 & \sqrt{1 + \epsilon^2 h^2} & \sqrt{1 + 4\epsilon^2 h^2} \\ \sqrt{1 + \epsilon^2 h^2} & 1 & \sqrt{1 + \epsilon^2 h^2} \\ \sqrt{1 + 4\epsilon^2 h^2} & \sqrt{1 + \epsilon^2 h^2} & 1 \end{bmatrix} \cdot \begin{bmatrix} \lambda_0 \\ \lambda_1 \\ \lambda_2 \end{bmatrix} = \begin{bmatrix} u_0 \\ u_1 \\ u_2 \end{bmatrix}. \tag{14}$$

By the closed-form expression for the RBF interpolant in [4],

$$I_2^R u(x) = \sum_{i=0}^{2} \frac{u_i}{\det(A)} \det(A_i(x)). \tag{15}$$

where $A_i(x)$, a 3×3 matrix, is obtained by replacing the ith row of A with the row vector

$$\left[\sqrt{1+\epsilon^2(x-x_0)^2} \quad \sqrt{1+\epsilon^2(x-x_1)^2} \quad \sqrt{1+\epsilon^2(x-x_2)^2}\right].$$

Differentiating the interpolant, we obtain the first-order derivative

$$\frac{d}{dx} I_2^R u(x) = \sum_{i=0}^{2} \frac{u_i}{\det(A)} \cdot \frac{d}{dx} \det(A_i(x)). \tag{16}$$

We then estimate the derivative of u at $x = x_1$ as we do in polynomial interpolation for the central difference formula:

$$\frac{d}{dx} I_2^R u(x_1) = \frac{\sqrt{1+4\epsilon^2h^2}+1}{4h\sqrt{1+\epsilon^2h^2}}(u_2-u_0). \tag{17}$$

By employing the Taylor expansion of the quotient on the right-hand side of (17), we have

$$\frac{d}{dx} I_2^R u(x_1) = \left[\frac{1}{2h} + \epsilon^2\frac{h}{4} + O(h^3)\right](u_2-u_0). \tag{18}$$

The main feature of the RBF method is that it contains a free parameter, ϵ, which we could make use of to further inhibit the errors. In the following section, we will show that using the parameter ϵ coupled with h^p terms, where $p \geqslant 2$, we can increase the order of local truncation error and further promote the order of global error by adopting the error inhibiting scheme.

Construction of the Adaptive Error Inhibiting Scheme

Following the main feature of the RBF method explained in the preceding section, we try to establish a similar explicit block one-step scheme that provides a higher order of convergence by adding one more block of the free parameters ϵ_1 and ϵ_2

coupled with h^p term. With $p = 3$, we have

$$V_{n+1} = \begin{bmatrix} 1-\upsilon & \upsilon \\ 1-\upsilon & \upsilon \end{bmatrix} V_n + h \begin{bmatrix} b_{11} & b_{12} \\ b_{21} & b_{22} \end{bmatrix} \begin{bmatrix} f_{n+1/2} \\ f_n \end{bmatrix} + h^3 \begin{bmatrix} 0 & \epsilon_1 \\ 0 & \epsilon_2 \end{bmatrix} \begin{bmatrix} f_{n+1/2} \\ f_n \end{bmatrix}. \tag{19}$$

We measure the one-step error normalized by the step size as in Sect. Expanding $u_{n+3/2}$, u_{n+1} and $u_{n+1/2}$ around $t = t_n$ in Taylor series again yields the local truncation error

$$\boldsymbol{\tau}_n = [\tau_{n+1/2}, \tau_n]^T,$$

where

$$\tau_{n+1/2} = \frac{1}{2}(2 - 2b_{11} - 2b_{12} + \upsilon)u_n' + \frac{1}{8}(8 - 4b_{11} + \upsilon)u_n''h + \left(-\epsilon_1 u_n' + \frac{1}{48}(26 - 6b_{11} + \upsilon)u_n^{(3)}\right)h^2 + \frac{1}{384}(80 - 8b_{11} + \upsilon)u_n^{(4)}h^3 + O(h^4), \tag{20}$$

$$\tau_n = \frac{1}{2}(1 - 2b_{21} - 2b_{22} + \upsilon)u_n' + \frac{1}{8}(3 - 4b_{21} + \upsilon)u_n''h + \left(-\epsilon_2 u_n' + \frac{1}{48}(7 - 6b_{21} + \upsilon)u_n^{(3)}\right)h^2 + \frac{1}{384}(15 - 8b_{21} + \upsilon)u_n^{(4)}h^3 + O(h^4). \tag{21}$$

Annihilating the first two terms in (20) and (21), and equating the quotient of the coefficient of the terms h^3 in (20) and (21) to $\frac{\upsilon}{\upsilon-1}$, we have the scheme

$$V_{n+1} = \frac{1}{7}\begin{bmatrix} -1 & 8 \\ -1 & 8 \end{bmatrix} V_n + \frac{h}{28}\begin{bmatrix} 64 & -20 \\ 29 & 1 \end{bmatrix} \begin{bmatrix} f_{n+1/2} \\ f_n \end{bmatrix} + h^3 \begin{bmatrix} 0 & \epsilon_1 \\ 0 & \epsilon_2 \end{bmatrix} \begin{bmatrix} f_{n+1/2} \\ f_n \end{bmatrix}. \tag{22}$$

We can easily check that the scheme (22) satisfies those four conditions in Sect. Further annihilating the coefficients of the term h^2, we get the optimal values of ϵ_1 and ϵ_2:

$$\epsilon_1 = \frac{47u_n^{(3)}}{168u_n'}, \tag{23}$$

$$\epsilon_2 = \frac{9u_n^{(3)}}{224u_n'}. \tag{24}$$

Our new scheme has the truncation error

$$\tau_n = \frac{55}{2688}\begin{bmatrix} 8 \\ 1 \end{bmatrix} u_n^{(4)} h^3 + O(h^4). \tag{25}$$

Note that in our new scheme, we need the value of $u_n^{(3)}$ at each step. This higher order derivative can be computed by repeated differentiation of the function f on the right-hand side of (1) twice. However, we choose to estimate the third-order derivative. For u_n', we use the given condition from (1), i.e. $u'(t) = f(t, u(t))$. For the third-order derivative $u_n^{(3)}$, we employ the second-order central difference formula for $f''(t, u(t))$ at $t = t_n$ as

$$u_n^{(3)} = f''(t_n, u_n) \approx \frac{4(f_{n+1/2} + 2f_n - f_{n-1/2})}{h^2}, \tag{26}$$

where $f_{n+1/2}$, f_n and $f_{n-1/2}$ are given values. For this computation, no additional conditions are necessary. The truncation error is still third order accurate, $O(h^3)$, as in (25), so by the error inhibiting strategy we end up with a global error that is $O(h^4)$, which will soon be confirmed in the following section.

We conclude this section with a comparison of three methods. For DIMSIM of type 3,

$$\begin{bmatrix} v_{n+2} \\ v_{n+1} \end{bmatrix} = \frac{1}{4}\begin{bmatrix} 7 & -3 \\ 7 & -3 \end{bmatrix}\begin{bmatrix} v_{n+1} \\ v_n \end{bmatrix} + \frac{h}{8}\begin{bmatrix} 9 & -7 \\ -3 & -3 \end{bmatrix}\begin{bmatrix} f_{n+1} \\ f_n \end{bmatrix},$$

two steps v_n and v_{n+1} are employed to update the step v_{n+1} and obtain the step v_{n+2}.

For error inhibiting scheme,

$$\begin{bmatrix} v_{n+3/2} \\ v_{n+1} \end{bmatrix} = \frac{1}{6}\begin{bmatrix} -1 & 7 \\ -1 & 7 \end{bmatrix}\begin{bmatrix} v_{n+1/2} \\ v_n \end{bmatrix} + \frac{h}{24}\begin{bmatrix} 55 & -17 \\ 25 & 1 \end{bmatrix}\begin{bmatrix} f_{n+1/2} \\ f_n \end{bmatrix},$$

two steps v_n and $v_{n+1/2}$ are involved to generate the next two steps v_{n+1} and $v_{n+3/2}$. For our method (if we utilize (26) and substitute (23), (24) for respective ϵ_1 and ϵ_2 in (22) to avoid the zero denominator),

$$\begin{bmatrix} v_{n+3/2} \\ v_{n+1} \\ v_{n+1/2} \end{bmatrix} = \frac{1}{7}\begin{bmatrix} -1 & 8 & 0 \\ -1 & 8 & 0 \\ 1 & 0 & 0 \end{bmatrix}\begin{bmatrix} v_{n+1/2} \\ v_n \\ v_{n-1/2} \end{bmatrix} + \frac{h}{168}\begin{bmatrix} 572 & -496 & 188 \\ 201 & -48 & 27 \\ 0 & 0 & 0 \end{bmatrix}\begin{bmatrix} f_{n+1/2} \\ f_n \\ f_{n-1/2} \end{bmatrix},$$

we use previous three steps $v_{n-1/2}$, v_n and $v_{n+1/2}$ to evolve the next two steps v_{n+1} and $v_{n+3/2}$. In [5] the stability analysis has been done for the adaptive radial basis function methods for IVPs and it has been shown that some adaptive methods have a better stability condition than the original ones. However, it seems that the adaptive error inhibiting method is more computationally expensive than the original one when the approximation of (26) is used.

Numerical Results

We start with the nonlinear first-order differential equation used in [3]

$$\begin{aligned} u' &= -u^2,\ t \geqslant 0 \\ u(0) &= 1. \end{aligned} \tag{27}$$

The exact solution of the example is $u(t) = 1/(t+1)$. The left figure of Fig. shows the global errors at the time $t = 1$ versus N, the number of steps, in logarithmic scale for the type-3 DIMSIM (blue), the original error inhibiting scheme (red) and our proposed method (green). As seen in the figure, our proposed method is the most accurate among those three methods and yields high order convergence which is 4th order. Table shows the convergence with N for (27). The type-3 DIMSIM yields the 2nd order accuracy, the original error inhibiting scheme yields the 3rd order accuracy and our proposed method yields the 4th order accuracy.

Next we consider the following problem used in [6] where the solution changes rapidly between $[-2, 2]$

$$\begin{aligned} u' &= -4t^3u^2,\ t \geqslant -10 \\ u(-10) &= 1/10001. \end{aligned} \tag{28}$$

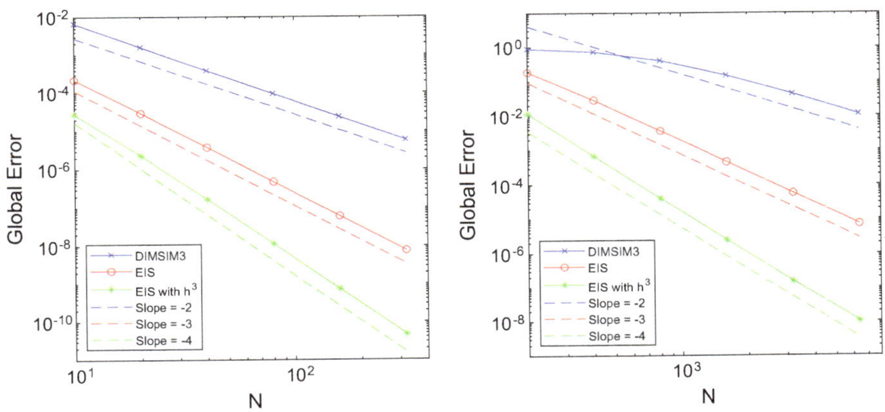

Fig. Global error versus N in logarithmic scale. Left: (27). Right: (28). Blue: DIMSIM (DIMSIM3) 2nd order. Red: error inhibiting scheme (EIS) 3rd order. Green: our proposed method (EIS with h^3) 4th order

Table Global error and order of convergence for $u' = -u^2$ with $u(0) = 1$

Method	N	Global error	order
DIMSIM type-3	10	6.60E−3	
	20	1.60E−3	2.0702
	40	3.82E−4	2.0402
	80	9.41E−5	2.0208
	160	2.34E−5	2.0105
	320	5.82E−6	2.0053
Error inhibiting scheme	10	2.17E−4	
	20	2.89E−5	2.9118
	40	3.73E−6	2.9536
	80	4.74E−7	2.9763
	160	5.97E−8	2.9880
	320	7.50E−9	2.9940
Error inhibiting scheme with h^3 term	10	2.71E−5	
	20	2.24E−6	3.5935
	40	1.64E−7	3.7781
	80	1.11E−8	3.8833
	160	7.22E−10	3.9400
	320	4.61E−11	3.9698

The exact solution is $u(t) = 1/(t^4 + 1)$. The right figure of Fig. shows the global errors at $t = 0$ versus N in logarithmic scale for the type-3 DIMSIM (blue), the original error inhibiting method (red) and our proposed method (green). We verify again that our proposed method is indeed the most accurate and yields the highest order of convergence. Table shows the convergence with N for (28). Although the type-3 DIMSIM does not reveal the 2nd order accuracy in the beginning, it eventually exhibits the order of accuracy as expected. The original error inhibiting scheme is 3rd order accurate and our proposed method 4th order accurate.

Conclusions

In this note, we modified and improved the original error inhibiting block one-step method proposed in [3] by introducing a free parameter. By exploiting the parameter, the local truncation error is further reduced resulting in higher order of the global error. It is numerically demonstrated that, with the proposed method, the local truncation error is of the 3rd order and the global error of the 4th order. As mentioned in Section, we will investigate the stability of the error inhibiting method and our proposed method as well as relaxing the fourth constraint in error inhibiting method in our future research.

Table Global error and order of convergence for $u' = -4t^3u^2$ with $u(-10) = 1/10001$

Method	N	Global error	Order
DIMSIM type-3	200	9.05E−1	
	400	7.24E−1	0.3221
	800	4.07E−1	0.8293
	1600	1.49E−1	1.4476
	3200	4.24E−2	1.8158
	6400	1.10E−2	1.9475
Error inhibiting scheme	200	1.86E−1	
	400	2.80E−2	2.7294
	800	3.60E−3	2.9639
	1600	4.50E−4	2.9965
	3200	5.63E−5	3.0002
	6400	7.03E−6	3.0005
Error inhibiting scheme with h^3 term	200	1.14E−2	
	400	6.57E−4	4.1132
	800	3.94E−5	4.0620
	1600	2.41E−6	4.0307
	3200	1.49E−7	4.0123
	6400	9.91E−9	3.9122

Acknowledgements The authors thank Adi Ditkowski for introducing the error inhibiting method to us and communicating with us on the subject. The research is partially supported by Ajou University.

References

1. Butcher, J.C.: Diagonally-implicit multi-stage integration method. Appl. Numer. Math. **11**, 347–363 (1993)
2. Ditkowski, A.: High order finite difference schemes for the heat equation whose convergence rates are higher than their truncation errors. In: R.M. Kirby et al. (eds.), Spectral and High Order Methods for Partial Differential Equations ICOSAHOM 2014, pp. 167–178. Springer, Switzerland (2015)
3. Ditkowski, A., Gottlieb, S.: Error inhibiting block one-step schemes for ordinary differential equations. J. Sci. Comput. **73**, 691–711 (2017)
4. Fornberg, B., Weight, G., Larsson, E.: Some observations regarding interpolants in the limit of flat radial basis functions. Comput. Math. Appl. **47**, 37–55 (2004)
5. Gu, J. Jung, J.-H.: Adaptive radial basis function methods for initial value problems. J. Sci. Comput. **82**, 47 (2020). https://doi.org/10.1007/s10915-020-01140-0
6. Sauer, T.: Numerical Analysis, 2nd edn. Pearson, New York (2012)
7. Shampine, L.F., Watts, H.A.: Block implicit one-step methods. Math. Comp. **23**, 731–740 (1969)

2

A Review of Regular Decompositions of Vector Fields: Continuous, Discrete, and Structure-Preserving

Ralf Hiptmair and Clemens Pechstein

Introduction

For a bounded Lipschitz domain $\Omega \subset \mathbb{R}^3$ recall the classical L^2-orthogonal *Helmholtz decompositions*

$$\mathbf{L}^2(\Omega) = \nabla H_0^1(\Omega) \oplus \mathbf{H}(\operatorname{div} 0, \Omega) = \nabla H^1(\Omega) \oplus \mathbf{H}_0(\operatorname{div} 0, \Omega)\,,$$

see, e.g.,. They can be used to derive decompositions of (subspaces of) $\mathbf{H}(\mathbf{curl}, \Omega)$:

$$\mathbf{H}_0(\mathbf{curl}, \Omega) = \nabla H_0^1(\Omega) \oplus \mathbf{X}_N(\Omega), \quad \mathbf{X}_N(\Omega) := \mathbf{H}_0(\mathbf{curl}, \Omega) \cap \mathbf{H}(\operatorname{div} 0, \Omega),$$
$$\mathbf{H}(\mathbf{curl}, \Omega) = \nabla H^1(\Omega) \oplus \mathbf{X}_T(\Omega), \quad \mathbf{X}_T(\Omega) := \mathbf{H}(\mathbf{curl}, \Omega) \cap \mathbf{H}_0(\operatorname{div} 0, \Omega)\,.$$

If the domain Ω is convex then the respective complementary space, $\mathbf{X}_N(\Omega)$ or $\mathbf{X}_T(\Omega)$, is continuously embedded in the space $\mathbf{H}^1(\Omega)$ of vector fields with Cartesian components in $H^1(\Omega)$, cf. [1]. Then one can, for instance, write any $\mathbf{u} \in \mathbf{H}(\mathbf{curl}, \Omega)$ as

$$\mathbf{u} = \nabla p + \mathbf{z}, \tag{1}$$

R. Hiptmair (✉)
Seminar for Applied Mathematics, ETH Zürich, Zürich, Switzerland
e-mail: hiptmair@sam.math.ethz.ch

C. Pechstein
Dassault Systèmes, Darmstadt, Germany
e-mail: clemens.pechstein@3ds.com

with $p \in H^1(\Omega)$ and $\mathbf{z} \in \mathbf{H}^1(\Omega)$. Since $\|\nabla p\|_{\mathbf{L}^2(\Omega)} \leq \|\mathbf{u}\|_{\mathbf{L}^2(\Omega)}$ one obtains (using the continuous embedding) the stability property[1]

$$\|\nabla p\|_{\mathbf{L}^2(\Omega)} + \|\mathbf{z}\|_{\mathbf{H}^1(\Omega)} \leq C\|\mathbf{z}\|_{\mathbf{H}(\mathbf{curl},\Omega)} . \tag{2}$$

A similar decomposition can be found for $\mathbf{u} \in \mathbf{H}_0(\mathbf{curl}, \Omega)$.

Generally, a decomposition of form (1) with the stability property (2) is called *regular decomposition*, even if $\mathbf{L}^2$-orthogonality does not hold. Actually, it turns out that (1)–(2) can be achieved even in cases where Ω is non-convex, in particular on non-smooth domains, or in cases where Ω or its boundary have non-trivial topology; only the $\mathbf{L}^2$-orthogonality has to be sacrificed, cf. [20].

Noting that $\nabla H^1(\Omega)$ is contained in the kernel of the **curl** operator and that—under mild smoothness assumptions on the domain—the whole kernel is spanned by $\nabla H^1(\Omega)$ plus a finite-dimensional *co-homology space* one can achieve a second decomposition,

$$\mathbf{u} = \mathbf{h} + \mathbf{z} , \tag{3}$$

with $\mathbf{h} \in \ker(\mathbf{curl}_{|\mathbf{H}(\mathbf{curl},\Omega)})$ and $\mathbf{z} \in \mathbf{H}^1(\Omega)$, where

$$\|\mathbf{h}\|_{\mathbf{L}^2(\Omega)} \leq C \|\mathbf{u}\|_{\mathbf{L}^2(\Omega)}, \qquad \|\mathbf{z}\|_{\mathbf{H}^1(\Omega)} \leq C \|\mathbf{curl}\ \mathbf{u}\|_{\mathbf{L}^2(\Omega)} . \tag{4}$$

The second stability estimate states that if $\mathbf{u}$ is already in the kernel of the **curl** operator, then $\mathbf{z}$ is zero. Hence, (1) the operator mapping $\mathbf{u}$ to $\mathbf{h}$ is a *projection* onto the kernel space and (2) the complement operator projects $\mathbf{u}$ to the function $\mathbf{z}$ of higher regularity $\mathbf{H}^1(\Omega)$. For trivial topology of Ω and $\partial\Omega$, the two decompositions (1)–(2) and (3)–(4) coincide.

As a few among many more, we would like to highlight two important applications of these regular decompositions.

1. The second form (3)–(4), in the sequel called *rotation-bounded decomposition*, can be used to show that the operator underlying a certain boundary value problem for Maxwell's equations is a *Fredholm operator*. The key point is that the complement space of the kernel (from the view of the mentioned projections) is $\mathbf{H}^1(\Omega)$ which is *compactly* embedded in $\mathbf{L}^2(\Omega)$, see e.g., [14, 16] and references therein.
2. The first form (1)–(2), in the sequel called *gradient-based decomposition*, has been used to generate stable three-term splittings of a finite element subspace of $\mathbf{H}(\mathbf{curl}, \Omega)$, cf. [19–21, 23], which allows the construction of so-called *fictitious* or *auxiliary* space preconditioners for the ill-conditioned system matrix underlying the discretized Maxwell equations.

[1]Here and below C stands for a positive "generic constant" that may depend only on Ω, unless specified otherwise.

In both applications, it is desirable to obtain the decompositions for minimal smoothness of the domain, e.g., Lipschitz domains, which are not necessarily convex. Moreover, it is also desirable to go beyond decompositions of the entire space $\mathbf{H}(\mathbf{curl}, \Omega)$ and extend them to subspaces for which the appropriate trace vanishes on a "Dirichlet part" Γ_D of the boundary. In this case traces of the two summands should also vanish on Γ_D.

In the present paper, we provide regular decompositions of both types for subspaces of $\mathbf{H}(\mathbf{curl}, \Omega)$ (in Section) and $\mathbf{H}(\mathrm{div}, \Omega)$ (in Sect.) comprising functions with vanishing trace on a part Γ_D of the boundary $\partial\Omega$ for Lipschitz domains Ω of *arbitrary topology*. In particular, Ω is allowed to have handles, and $\partial\Omega$ and Γ_D may have several connected components. The Dirichlet boundary Γ_D must satisfy a certain smoothness assumption that we shall introduce in Section. In addition to the stability estimates (2) a n d (4), we show that the decompositions are stable even in $\mathbf{L}^2(\Omega)$.

In the final part of the manuscript, in Section, we establish regular decompositions of spaces of Whitney forms, which are lowest-order conforming finite element subspaces of $\mathbf{H}(\mathbf{curl}, \Omega)$ and $\mathbf{H}(\mathrm{div}, \Omega)$, respectively, built upon simplicial trian-gulations of Ω.

This note is based on [17] and is an abridged version of [18]. Please refer to this latter preprint for complete proofs of the results quoted below.

Preliminaries

Since subtle geometric arguments will play a major role for parts of the theory, we start with a precise characterization of the geometric setting: Let $\Omega \subset \mathbb{R}^3$ be an open, bounded, connected Lipschitz domain.[2] We write $\mathrm{d}(\Omega)$ for its diameter. Its boundary $\Gamma := \partial\Omega$ is partitioned according to $\Gamma = \Gamma_D \cup \Sigma \cup \Gamma_N$, with relatively open sets Γ_D and Γ_N. We assume that this provides a *piecewise C^1 dissection* of $\partial\Omega$ in the sense of [12, Definition 2.2]. Sloppily speaking, this means that Σ is the union of closed curves that are piecewise C^1.

Under the above assumptions on Ω and Γ_D, [12, Lemma 4.4] guarantees the existence of an open Lipschitz neighborhood Ω_Γ ("Lipschitz collar") of Γ and of a "bulge" $\Upsilon_D \subset \Omega_\Gamma \setminus \Omega$. We recall the properties of bulge domains from [12, Sect, Thm. 2.3], also stated in [17, Thm. 2.2]:

Theorem 1 (Bulge-Augmented Domain) *There exists a* Lipschitz domain $\Upsilon_D \subset \mathbb{R}^3 \setminus \overline{\Omega}$, *such that* $\overline{\Upsilon}_D \cap \overline{\Omega} = \Gamma_D$, $\Omega^e := \Upsilon_D \cup \Gamma_D \cup \Omega$ *is Lipschitz,* $\mathrm{d}(\Omega^e) \leq 2\,\mathrm{d}(\Omega)$, *and* $\overline{\Upsilon}_D \subset \Omega_\Gamma$. *Moreover, each connected component* $\Gamma_{D,k}$ *of* Γ_D *corresponds to a connected component* $\Upsilon_{D,k}$ *of* Υ_D, *and these have positive distance from each other.*

[2] Strongly Lipschitz, in the sense that the boundary is locally the graph of a Lipschitz continuous function.

Let

$$\begin{aligned} H^1_{\Gamma_D}(\Omega) &:= \{u \in H^1(\Omega)\colon (\gamma u)_{|\Gamma_D} = 0\}, \\ \mathbf{H}_{\Gamma_D}(\mathbf{curl}, \Omega) &:= \{\mathbf{u} \in \mathbf{H}(\mathbf{curl}, \Omega)\colon (\boldsymbol{\gamma}_\tau \mathbf{u})_{|\Gamma_D} = 0\}, \\ \mathbf{H}_{\Gamma_D}(\mathrm{div}, \Omega) &:= \{\mathbf{u} \in \mathbf{H}(\mathrm{div}, \Omega)\colon (\gamma_n \mathbf{u})_{|\Gamma_D} = 0\}, \end{aligned}$$

denote the standard Sobolev spaces where the distributional gradient, curl, or divergence is in L^2 and where the pointwise trace γu, the tangential trace $\gamma_\tau \mathbf{u}$, or the normal trace $\gamma_n \mathbf{u}$, respectively, vanishes on the Dirichlet boundary Γ_D, see e.g. [3, 6, 26]. These space are linked via the *de Rham complex*,

$$\mathcal{K}_{\Gamma_D}(\Omega) \xrightarrow{\mathrm{id}} H^1_{\Gamma_D}(\Omega) \xrightarrow{\nabla} \mathbf{H}_{\Gamma_D}(\mathbf{curl}, \Omega) \xrightarrow{\mathbf{curl}} \mathbf{H}_{\Gamma_D}(\mathrm{div}, \Omega) \xrightarrow{\mathrm{div}} L^2(\Omega), \tag{5}$$

where

$$\mathcal{K}_{\Gamma_D}(\Omega) := \{v \in H^1_{\Gamma_D}(\Omega)\colon v = \mathrm{const}\} = \begin{cases} \mathrm{span}\{1\}, & \text{if } \Gamma_D = \emptyset, \\ \{0\}, & \text{otherwise.} \end{cases}$$

The range of each operator in (5) lies in the kernel space of the succeeding one, cf. [3, Lemma 2.2]. We define

$$\begin{aligned} \mathbf{H}_{\Gamma_D}(\mathbf{curl}\,0, \Omega) &:= \{\mathbf{v} \in \mathbf{H}_{\Gamma_D}(\mathbf{curl}, \Omega)\colon\ \mathbf{curl}\,\mathbf{v} = 0\}, \\ \mathbf{H}_{\Gamma_D}(\mathrm{div}\,0, \Omega) &:= \{\mathbf{v} \in \mathbf{H}_{\Gamma_D}(\mathrm{div}, \Omega)\colon\ \mathrm{div}\,\mathbf{v} = 0\}. \end{aligned} \tag{6}$$

Barring topological obstructions these kernels can be represented through potentials: Let $\beta_1(\Omega)$ denote the *first Betti number* of Ω (the number of "handles") and $\beta_2(\Omega)$ the *second Betti number* (the number of connected components of $\partial\Omega$ minus one). By the very definition of the Betti numbers as dimensions of co-homology spaces we have

$$\beta_1(\Omega) = 0 \quad \Longrightarrow \quad \mathbf{H}(\mathbf{curl}\,0, \Omega) = \nabla H^1(\Omega), \tag{7}$$

$$\beta_2(\Omega) = 0 \quad \Longrightarrow \quad \mathbf{H}(\mathrm{div}\,0, \Omega) = \mathbf{curl}\,\mathbf{H}(\mathbf{curl}, \Omega), \tag{8}$$

cf. [26]. We call Ω topologically trivial if $\beta_1(\Omega) = \beta_2(\Omega) = 0$.

Regular Decompositions and Potentials Related to H(curl)

Throughout we rely on the properties of Ω and Γ_D as introduced in Section and use the notations from Theorem 1. We write C for positive "generic constants" and say that a constant "depends only on the shape of Ω and Γ_D", if it depends on the geometric setting alone, but is invariant with respect to similarity transformations. To achieve this the diameter of Ω will have to enter the estimates; we denote it by $\mathrm{d}(\Omega)$.

Gradient-Based Regular Decomposition of H(curl)

The following theorem is essentially [17, Thm. 2.1].

Theorem 2 (Gradient-Based Regular Decomposition of H(curl)) *Let* (Ω, Γ_D) *satisfy the assumptions of Section. Then for each* $\mathbf{u} \in \mathbf{H}_{\Gamma_D}(\mathbf{curl}, \Omega)$ *there exist* $\mathbf{z} \in \mathbf{H}^1_{\Gamma_D}(\Omega)$ *and* $p \in H^1_{\Gamma_D}(\Omega)$ *depending linearly on* $\mathbf{u}$ *such that*

(i)
$$\mathbf{u} = \mathbf{z} + \nabla p,$$

(ii)
$$\|\mathbf{z}\|_{0,\Omega} + \|\nabla p\|_{0,\Omega} \leq C\,\|\mathbf{u}\|_{0,\Omega},$$

(iii)
$$\|\nabla \mathbf{z}\|_{0,\Omega} + \frac{1}{\mathrm{d}(\Omega)}\|\mathbf{z}\|_{0,\Omega} \leq C\|\,\mathbf{curl}\,\mathbf{u}\|_{0,\Omega} + \frac{1}{\mathrm{d}(\Omega)}\|\mathbf{u}\|_{0,\Omega}\,,$$

with constants depending only on the shape of Ω *and* Γ_D*, but not on* $\mathrm{d}(\Omega)$.

Remark 1 An early decomposition of a subspace of $\mathbf{H}(\mathbf{curl}, \Omega) \cap \mathbf{H}(\mathrm{div}, \Omega)$ into a regular part in $\mathbf{H}^1(\Omega)$ and a singular part in $\nabla H^1(\Omega)$ can be found in [4] and in [5, Proposition 5.1], see also [7, Section] and references therein. Theorem 2 was proved in [14, Lemma 2.4] for the case of $\Gamma_D = \partial\Omega$ and without the $\mathbf{L}^2$-stability estimate, following [5, Proposition 5.1]. Pasciak and Zhao [28, Lemma 2.2] provided a version for simply connected Ω and the case $\Gamma_D = \partial\Omega$ *with* pure $\mathbf{L}^2$-stability, but p is only constant on each connected component of $\partial\Omega$ (see also Theorem 5 and Remark 3). This result was refined in [24, Thm. 3.1]. For the case $\Gamma_D = \emptyset$, [14, Lemma 2.4] gives a similar decomposition but ∇p must be replaced by an element from $\mathbf{H}(\mathbf{curl}\,0, \Omega)$ in general. Finally, Theorem 2 without the pure $\mathbf{L}^2$-stability was proved in [20, Thm. 5.2].[3]

Remark 2 The constant C in Theorem 2 depends mainly on the stability constants of key extension operators. If the bulge Υ_D has multiple components $\Upsilon_{D,k}$, the final estimate will depend on the relative distances between $\Upsilon_{D,k}$, $\Upsilon_{D,\ell}$, $k \neq \ell$ and the ratios $\mathrm{d}(\Upsilon_{D,k})/\,\mathrm{d}(\Omega)$.

[3] This reference contains a typo which is easily identified when inspecting the proof: In general, $\mathbf{z}$ cannot be estimated in terms of $\|\,\mathbf{curl}\,\mathbf{u}\|_{0,\Omega}$ but one must use the full $\mathbf{H}(\mathbf{curl})$ norm.

Remark 3 If $\Gamma_D = \partial\Omega$, one obtains only $p \in H^1(\Omega)$ being constant on each connected component of Γ_D but the improved bound

$$\| \nabla \mathbf{z}\|_{0,\Omega} + \mathrm{d}(\Omega)^{-1} \|\mathbf{z}\|_{0,\Omega} \leq C \, \| \mathbf{curl}\, \mathbf{u}\|_{0,\Omega} \,.$$

Results on regular decompositions in this special case can be found in [24, 28].

Regular Potentials for Some Divergence-Free Functions

Let the domain Ω and the Dirichlet boundary part Γ_D be as introduced in Section. Let Γ_i, $i = 0, \ldots, \beta_2(\Omega)$, denote the connected components of $\partial\Omega$, where $\beta_2(\Omega)$ is the second Betti number of Ω.

We define the space[4]

$$\mathbf{H}_{\Gamma_D}(\mathrm{div}\, 00, \Omega) := \left\{ \mathbf{q} \in \mathbf{H}_{\Gamma_D}(\mathrm{div}\, 0, \Omega) \colon \langle \gamma_n \mathbf{q}, 1 \rangle_{\Gamma_i} = 0, \ i = 0, \ldots, \beta_2(\Omega) \right\}. \tag{9}$$

Above γ_n denotes the normal trace operator, and the duality pairing is that between $H^{-1/2}(\Gamma_i)$ and $H^{1/2}(\Gamma_i)$. If $\Gamma_D = \emptyset$ we simply drop the subscript Γ_D. Obviously,

$$\mathbf{H}_{\Gamma_D}(\mathrm{div}\, 00, \Omega) \subset \mathbf{H}(\mathrm{div}\, 00, \Omega) \,.$$

The next result identifies the above space as the range of the curl operator.

Theorem 3 (Regular Potential of Range(**curl**)) *Let (Ω, Γ_D) be as in Sect. and assume in addition that each connected component $\Upsilon_{D,k}$ of the bulge has vanishing first Betti number, $\beta_1(\Upsilon_{D,k}) = 0$. Then*

$$\mathbf{H}_{\Gamma_D}(\mathrm{div}\ 00, \Omega) = \mathbf{curl}\ \mathbf{H}_{\Gamma_D}(\mathbf{curl}, \Omega) = \mathbf{curl}\ \mathbf{H}^1_{\Gamma_D}(\Omega) \,,$$

and for each $\mathbf{q} \in \mathbf{H}_{\Gamma_D}(\mathrm{div}\ 00, \Omega)$ there exists $\boldsymbol{\psi} \in \mathbf{H}^1_{\Gamma_D}(\Omega)$ depending linearly on $\mathbf{q}$ such that

$$\mathbf{curl}\ \boldsymbol{\psi} = \mathbf{q} \quad \textit{and} \quad \| \nabla \boldsymbol{\psi} \|_{0,\Omega} + \frac{1}{\mathrm{d}(\Omega)} \|\boldsymbol{\psi}\|_{0,\Omega} \leq C \, \|\mathbf{q}\|_{0,\Omega} \,,$$

where C depends only on the shape of Ω and Γ_D, but not on $\mathrm{d}(\Omega)$

[4] Alternatively we can define $\mathbf{H}_{\Gamma_D}(\mathrm{div}\, 00, \Omega)$ as the functions in $\mathbf{H}_{\Gamma_D}(\mathrm{div}\, 0, \Omega)$ orthogonal to the *harmonic Dirichlet fields* $\mathbf{H}(\mathrm{div}\, 0, \Omega) \cap \mathbf{H}_0(\mathbf{curl}\, 0, \Omega)$.

Remark 4 For the case that $\Gamma_D = \emptyset$, we reproduce the classical result

$$\mathbf{H}(\text{div}\,00, \Omega) = \mathbf{curl}\ \mathbf{H}(\mathbf{curl}, \Omega) = \mathbf{curl}\ \mathbf{H}^1(\Omega),$$

see [11, Thm. 3.4]. In that case, Step 4 of the proof can be left out and $\boldsymbol{\psi} = \mathbf{w}_1$ which is why div $\boldsymbol{\psi} = 0$ in Ω. This property, however, is lost in the general case.

Rotation-Bounded Regular Decomposition of H(curl)

We can now formulate another *new* variety of regular decompositions, for which the $\mathbf{H}^1$-component will vanish for curl-free fields.

Theorem 4 (Rotation-Bounded Regular Decomposition of $\mathbf{H}(\mathbf{curl})$ (I)) *Let (Ω, Γ_D) be as in Sect. and assume, in addition, that each connected component $\Upsilon_{D,k}$ of the bulge has vanishing first Betti number, $\beta_1(\Upsilon_{D,k}) = 0$. Then, for each $\mathbf{u} \in \mathbf{H}_{\Gamma_D}(\mathbf{curl}, \Omega)$ there exist $\mathbf{z} \in \mathbf{H}^1_{\Gamma_D}(\Omega)$ and a curl-free vector field $\mathbf{h} \in \mathbf{H}_{\Gamma_D}(\mathbf{curl}\,0, \Omega)$, depending linearly on $\mathbf{u}$ such that*

$$\mathbf{u} = \mathbf{z} + \mathbf{h},$$

$$\|\mathbf{h}\|_{0,\Omega} \leq \|\mathbf{u}\|_{0,\Omega} + C\ \mathrm{d}(\Omega)\,\|\ \mathbf{curl}\ \mathbf{u}\|_{0,\Omega},$$

$$\|\ \nabla\,\mathbf{z}\|_{0,\Omega} + \frac{1}{\mathrm{d}(\Omega)}\|\mathbf{z}\|_{0,\Omega} \leq C\,\|\ \mathbf{curl}\ \mathbf{u}\|_{0,\Omega}\ ,$$

where C depends only on the shape of Ω and Γ_D, but not on $\mathrm{d}(\Omega)$.

Remark 5 The constant C in Theorem 4 depends essentially on the stability constants of the divergence-free extension operator $E^{\text{div},0}_{\Omega^e}$ and the (adapted) Stein extension operator $E^{\nabla,\text{Stein}}_{\Upsilon_D}$.

Another *stronger* version of the rotation-bounded regular decomposition of $\mathbf{H}(\mathbf{curl})$ gets rid of the assumptions on the topology of the Dirichlet boundary and has improved stability properties (though with less explicit constants).

Theorem 5 (Rotation-Bounded Regular Decomposition of $\mathbf{H}(\mathbf{curl})$ (II)) *Let (Ω, Γ_D) be as in Sect.. Then for each $\mathbf{u} \in \mathbf{H}_{\Gamma_D}(\mathbf{curl}, \Omega)$ there exist $\mathbf{z} \in \mathbf{H}^1_{\Gamma_D}(\Omega)$ and a curl-free $\mathbf{h} \in \mathbf{H}_{\Gamma_D}(\mathbf{curl}\,0, \Omega)$ depending linearly on $\mathbf{u}$ such that*

$$\mathbf{u} = \mathbf{z} + \mathbf{h}\,,$$

$$\|\mathbf{z}\|_{0,\Omega} + \|\mathbf{h}\|_{0,\Omega} \leq C\,\|\mathbf{u}\|_{0,\Omega}\ ,$$

$$\|\ \nabla\,\mathbf{z}\|_{0,\Omega} + \mathrm{d}(\Omega)^{-1}\|\mathbf{z}\|_{0,\Omega} \leq C\,\|\ \mathbf{curl}\ \mathbf{u}\|_{0,\Omega}\ ,$$

where C depends only on the shape of Ω and Γ_D, but not on $\mathrm{d}(\Omega)$.

Remark 6 For the case $\Gamma_D = \partial\Omega$ the result of the theorem is already proved by Remark 3 since we obtain $\mathbf{u} = \mathbf{z} + \nabla p$ with $\nabla p \in \nabla H^1_{0,\mathrm{const}}(\Omega) = \mathbf{H}_0(\mathbf{curl}, \Omega)$.

Remark 7 We would like to emphasize that both in Theorems 2 and 5, the domain Ω may be non-convex, non-smooth, and may have non-trivial topology: it may have handles and its boundary may have multiple components. Also the Dirichlet boundary Γ_D may have multiple components, each of which with non-trivial topology. Moreover, we have the pure $\mathbf{L}^2(\Omega)$-stability in both theorems. In this sense, the results of Theorems 2 and 5 are superior to those found, e.g., in [7, Thm 3.4], [19] or the more recent ones in [8, Thm. 2.3], [22].

Remark 8 If Ω has vanishing first Betti number, $\beta_1(\Omega) = 0$, then $\mathbf{H}_{\Gamma_D}(\mathbf{curl}\,0, \Omega) = \nabla H^1_{\Gamma_D,\mathrm{const}}(\Omega)$. Hence, we can split each $\mathbf{u} \in \mathbf{H}_{\Gamma_D}(\mathbf{curl}, \Omega)$ into $\mathbf{z} \in \mathbf{H}^1_{\Gamma_D}(\Omega)$ and ∇p with $p \in H^1(\Omega)$ being constant on each connected component of Γ_D. If Γ_D is connected, then $p \in H^1_{\Gamma_D}(\Omega)$. Summarizing, if Ω has no handles and if Γ_D is connected, then we have the combined features of Theorems 2 and 5.

Finally, we mention that the regular decomposition theorems spawn projection operators that play a fundamental role in the analysis of weak formulations of Maxwell's equations in frequency domain.

Corollary 1 *Let (Ω, Γ_D) be as in Sect. Then there exist continuous projection operators* $\mathsf{R}\colon \mathbf{H}_{\Gamma_D}(\mathbf{curl}, \Omega) \to \mathbf{H}^1_{\Gamma_D}(\Omega)$ *and* $\mathsf{N}\colon \mathbf{H}_{\Gamma_D}(\mathbf{curl}, \Omega) \to \mathbf{H}_{\Gamma_D}(\mathbf{curl}\,0, \Omega)$ *such that* $\mathsf{R} + \mathsf{N} = \mathsf{id}$ *and*

$$\|\mathsf{R}\mathbf{v}\|_{\mathbf{H}^1(\Omega)} + \|\mathsf{N}\mathbf{v}\|_{\mathbf{L}^2(\Omega)} \le C\,\|\mathbf{v}\|_{\mathbf{H}(\mathbf{curl},\Omega)} \qquad \forall \mathbf{v} \in \mathbf{H}(\mathbf{curl}, \Omega),$$

where C is a constant independent of $\mathbf{v}$. *Moreover,* $\mathsf{F}\colon \mathbf{H}_{\Gamma_D}(\mathbf{curl}, \Omega) \to \mathbf{H}_{\Gamma_D}(\mathbf{curl}, \Omega)$ *defined by* $\mathsf{F}\mathbf{v} := \mathsf{R}\mathbf{v} - \mathsf{N}\mathbf{v}$ *is an isomorphism.*

Remark 9 The $\mathbf{L}^2$-estimates from Theorem 4 then show that the corresponding operator R can be extended to a continuous operator mapping from $\mathbf{L}^2(\Omega)$ to $\mathbf{L}^2(\Omega)$.

Regular Decompositions and Potentials Related to $\mathbf{H}(\mathrm{div})$

The developments of this section are largely parallel to those of Sect. with some new aspects concerning extensions and topological considerations.

Rotation-Based Regular Decomposition of $\mathbf{H}(\mathrm{div})$

The following theorem is the $\mathbf{H}(\mathrm{div})$-counterpart of Theorem 2.

Theorem 6 (Rotation-Based Regular Decomposition of $\mathbf{H}(\mathrm{div})$) *Let (Ω, Γ_D) satisfy the assumptions made in Sect. Then for each $\mathbf{v} \in \mathbf{H}_{\Gamma_D}(\mathrm{div}, \Omega)$ there exist $\mathbf{z} \in \mathbf{H}^1_{\Gamma_D}(\Omega)$ and $\mathbf{q} \in \mathbf{H}^1_{\Gamma_D}(\Omega)$ depending linearly on $\mathbf{v}$ such that*

$$\mathbf{v} = \mathbf{z} + \mathbf{curl}\, \mathbf{q},$$

$$\|\mathbf{z}\|_{0,\Omega} + \|\, \mathbf{curl}\, \mathbf{q}\|_{0,\Omega} + \frac{1}{\mathrm{d}(\Omega)} \|\mathbf{q}\|_{0,\Omega} \leq C\, \|\mathbf{v}\|_{0,\Omega}\,,$$

$$\|\, \nabla\, \mathbf{z}\|_{0,\Omega} + \frac{1}{\mathrm{d}(\Omega)} \|\mathbf{z}\|_{0,\Omega} + \frac{1}{\mathrm{d}(\Omega)} \|\, \nabla\, \mathbf{q}\|_{0,\Omega} \leq C\big(\|\, \mathbf{curl}\ \mathbf{v}\|_{0,\Omega} + \frac{1}{\mathrm{d}(\Omega)} \|\mathbf{v}\|_{0,\Omega}\big)\,,$$

with constant C depending only on the shape of Ω and Γ_D, but not on $\mathrm{d}(\Omega)$.

Regular Potential with Prescribed Divergence

The next result carries Theorem 3 over to $\mathbf{H}(\mathrm{div})$.

Theorem 7 (Regular Potentials for the Image Space of div) *Let (Ω, Γ_D) be as in Sect. and, in addition, assume that each connected component $\Upsilon_{D,k}$ of the bulge has a connected boundary, i.e., $\beta_2(\Upsilon_{D,k}) = 0$. Then*

$$L^2(\Omega) = \mathrm{div}\, \mathbf{H}_{\Gamma_D}(\mathrm{div}, \Omega) = \mathrm{div}\, \mathbf{H}^1_{\Gamma_D}(\Omega).$$

Moreover, for each $v \in L^2(\Omega)$ there exists $\mathbf{q} \in \mathbf{H}^1_{\Gamma_D}(\Omega)$ depending linearly on v such that, with a constant C depending on Ω and Γ_D but not on $\mathrm{d}(\Omega)$,

$$\mathrm{div}\, \mathbf{q} = v \quad \textit{and} \quad \|\, \nabla\, \mathbf{q}\|_{0,\Omega} + \frac{1}{\mathrm{d}(\Omega)} \|\mathbf{q}\|_{0,\Omega} \leq C\, \|v\|_{0,\Omega}\,.$$

Divergence-Bounded Regular Decompositions of $\mathbf{H}(\mathbf{div})$

We can now formulate other variants of regular decompositions of $\mathbf{H}(\mathrm{div})$ in analogy to what we did in other section.

Theorem 8 (Divergence-Bounded Regular Decomposition of $\mathbf{H}(\mathrm{div})$ (I)) *Let (Ω, Γ_D) be as in Sect. In addition, assume that each connected component $\Upsilon_{D,k}$ of the bulge has a connected boundary, i.e., $\beta_2(\Upsilon_{D,k}) = 0$. Then, for each $\mathbf{v} \in \mathbf{H}_{\Gamma_D}(\mathrm{div}, \Omega)$ there exists $\mathbf{z} \in \mathbf{H}^1_{\Gamma_D}(\Omega)$ and a divergence-free vector field*

$\mathbf{h} \in \mathbf{H}_{\Gamma_D}(\operatorname{div} 0, \Omega)$ *depending linearly on* $\mathbf{v}$ *such that*

$$\mathbf{v} = \mathbf{z} + \mathbf{h}, \tag{10}$$

$$\|\mathbf{h}\|_{0,\Omega} \le \|\mathbf{v}\|_{0,\Omega} + C\,\mathrm{d}(\Omega)\|\operatorname{div}\mathbf{v}\|_{0,\Omega}\,, \tag{11}$$

$$\|\nabla\mathbf{z}\|_{0,\Omega} + \frac{1}{\mathrm{d}(\Omega)}\|\mathbf{z}\|_{0,\Omega} \le C\,\|\operatorname{div}\mathbf{v}\|_{0,\Omega}\,, \tag{12}$$

where C *depends only on the shape of* Ω *and* Γ_D*, but not on* $\mathrm{d}(\Omega)$.

The last variant of $\mathbf{H}(\operatorname{div})$ regular decomposition of $\mathbf{H}(\operatorname{div})$ dispenses with the assumptions on the topology of the Dirichlet boundary and has better stability properties than the splitting from Theorem 8 (though with less explicit constants).

Theorem 9 (Divergence-Bounded Regular Decomposition of $\mathbf{H}(\operatorname{div})$ (II)) *Let* (Ω, Γ_D) *be as in Sect. Then, for each* $\mathbf{v} \in \mathbf{H}_{\Gamma_D}(\operatorname{div}, \Omega)$ *there exists* $\mathbf{z} \in \mathbf{H}^1_{\Gamma_D}(\Omega)$ *and a divergence-free vector field* $\mathbf{h} \in \mathbf{H}_{\Gamma_D}(\operatorname{div} 0, \Omega)$ *depending linearly on* $\mathbf{v}$ *such that*

$$\mathbf{v} = \mathbf{z} + \mathbf{h}, \tag{13}$$

$$\|\mathbf{z}\|_{0,\Omega} + \|\mathbf{h}\|_{0,\Omega} \le \|\mathbf{v}\|_{0,\Omega}\,, \tag{14}$$

$$\|\nabla\mathbf{z}\|_{0,\Omega} + \frac{1}{\mathrm{d}(\Omega)}\|\mathbf{z}\|_{0,\Omega} \le C\,\|\operatorname{div}\mathbf{v}\|_{0,\Omega}\,, \tag{15}$$

where C *depends only on the shape of* Ω *and* Γ_D*, but not on* $\mathrm{d}(\Omega)$.

Discrete Counterparts of the Regular Decompositions

The discrete setting to which we want to extend the concept of regular decompositions is provided by finite element exterior calculus (FEEC, [2]) which introduces finite element subspaces of $\mathbf{H}(\mathbf{curl})$ and $\mathbf{H}(\operatorname{div})$ as special instances of spaces of discrete differential forms. In this section we confine ourselves to the lowest-order case of piecewise linear finite element functions.

Throughout, we assume that (Ω, Γ_D) is as in Sect., and, additionally, that Ω is a polyhedron and that $\partial\Gamma_D$ consists of straight line segments. All considerations take for granted a shape-regular family of meshes $\{\mathcal{T}^h\}_h$ of Ω, consisting of tetrahedral elements, and resolving Γ_D in the sense that Γ_D is a union of faces of some of the tetrahedra.

The following finite element spaces will be relevant:

- the space $\mathcal{W}^0_{h,\Gamma_D}(\Omega)$ of $H^1_{\Gamma_D}(\Omega)$-conforming piecewise linear Lagrangian finite element functions,
- the space $\boldsymbol{\mathcal{W}}^1_{h,\Gamma_D}(\Omega)$ of $\mathbf{H}_{\Gamma_D}(\mathbf{curl}, \Omega)$-conforming lowest order Nédélec elements, also known as *edge elements*,

- the space $\mathcal{W}^2_{h,\Gamma_D}(\Omega)$ of $\mathbf{H}_{\Gamma_D}(\mathrm{div},\Omega)$-conforming lowest order tetrahedral Raviart-Thomas finite elements, aka, *face elements*,
- the space $\boldsymbol{\mathcal{W}}^0_{h,\Gamma_D}(\Omega) := [\mathcal{W}^0_{h,\Gamma_D}(\Omega)]^3$ of piecewise linear globally continuous vector fields vanishing on Γ_D.

Functions in $\mathcal{W}^\ell_{h,\Gamma_D}(\Omega)$, $\ell = 1, 2, 3$, are so-called Whitney forms, lowest-order discrete differential forms of the first family as introduced.

Discrete Regular Decompositions for Edge Elements

Commuting projectors, also known as co-chain projectors, are the linchpin of FEEC theory and it is not different with our developments. Thus, let

$$R^0_{h,\Gamma_D} : H^1_{\Gamma_D}(\Omega) \to \mathcal{W}^0_{h,\Gamma_D}(\Omega)$$

$$\text{and}\quad \mathbf{R}^1_{h,\Gamma_D} : \mathbf{H}_{\Gamma_D}(\mathbf{curl},\Omega) \to \mathcal{W}^1_{h,\Gamma_D}(\Omega)$$

denote the *continuous*, *boundary-aware* cochain projectors from which extend the pioneering work [10] by Falk and Winther. These two linear operators are projectors onto their ranges, they fulfill the commuting property

$$\nabla(R^0_{h,\Gamma_D}\varphi) = \mathbf{R}^1_{h,\Gamma_D}(\nabla\varphi) \qquad \forall\varphi \in H^1_{\Gamma_D}(\Omega)\,, \tag{16}$$

and local stability estimates

Theorem 10 ([17, Thm. 1.2]) *For each $\mathbf{v}_h \in \mathcal{W}^1_{0,\Gamma_D}(\Omega)$ there exists a continuous and piecewise linear vector field $\mathbf{z}_h \in \boldsymbol{\mathcal{W}}^0_{h,\Gamma_D}(\Omega)$, a continuous and piecewise linear scalar function $p_h \in \mathcal{W}^0_{h,\Gamma_D}(\Omega)$, and a remainder $\widetilde{\mathbf{v}}_h \in \mathcal{W}^1_{0,\Gamma_D}(\Omega)$, all depending linearly on $\mathbf{v}_h$, providing the discrete regular decomposition*

$$\mathbf{v}_h = \mathbf{R}^1_{h,\Gamma_D}\mathbf{z}_h + \widetilde{\mathbf{v}}_h + \nabla p_h$$

and satisfying the stability estimates

$$\|\mathbf{z}_h\|_{0,\Omega} + \|\nabla p_h\|_{0,\Omega} + \|\widetilde{\mathbf{v}}_h\|_{0,\Omega} \le C\,\|\mathbf{v}_h\|_{0,\Omega}\,, \tag{17}$$

$$\|\boldsymbol{\nabla}\mathbf{z}_h\|_{0,\Omega} + \|h^{-1}\widetilde{\mathbf{v}}_h\|_{0,\Omega} \le C\big(\|\mathbf{curl}\,\mathbf{v}_h\|_{0,\Omega} + \tfrac{1}{\mathrm{d}(\Omega)}\|\mathbf{v}_h\|_{0,\Omega}\big)\,, \tag{18}$$

where C is a generic constant that depends only on the shape of (Ω, Γ_D), but not on $\mathrm{d}(\Omega)$, and on the shape regularity constant of $\mathcal{T}^h(\Omega)$. Above, h^{-1} is the piecewise constant function that is equal to h_T^{-1} on every element T.

Obviously, this is a discrete counterpart of the regular decomposition of $\mathbf{H}(\mathbf{curl})$ from Theorem 2. The following theorem appears to be new and it corresponds to the rotation-bounded regular decomposition of Theorem 5. For the sake of brevity define the discrete nullspace of the curl operator

$$\mathcal{N}_h^1 := \{\mathbf{v}_h \in \mathcal{W}^1_{h,\Gamma_D}(\Omega) \colon \ \mathbf{curl}\,\mathbf{v}_h = 0\}\,. \tag{19}$$

If Ω and Γ_D have simple topology, $\mathcal{X}_h = \nabla\, \mathcal{W}^0_{h,\Gamma_D}(\Omega)$, but if the first Betti number of Ω is non-zero, or if Γ_D has multiple components, then a finite-dimensional co-homology space has to be added.

Theorem 11 (Rotation-Bounded Discrete Regular Decomposition for Edge Elements) *For each* $\mathbf{v}_h \in \mathcal{W}^1_{0,\Gamma_D}(\Omega)$ *there exists a continuous and piecewise linear vector field* $\mathbf{z}_h \in \mathcal{W}^0_{h,\Gamma_D}(\Omega)$, *an* curl-free *edge element function* $\mathbf{h}_h \in \mathcal{N}_h^1$, *and a remainder* $\widetilde{\mathbf{v}}_h \in \mathcal{W}^1_{0,\Gamma_D}(\Omega)$, *all depending linearly on* $\mathbf{v}_h$, *providing the discrete regular decomposition*

$$\mathbf{v}_h = \mathbf{R}^1_{h,\Gamma_D}\mathbf{z}_h + \widetilde{\mathbf{v}}_h + \mathbf{h}_h$$

and satisfying the stability bounds

$$\left.\begin{array}{l}\|\mathbf{z}_h\|_{0,\Omega}\\ \|\mathbf{h}_h\|_{0,\Omega}\\ \|\widetilde{\mathbf{v}}_h\|_{0,\Omega}\end{array}\right\} \le C\,\|\mathbf{v}_h\|_{0,\Omega}\,, \qquad \left.\begin{array}{l}\|\nabla\,\mathbf{z}_h\|_{0,\Omega}\\ \|h^{-1}\widetilde{\mathbf{v}}_h\|_{0,\Omega}\end{array}\right\} \le C\,\|\,\mathbf{curl}\,\mathbf{v}_h\|_{0,\Omega}\,,$$

where C *is a uniform constant that depends only on the shape of* (Ω, Γ_D), *but not on* $\mathrm{d}(\Omega)$, *and on the shape regularity constant of* $\mathcal{T}^h(\Omega)$.

We stress that the statements of Theorems 10 and 11 do not hinge on *any* assumptions on the topological properties of Ω and Γ_D.

Discrete Regular Decompositions for Face Elements

For face elements, the construction of a boundary-aware co-chain projection operator

$$\mathbf{R}^2_{h,\Gamma_D} \colon \mathbf{H}_{\Gamma_D}(\mathrm{div}, \Omega) \to \mathcal{W}^2_{h,\Gamma_D}(\Omega)$$

that commutes with $\mathbf{R}^1_{h,\Gamma_D}$ and the **curl**-operator has not yet been accomplished. Fortunately, in the case $\Gamma_D = \emptyset$, this operator is available from [10]. Thus, in the following, we treat only the case $\Gamma_D = \emptyset$ and just omit the subscript Γ_D. Then, from [10] we can borrow a linear operator $\mathbf{R}^2_h \colon \mathbf{H}(\mathrm{div}, \Omega) \to \mathcal{W}^2_h(\Omega)$ such that

$$\mathbf{curl}\,\mathbf{R}^1_h\mathbf{u} = \mathbf{R}^2_h\,\mathbf{curl}\,\mathbf{u} \qquad \forall \mathbf{u} \in \mathbf{H}(\mathbf{curl}, \Omega)\,. \tag{20}$$

The next result takes Theorem 6 to the discrete setting.

Theorem 12 (Discrete Regular Decomposition of $\mathcal{W}_h^2(\Omega)$) *For each vector field $\mathbf{v}_h$ in the lowest-order Raviart-Thomas space $\mathcal{W}_h^2(\Omega)$, there exists a continuous and piecewise linear vector field $\mathbf{z}_h \in \mathcal{W}_h^0(\Omega)$, a vector field $\mathbf{q}_h$ in the lowest-order Nédélec space $\mathcal{W}_h^1(\Omega)$, and a remainder $\widetilde{\mathbf{v}}_h \in \mathcal{W}_h^2(\Omega)$, all depending linearly on $\mathbf{v}_h$, providing the discrete regular decomposition*

$$\mathbf{v}_h = \mathbf{R}_h^2 \mathbf{z}_h + \widetilde{\mathbf{v}}_h + \mathbf{curl}\, \mathbf{q}_h ,$$

and the stability estimates

$$\left.\begin{array}{r} \|\mathbf{z}_h\|_{0,\Omega} \\ \|\,\mathbf{curl}\,\mathbf{q}_h\|_{0,\Omega} + \frac{1}{\mathrm{d}(\Omega)}\|\mathbf{q}_h\|_{0,\Omega} \\ \|\widetilde{\mathbf{v}}_h\|_{0,\Omega} \end{array}\right\} \leq C\, \|\mathbf{v}_h\|_{0,\Omega} ,$$

$$\left.\begin{array}{r} \|\,\nabla\, \mathbf{z}_h\|_{0,\Omega} \\ \|h^{-1}\widetilde{\mathbf{v}}_h\|_{0,\Omega} \end{array}\right\} \leq C \|\,\mathrm{div}\,\mathbf{v}_h\|_{0,\Omega} + \frac{1}{\mathrm{d}(\Omega)}\|\mathbf{v}_h\|_{0,\Omega} .$$

The constant C depends only on the shape of Ω, but not on $\mathrm{d}(\Omega)$, *and the shape-regularity of $\mathcal{T}^h(\Omega)$.*

Finally, we present a counterpart to the divergence-bounded regular decomposition of Theorem 9. For convenience we introduce the space of divergence-free face element functions

$$\mathcal{N}_h^2 := \{\mathbf{q}_h \in \mathcal{W}_h^2(\Omega) \colon \ \mathrm{div}\,\mathbf{q}_h = 0\} . \tag{21}$$

Theorem 13 (Divergence-Bounded Discrete Regular Decomposition of $\mathcal{W}_h^2(\Omega)$) *For each vector field $\mathbf{v}_h$ in the lowest-order Raviart-Thomas space $\mathcal{W}_h^2(\Omega)$, there exists a continuous and piecewise linear vector field $\mathbf{z}_h \in \mathcal{W}_h^0(\Omega)$, an element $\mathbf{h}_h$ in the discrete divergence-free subspace $\mathcal{N}_h^2$, and a remainder $\widetilde{\mathbf{v}}_h \in \mathcal{W}_h^2(\Omega)$, all depending linearly on $\mathbf{v}_h$, providing the discrete regular decomposition*

$$\mathbf{v}_h = \mathbf{R}_h^2 \mathbf{z}_h + \widetilde{\mathbf{v}}_h + \mathbf{h}_h$$

and the stability estimates

$$\left.\begin{array}{r} \|\mathbf{z}_h\|_{0,\Omega} \\ \|\widetilde{\mathbf{v}}_h\|_{0,\Omega} \\ \|\mathbf{h}_h\|_{0,\Omega} \end{array}\right\} \leq C\, \|\mathbf{v}_h\|_{0,\Omega} , \qquad \left.\begin{array}{r} \|\,\nabla\, \mathbf{z}_h\|_{0,\Omega} \\ \|h^{-1}\widetilde{\mathbf{v}}_h\|_{0,\Omega} \end{array}\right\} \leq C \,\|\,\mathrm{div}\,\mathbf{v}_h\|_{0,\Omega} .$$

The constants C depend only on the shape of Ω, but not on $\mathrm{d}(\Omega)$, *and the shape regularity of $\mathcal{T}^h(\Omega)$.*

Remark 10 The result of Theorem 13 can be viewed as an improvement of the decompositions in [25] which are elaborated for the case of essential boundary conditions on $\partial\Omega$.

Corollary 2 *If the second Betti number of Ω vanishes, that is, if $\partial\Omega$ is connected, then $\mathbf{h}_h$ in Theorem 13 can be chosen as $\mathbf{h}_h = \mathbf{curl}\,\mathbf{q}_h$ with $\mathbf{q}_h \in \mathcal{W}_h^1(\Omega)$ such that*

$$\mathbf{v}_h = \mathbf{R}_h^2\mathbf{z} + \widetilde{\mathbf{v}}_h + \mathbf{curl}\,\mathbf{q}_h\,,$$

with the bounds

$$\left.\begin{array}{r}\|\mathbf{z}_h\|_{0,\Omega}\\ \|\widetilde{\mathbf{v}}_h\|_{0,\Omega}\\ \|\,\mathbf{curl}\,\mathbf{q}_h\|_{0,\Omega}\\ \mathrm{d}(\Omega)^{-1}\|\mathbf{q}_h\|_{0,\Omega}\end{array}\right\} \leq C\,\|\mathbf{v}_h\|_{0,\Omega}\,, \qquad \left.\begin{array}{r}\|\,\nabla\,\mathbf{z}_h\|_{0,\Omega}\\ \|h^{-1}\widetilde{\mathbf{v}}_h\|_{0,\Omega}\end{array}\right\} \leq C\,\|\,\mathrm{div}\,\mathbf{v}_h\|_{0,\Omega}\,.$$

Remark 11 The result of Corollary 2 is an improvement of [19, Lemma 5.2] which assumes a domain Ω that is smooth enough to allow H^2-regularity of the Laplace problem (2-regular case, for details see [19, Section). This lemma is used in [27] in a domain decomposition framework, where convex subdomains are assumed. With our improved version, this assumption can be weakened considerably.

Acknowledgements The second author would like to thank Dirk Pauly (Essen) for enlightening discussions.

References

1. Amrouche, C., Bernardi, C., Dauge, M., Girault, V.: Vector potentials in three-dimensional non-smooth domains. Math. Meth. Appl. Sci. **21**(9), 823–864 (1998)
2. Arnold, D., Falk, R., Winther, R.: Finite element exterior calculus, homological techniques, and applications. Acta Numer. **15**, 1–155 (2006)
3. Bauer, S., Pauly, D., Schomburg, M.: The Maxwell compactness property in bounded weak Lipschitz domains with mixed boundary conditions. SIAM J. Math. Anal. **48**(4), 2912–2943 (2016)
4. Birman, M.S., Solomyak, M.Z.: L^2-theory of the Maxwell operator in arbitrary domains. Usp. Mat. Nauk. **42**(6), 61–76 (1987). English Transliterated, Russian Math. Surveys **42**(6), 75–96 (1987)
5. Bonnet-Ben Dhia, A., Hazard, C., Lohrengel, S.: A singular field method for the solution of Maxwell's equations in polyhedral domains. SIAM J. Appl. Math. **59**(6), 2028–2044 (1999)
6. Buffa, A., Costabel, M., Sheen, D.: On traces for $\mathbf{H}(\mathbf{curl}, \Omega)$ in Lipschitz domains. J. Math. Anal. Appl. **276**, 845–867 (2002)
7. Costabel, M., Dauge, M., Nicaise, S.: Singularities of Maxwell interface problems. Model. Math. Anal. Numer. **33**(2), 627–649 (1999)
8. Creusé, E., Nicaise, S., Tang, Z.: Helmholtz decomposition of vector fields with mixed boundary conditions and an application to a posteriori finite element analysis of the Maxwell system. Math. Methods Appl. Sci. **38**(4), 738–750 (2015)

9. Dautray, R., Lions, J.: Mathematical Analysis and Numerical Methods for Science and Technology. Springer, Berlin (2000). Original French edition published by Masson, S.A., Paris (1984)
10. Falk, R.S., Winther, R.: Local bounded cochain projections. Math. Comp. **83**(290), 2631–2656 (2014)
11. Girault, V., Raviart, P.A.: Finite Element Methods for Navier–Stokes Equations. Springer, New York (1986)
12. Gopalakrishnan, J., Qiu, W.: Partial expansion of a Lipschitz domain and some applications. Front. Math. China **7**(2), 249–272 (2012)
13. Hiptmair, R.: Canonical construction of finite elements. Math. Comp. **68**, 1325–1346 (1999)
14. Hiptmair, R.: Finite elements in computational electromagnetism. Acta Numer. **11**, 237–339 (2002)
15. Hiptmair, R.: Boundary element methods for eddy current computation. In: Schanz, M., Steinbach, O. (eds.) Boundary Element Analysis: Mathematical Aspects and Applications. Lecture Notes in Applied and Computational Mechanics, vol. 29, pp. 213–248. Springer, Heidelberg (2007)
16. Hiptmair, R.: Maxwell's equations: Continuous and discrete. In: Bermúdez de Castro, A., Valli, A. (eds.) Computational Electromagnetism. Lecture Notes in Mathematics, vol. 2148. Springer, Cham (2015)
17. Hiptmair, R., Pechstein, C.: Discrete regular decompositions of tetrahedral discrete 1-forms. In: Langer, U., Pauly, D., Repin, S. (eds.) Maxwell's Equations. Radon Series on Computational and Applied Mathematics, vol. 24, pp. 199–258. De Gruyter, Stuttgart (2019)
18. Hiptmair, R., Pechstein, C.: Regular decompositions of vector fields - continuous, discrete, and structure-preserving. Technical Report 2019-18, Seminar for Applied Mathematics, ETH Zürich (2019). https://www.sam.math.ethz.ch/sam_reports/reports_final/reports2019/2019-18.pdf
19. Hiptmair, R., Xu, J.: Nodal auxiliary space preconditioning in H(curl) and H(div) spaces. SIAM J. Numer. Anal. **45**, 2483–2509 (2007)
20. Hiptmair, R., Zheng, W.: Local multigrid in H(curl). J. Comput. Math. **27**(5), 573–603 (2009)
21. Hiptmair, R., Widmer, G., Zou, J.: Auxiliary space preconditioning in $H_0(\mathrm{curl};\,\Omega)$. Numer. Math. **103**, 435–459 (2006)
22. Hu, Q.: Convergence of HX preconditioner for Maxwell's equations with jump coefficients (i): various extensions of the regular Helmholtz decomposition. Technical Report arXiv:1708.05850v2 [math.AP] (2018)
23. Kolev, T.V., Pasciak, J.E., Vassilevski, P.S.: H(curl) auxiliary mesh preconditioning. Numer. Linear Algebr. Appl. **15**(5), 455–471 (2008)
24. Kolev, T.V., Vassilevski, P.S.: Parallel auxiliary space AMG for **H**(curl) problems. J. Comput. Math. **27**(5), 604–623 (2009)
25. Kolev, T.V., Vassilevski, P.S.: Parallel auxiliary space AMG for **H**(div) problems. SIAM J. Sci. Comput. **34**(6), A3079–A3098 (2012)
26. Monk, P.: Finite Element Methods for Maxwell's Equations. Numerical Mathematics and Scientific Computation. Oxford University Press, Oxford (2003)
27. Oh, D., Widlund, O.B., Zampini, S., Dohrmann, C.R.: BDDC algorithms with deluxe scaling and adaptive selection of primal constraints for Raviart–Thomas vector fields. Math. Comp. **87**, 659–692 (2018)
28. Pasciak, J.E., Zhao, J.: Overlapping Schwarz methods in H(curl) on polyhedral domains. Numer. Math. **10**(3), 211–234 (2002)

3

Recurrence Relations for a Family of Orthogonal Polynomials on a Triangle

Sheehan Olver, Alex Townsend, and Geoffrey M. Vasil

Introduction

In 1975, Koornwinder described a general procedure for constructing multivariate orthogonal polynomials from univariate ones [4, §3.7.2]. The procedure allows for the construction of seven classes of bivariate orthogonal polynomials from Jacobi polynomials, some of which were previously known [9]. In this paper, we consider a four-parameter variant of Koornwinder's class IV polynomials (the four-parameter variant was not constructed by Koornwinder) defined as [2]

$$\begin{aligned} P_{n,k}^{(a,b,c,d)}(x,y) &= P_{n-k}^{(2k+b+c+d+1,a)}(2x-1)(1-x)^k P_k^{(c,b)}\left(-1+\tfrac{2y}{1-x}\right) \\ &= \tilde{P}_{n-k}^{(2k+b+c+d+1,a)}(x)(1-x)^k \tilde{P}_k^{(c,b)}\left(\tfrac{y}{1-x}\right), \end{aligned} \tag{1}$$

where $a, b, c > -1$, n and k are integers such that $n \geq k \geq 0$, $P_k^{(a,b)}(x)$ is the Jacobi polynomial of degree k [7, Table], and $\tilde{P}_k^{(a,b)}$ is the Jacobi polynomial of degree k shifted to have support on $(0, 1)$. Koornwinder's construction derives the polynomials with $d = 0$, which we denote by $P_{n,k}^{(a,b,c)}$. The polynomials in (1)

S. Olver (✉)
Department of Mathematics, Imperial College, London, UK
e-mail: s.olver@imperial.ac.uk

A. Townsend
Department of Mathematics, Cornell University, Ithaca, NY, USA
e-mail: townsend@cornell.edu

G. M. Vasil
School of Mathematics and Statistics, University of Sydney, Sydney, NSW, Australia
e-mail: geoffrey.vasil@sydney.edu.au

are orthogonal on the right-angled triangle $\{(x, y) : 0 < x < 1, 0 < y < 1 - x\}$ with respect to the weight function $w_{a,b,c,d}(x, y) = x^a y^b (1 - x - y)^c (1 - x)^d$.

The basis $P_{n,k}^{(a,b,c)}$ has been used extensively by the spectral element community, see the overview in [3]. The recurrence relations we derive can be employed to reduce partial differential operators to sparse matrices, enabling efficient solution of linear partial differential equations defined on triangles, which will be the topic of a future paper. This is analogous to the ultraspherical spectral method for solving ordinary differential equations on bounded intervals [8]. A similar idea using a hierarchy of Zernike polynomials, which are bivariate orthogonal polynomials on the unit disk, is used in [13] to develop a sparse spectral method for solving partial differential equations defined on the disk [13]. On the disk, polar coordinates allow for radially symmetric partial differential operators to be reduced to ordinary differential operators acting on Jacobi polynomials [13]. This simplification does not translate to non-radially symmetric partial differential operators on the disk, nor partial differential operators on the triangle.

Several of the formulae in this paper have already be derived by directly employing recurrence relations satisfied by Jacobi polynomials [14]. Our approach via ladder operators is a more systematic study that derives previously unreported recurrence relations for $P_{n,k}^{(a,b,c)}$. We also hope to use ladder operators to derive sparse recurrence relations for multivariate orthogonal polynomials built from Jacobi polynomials on higher-dimensional simplices.

Throughout this paper, the recurrence relations hold for choices of the parameters n, k, a, b, c, and d that make the Jacobi polynomials well-defined. Moreover, we take $P_{-1}^{(a,b)}(x) = 0$. Also, note that orthogonal polynomials remain orthogonal after an affine transformation so the recurrence relations in this paper for (1) on a right-angled triangle can be extended to any triangle, including triangles with the corners permuted.

The paper is structured as follows. In the next section, we give 12 ladder operators for Jacobi polynomials and use them to derive sparse recurrence relations for $P_n^{(a,b)}$. In Sect. we give 24 ladder operators for (1) and write down the corresponding sparse recurrence relations for $P_{n,k}^{(a,b,c,d)}$ in other sec. we use the ladder operators to derive a collection of sparse recurrence relations for differentiation, conversion, and multiplication that are satisfied by $P_{n,k}^{(a,b,c)}$. Section applies these sparse recurrence relations to efficiently calculating Laplacians of functions on the triangle.

Ladder Operators for Jacobi Polynomials

We give 12 ordinary differential operators that increment or decrement the parameters and degree of Jacobi polynomials by zero or one. Each ladder operator maps $P_n^{(a,b)}(x)$ to $P_{\tilde{n}}^{(\tilde{a},b)}(x)$, where $|\tilde{n} - n| \leq 1$, $|\tilde{a} - a| \leq 1$, and $|\tilde{b} - b| \leq 1$.

Definition 1 The following operators are ladder operators for Jacobi polynomials:

$$\mathcal{L}_1 u = \tfrac{du}{dx}$$
$$\mathcal{L}_1^\dagger u = ((1+x)a - (1-x)b)u - (1-x^2)\tfrac{du}{dx}$$
$$\mathcal{L}_2 u = (a+b+n+1)u + (1+x)\tfrac{du}{dx}$$
$$\mathcal{L}_2^\dagger u = (2a + (1-x)n)u - (1-x^2)\tfrac{du}{dx}$$
$$\mathcal{L}_3 u = (a+b+n+1)u - (1-x)\tfrac{du}{dx}$$
$$\mathcal{L}_3^\dagger u = (2b + (1+x)n)u + (1-x^2)\tfrac{du}{dx}$$
$$\mathcal{L}_4 u = ((1+x)a - (1-x)(b+n+1))u - (1-x^2)\tfrac{du}{dx} \qquad \mathcal{L}_4^\dagger u = -nu + (1+x)\tfrac{du}{dx}$$
$$\mathcal{L}_5 u = ((1+x)(a+n+1) - (1-x)b)u - (1-x^2)\tfrac{du}{dx} \qquad \mathcal{L}_5^\dagger u = nu + (1-x)\tfrac{du}{dx}$$
$$\mathcal{L}_6 u = bu + (1+x)\tfrac{du}{dx} \qquad \mathcal{L}_6^\dagger u = au - (1-x)\tfrac{du}{dx}.$$

The notation for the ladder operators is chosen so that $\mathcal{L}_s^\dagger \mathcal{L}_s P_n^{(a,b)}$ and $\mathcal{L}_s \mathcal{L}_s^\dagger P_n^{(a,b)}$ are scalar multiples of $P_n^{(a,b)}$ for $1 \le s \le 6$. These ladder operators are carefully constructed to give rise to sparse recurrence relations for Jacobi polynomials.

Lemma 1 *The ladder operators give sparse recurrence relations for Jacobi polynomials:*

$$\mathcal{L}_1 P_n^{(a,b)} = \tfrac{1}{2}(n+a+b+1)P_{n-1}^{(a+1,b+1)} \qquad \mathcal{L}_1^\dagger P_n^{(a,b)} = 2(n+1)P_{n+1}^{(a-1,b-1)}$$
$$\mathcal{L}_2 P_n^{(a,b)} = (n+a+b+1)P_n^{(a+1,b)} \qquad \mathcal{L}_2^\dagger P_n^{(a,b)} = 2(n+a)P_n^{(a-1,b)}$$
$$\mathcal{L}_3 P_n^{(a,b)} = (n+a+b+1)P_n^{(a,b+1)} \qquad \mathcal{L}_3^\dagger P_n^{(a,b)} = 2(n+b)P_n^{(a,b-1)}$$
$$\mathcal{L}_4 P_n^{(a,b)} = 2(n+1)P_{n+1}^{(a-1,b)} \qquad \mathcal{L}_4^\dagger P_n^{(a,b)} = (n+b)P_{n-1}^{(a+1,b)}$$
$$\mathcal{L}_5 P_n^{(a,b)} = 2(n+1)P_{n+1}^{(a,b-1)} \qquad \mathcal{L}_5^\dagger P_n^{(a,b)} = (n+a)P_{n-1}^{(a,b+1)}$$
$$\mathcal{L}_6 P_n^{(a,b)} = (n+b)P_n^{(a+1,b-1)} \qquad \mathcal{L}_6^\dagger P_n^{(a,b)} = (n+a)P_n^{(a-1,b+1)}.$$

Proof The relationship for $\mathcal{L}_1$ is a formula for the derivative of $P_n^{(a,b)}(x)$ [7, 18.9.15] and relationship $\mathcal{L}_1^\dagger$ is equivalent to [7, 18.9.16]. Six more follow from expressing the left- and right-hand sides in terms of ${}_2F_1$ functions using [7, 18.5.7] and the reflection formula $P_n^{(a,b)}(x) = (-1)^n P_n^{(b,a)}(-x)$: $\mathcal{L}_4^\dagger$ and $\mathcal{L}_4^\dagger$ are equivalent to [7, 15.5.3], $\mathcal{L}_6^\dagger$ is equivalent to [7, 15.5.4], $\mathcal{L}_4$ and $\mathcal{L}_5$ are equivalent to [7, 15.5.5], and $\mathcal{L}_6$ is equivalent to [7, 15.5.6].

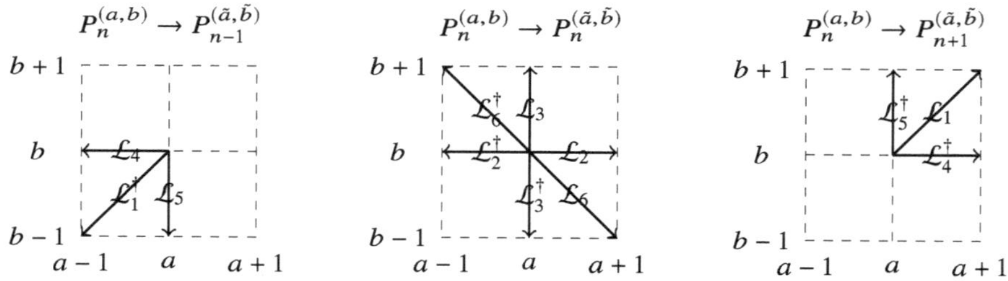

Fig. Illustration of the 12 ladder operators for Jacobi polynomials in Definition 1

The relationship for $\mathcal{L}_2$ follows from combining [7, 18.9.5] and [7, 18.9.6]. The relationships for $\mathcal{L}_2^\dagger$ follows by writing

$$\mathcal{L}_2^\dagger = \mathcal{L}_1^\dagger + (n + a + b)(1 - x)$$

and then using [7, 18.9.5] and [7, 18.9.6]. Finally, $\mathcal{L}_3$ and $\mathcal{L}_3^\dagger$ follow just as $\mathcal{L}_2$ and $\mathcal{L}_2^\dagger$, using the reflection formula.

Remark 1 We note that the first-order differential operators occurring in Lemma 1 form together an action of the Lie algebra sl(4) [1, 5, 6].

Figure illustrates the ladder operators and how they increment or decrement the parameters associated to a Jacobi polynomial.

The ladder operators can be easily adapted to the shifted Jacobi polynomials, denoted by $\tilde{P}_k^{(a,b)}$, which are supported on $(0, 1)$, with x in place of $1 + x$, and no factors of 2. The corresponding recurrence relations for $\tilde{P}_k^{(a,b)}$ are the same as in Lemma 1, except the multiplicative factors of $\frac{1}{2}$ and 2 are replaced by 1.

Ladder Operators for $P_{n,k}^{(a,b,c,d)}$

The 12 ladder operators for the Jacobi polynomials in Sect. allow us to derive 24 ladder operators for $P_{n,k}^{(a,b,c,d)}$. The ladder operators are carefully defined so that they map $P_{n,k}^{(a,b,c,d)}$ to a scalar multiple of $P_{\tilde{n},\tilde{k}}^{(\tilde{a},\tilde{b},\tilde{c},\tilde{d})}$, where the new parameters in $P_{\tilde{n},\tilde{k}}^{(\tilde{a},\tilde{b},\tilde{c},\tilde{d})}$ are n, k, a, b, c or d, respectively, incremented or decremented by 0 or 1.

To highlight the symmetries of the right-angled triangle and make the recurrences more convenient to write down, we define

$$z := 1 - x - y \qquad \text{and} \qquad \frac{\partial}{\partial z} := \frac{\partial}{\partial y} - \frac{\partial}{\partial x},$$

as in [14]. Now, the variables x, y, and z have the convenient property that any affine transformation that maps the triangle onto itself has the effect of exchanging the roles of x, y, and z.

Definition 2 The following operators are ladder operators for $P_{n,k}^{(a,b,c,d)}$. The first set of 12 are:

$$\mathcal{M}_{0,1}u = \frac{\partial u}{\partial y} \qquad \mathcal{M}_{0,1}^{\dagger}u = (yc - zb)u - yz\frac{\partial u}{\partial y}$$

$$\mathcal{M}_{0,2}u = (k+b+c+1)u + y\frac{\partial u}{\partial y} \qquad \mathcal{M}_{0,2}^{\dagger}u = \left(c + k - \frac{yk}{1-x}\right)u - \frac{y}{1-x}z\frac{\partial u}{\partial y}$$

$$\mathcal{M}_{0,3}u = (k+b+c+1)u - x\frac{\partial u}{\partial y} \qquad \mathcal{M}_{0,3}^{\dagger}u = \left(b + \frac{ky}{1-x}\right)u + \frac{y}{1-x}z\frac{\partial u}{\partial y}$$

$$\mathcal{M}_{0,4}u = (yc - z(b+k+1))u - yz\frac{\partial u}{\partial y} \qquad \mathcal{M}_{0,4}^{\dagger}u = -\frac{k}{1-x}u + \frac{y}{1-x}\frac{\partial u}{\partial y}$$

$$\mathcal{M}_{0,5}u = (y(c+k+1))u - zb - yz\frac{\partial u}{\partial y} \qquad \mathcal{M}_{0,5}^{\dagger}u = \frac{k}{1-x}u + \left(1 - \frac{y}{1-x}\right)\frac{\partial u}{\partial y}$$

$$\mathcal{M}_{0,6}u = cu - z\frac{\partial u}{\partial y} \qquad \mathcal{M}_{0,6}^{\dagger}u = bu + y\frac{\partial u}{\partial y}.$$

The second set of 12 are:

$$\mathcal{M}_{1,0}u = \frac{k}{1-x}u + \frac{\partial u}{\partial x} - \frac{y}{1-x}\frac{\partial u}{\partial y},$$

$$\mathcal{M}_{1,0}^{\dagger}u = (x(k+a+b+c+d+1) - a)u - x(1-x)\frac{\partial u}{\partial x} + xy\frac{\partial u}{\partial y}$$

$$\mathcal{M}_{2,0}u = (n+k+a+b+c+d+2)u + \frac{xk}{1-x}u + x\frac{\partial u}{\partial x} - \frac{xy}{1-x}\frac{\partial u}{\partial y}$$

$$\mathcal{M}_{2,0}^{\dagger}u = (n+k+b+c+d+1-xn)u - x(1-x)\frac{\partial u}{\partial x} + xy\frac{\partial u}{\partial y}$$

$$\mathcal{M}_{3,0}u = (n+a+b+c+d+2)u - (1-x)\frac{\partial u}{\partial x} + y\frac{\partial u}{\partial y}$$

$$\mathcal{M}_{3,0}^{\dagger}u = (a+xn)u + x(1-x)\frac{\partial u}{\partial x} - xy\frac{\partial u}{\partial y}$$

$$\mathcal{M}_{4,0}u = (x(n+a+b+c+d+2) - a - n + k - 1)u - x(1-x)\frac{\partial u}{\partial x} + xy\frac{\partial u}{\partial y}$$

$$\mathcal{M}_{4,0}^{\dagger}u = \frac{k}{1-x}u - nu + x\frac{\partial u}{\partial x} - \frac{xy}{1-x}\frac{\partial u}{\partial y}, \qquad \mathcal{M}_{5,0}u = nu + (1-x)\frac{\partial u}{\partial x} - y\frac{\partial u}{\partial y}$$

$$\mathcal{M}_{5,0}^{\dagger}u = x(n+a+b+c+d+2)u - au - x(1-x)\frac{\partial u}{\partial x} + xy\frac{\partial u}{\partial y}$$

$$\mathcal{M}_{6,0}u = au + \frac{xk}{1-x}u + x\frac{\partial u}{\partial x} - \frac{xy}{1-x}\frac{\partial u}{\partial y},$$

$$\mathcal{M}_{6,0}^{\dagger}u = (k+b+c+d+1)u - (1-x)\frac{\partial u}{\partial x} + y\frac{\partial u}{\partial y}.$$

The notation for the ladder operators is chosen so that the recurrence relations in Theorem 1 are derived for $\mathcal{M}_{s,0}$ (resp. $\mathcal{M}_{0,s}$) by applying $\mathcal{L}_s$ or $\mathcal{L}_s^{\dagger}$ to the first (resp. second) Jacobi polynomial in $P_{n,k}^{(a,b,c,d)}(x,y)$ for $1 \le s \le 6$. Moreover, we know that $\mathcal{M}_{s,0}^{\dagger}\mathcal{M}_{s,0}P_{n,k}^{(a,b,c,d)}$, $\mathcal{M}_{0,s}^{\dagger}\mathcal{M}_{0,s}P_{n,k}^{(a,b,c,d)}$, $\mathcal{M}_{s,0}\mathcal{M}_{s,0}^{\dagger}P_{n,k}^{(a,b,c,d)}$, and $\mathcal{M}_{0,s}\mathcal{M}_{0,s}^{\dagger}P_{n,k}^{(a,b,c,d)}$ are scalar multiples of $P_{n,k}^{(a,b,c,d)}$ for $1 \le s \le 6$.

The ladder operators in Definition 2 correspond to 24 sparse recurrence relations for $P_{n,k}^{(a,b,c,d)}$. To derive these recurrences, we first express the partial derivatives of $P_{n,k}^{(a,b,c,d)}$ as derivatives of shifted Jacobi polynomials.

Proposition 1 *The following relationships hold:*

$$\tilde{P}_{n-k}^{(2k+b+c+d+1,a)}(x)(1-x)^k\left[\tilde{P}_k^{(c,b)}\right]'\left(\tfrac{y}{1-x}\right) = (1-x)\tfrac{\partial}{\partial y}P_{n,k}^{(a,b,c,d)}(x,y), \tag{2}$$

$$\left[\tilde{P}_{n-k}^{(2k+b+c+d+1,a)}\right]'(x)(1-x)^{k+1}\tilde{P}_k^{(c,b)}\left(\tfrac{y}{1-x}\right) = \left(k+(1-x)\tfrac{\partial}{\partial x} - y\tfrac{\partial}{\partial y}\right)P_{n,k}^{(a,b,c,d)}(x,y). \tag{3}$$

Proof The first relationship is immediate. The second relationship follows from the chain-rule:

$$(1-x)\tfrac{\partial}{\partial x}\left(f(x)(1-x)^k g\left(\tfrac{y}{1-x}\right)\right) = f'(x)(1-x)^{k+1}g\left(\tfrac{y}{1-x}\right)$$
$$- kf(x)(1-x)^k g(x) + yf(x)(1-x)^{k-1}g'\left(\tfrac{y}{1-x}\right)$$

and an application of (2) to simplify the last term.

The 24 sparse recurrence relations for $P_{n,k}^{(a,b,c,d)}$ are given in the following theorem.

Theorem 1 *Let $t = a+b+c+d$. The first set of 12 are:*

$$\begin{aligned}
\mathcal{M}_{0,1}P_{n,k}^{(a,b,c,d)} &= (k+b+c+1)P_{n-1,k-1}^{(a,b+1,c+1,d)} & \mathcal{M}_{0,1}^{\dagger}P_{n,k}^{(a,b,c,d)} &= (k+1)P_{n+1,k+1}^{(a,b-1,c-1,d)}\\
\mathcal{M}_{0,2}P_{n,k}^{(a,b,c,d)} &= (k+b+c+1)P_{n,k}^{(a,b,c+1,d-1)} & \mathcal{M}_{0,2}^{\dagger}P_{n,k}^{(a,b,c,d)} &= (k+c)P_{n,k}^{(a,b,c-1,d+1)}\\
\mathcal{M}_{0,3}P_{n,k}^{(a,b,c,d)} &= (k+b+c+1)P_{n,k}^{(a,b+1,c,d-1)} & \mathcal{M}_{0,3}^{\dagger}P_{n,k}^{(a,b,c,d)} &= (k+b)P_{n,k}^{(a,b-1,c,d+1)}\\
\mathcal{M}_{0,4}P_{n,k}^{(a,b,c,d)} &= (k+1)P_{n+1,k+1}^{(a,b,c-1,d-1)} & \mathcal{M}_{0,4}^{\dagger}P_{n,k}^{(a,b,c,d)} &= (k+b)P_{n-1,k-1}^{(a,b,c+1,d+1)}\\
\mathcal{M}_{0,5}P_{n,k}^{(a,b,c,d)} &= (k+1)P_{n+1,k+1}^{(a,b-1,c,d-1)} & \mathcal{M}_{0,5}^{\dagger}P_{n,k}^{(a,b,c,d)} &= (k+c)P_{n-1,k-1}^{(a,b+1,c,d+1)}\\
\mathcal{M}_{0,6}P_{n,k}^{(a,b,c,d)} &= (k+c)P_{n,k}^{(a,b+1,c-1,d)} & \mathcal{M}_{0,6}^{\dagger}P_{n,k}^{(a,b,c,d)} &= (k+b)P_{n,k}^{(a,b-1,c+1,d)}.
\end{aligned}$$

The second set of 12 are:

$$\begin{aligned}
\mathcal{M}_{1,0}P_{n,k}^{(a,b,c,d)} &= (n+k+t+2)P_{n-1,k}^{(a+1,b,c,d+1)}\\
\mathcal{M}_{1,0}^{\dagger}P_{n,k}^{(a,b,c,d)} &= (n-k+1)P_{n+1,k}^{(a-1,b,c,d-1)}\\
\mathcal{M}_{2,0}P_{n,k}^{(a,b,c,d)} &= (n+k+t+2)P_{n,k}^{(a,b,c,d+1)}\\
\mathcal{M}_{2,0}^{\dagger}P_{n,k}^{(a,b,c,d)} &= (n+k+t-a+1)P_{n,k}^{(a,b,c,d-1)}
\end{aligned}$$

$$
\begin{aligned}
\mathcal{M}_{3,0} P_{n,k}^{(a,b,c,d)} &= (n+k+t+2) P_{n,k}^{(a+1,b,c,d)} \\
\mathcal{M}_{3,0}^{\dagger} P_{n,k}^{(a,b,c,d)} &= (n-k+a) P_{n,k}^{(a-1,b,c,d)} \\
\mathcal{M}_{4,0} P_{n,k}^{(a,b,c,d)} &= (n-k+1) P_{n+1,k}^{(a,b,c,d-1)} \\
\mathcal{M}_{4,0}^{\dagger} P_{n,k}^{(a,b,c,d)} &= (n-k+a) P_{n-1,k}^{(a,b,c,d+1)} \\
\mathcal{M}_{5,0} P_{n,k}^{(a,b,c,d)} &= (n+k+t-a+1) P_{n-1,k}^{(a+1,b,c,d)} \\
\mathcal{M}_{5,0}^{\dagger} P_{n,k}^{(a,b,c,d)} &= (n-k+1) P_{n+1,k}^{(a-1,b,c,d)} \\
\mathcal{M}_{6,0} P_{n,k}^{(a,b,c,d)} &= (n-k+a) P_{n,k}^{(a-1,b,c,d+1)} \\
\mathcal{M}_{6,0}^{\dagger} P_{n,k}^{(a,b,c,d)} &= (n+k+t-a+1) P_{n,k}^{(a+1,b,c,d-1)}.
\end{aligned}
$$

Proof We present the proof of $\mathcal{M}_{0,1} P_{n,k}^{(a,b,c,d)} = (k+b+c+1) P_{n-1,k-1}^{(a,b+1,c+1,d)}$. By the definition of $P_{n,k}^{(a,b,c,d)}$ in (1), the chain rule, and the relationship in (2), we have

$$
\begin{aligned}
\frac{\partial}{\partial y} P_{n,k}^{(a,b,c,d)}(x,y) &= \tilde{P}_{n-k}^{(2k+b+c+d+1,a)}(x)(1-x)^{k-1}[\tilde{P}_k^{(c,b)}]'\left(\tfrac{y}{1-x}\right) \\
&= (k+c+b+1) \tilde{P}_{n-k}^{(2k+b+c+d+1,a)}(x)(1-x)^{k-1} \tilde{P}_{k-1}^{(c+1,b+1)}\left(\tfrac{y}{1-x}\right),
\end{aligned} \tag{4}
$$

where the last equality comes from applying $\mathcal{L}_1$ in Definition 1. The final expression in (4) is equivalent to $(k+b+c+1) P_{n-1,k-1}^{(a,b+1,c+1,d)}$. The manipulations for the remaining recurrence relations are similar, except with different choices of the operators $\mathcal{L}_s$ or $\mathcal{L}_s^{\dagger}$ and combinations of (2) and (3).

Sparse Recurrence Relations for $P_{n,k}^{(a,b,c)}$

We can combine the ladder operators in Sect. to derive sparse recurrence relations between $P_{n,k}^{(a,b,c)}$ polynomials with different parameters and their partial derivatives. These recurrence relations are analogous to many of the sparse recurrence relations for Jacobi polynomials [7, §18.9].

Differentiation

The partial derivatives of $P_{n,k}^{(a,b,c)}$ can be written in terms of Jacobi polynomials on the triangle with incremented parameters, which is analogous to a recurrence relation for the derivative of a Jacobi polynomial [7, 18.9.15]. A similar recurrence for $P_{n,k}^{(a,b,c)}$ can be found in [14, Prop. 4.6, 4.7, & 4.8].

Corollary 1 *The following recurrence relations hold:*

$$(2k+b+c+1)\tfrac{\partial}{\partial x}P_{n,k}^{(a,b,c)} = (n+k+a+b+c+2)(k+b+c+1)P_{n-1,k}^{(a+1,b,c+1)} + (k+b)(n+k+b+c+1)P_{n-1,k-1}^{(a+1,b,c+1)}, \tag{5}$$

$$\tfrac{\partial}{\partial y}P_{n,k}^{(a,b,c)} = (k+b+c+1)P_{n-1,k-1}^{(a,b+1,c+1)}, \tag{6}$$

$$(2k+b+c+1)\tfrac{\partial}{\partial z}P_{n,k}^{(a,b,c)} = -(n+k+a+b+c+2)(k+b+c+1)P_{n-1,k}^{(a+1,b+1,c)} + (k+c)(n+k+b+c+1)P_{n-1,k-1}^{(a+1,b+1,c)}. \tag{7}$$

Proof The recurrence (5) follows from the fact that $(\mathcal{M}_{1,0}\mathcal{M}_{0,2} + \mathcal{M}_{0,4}^{\dagger}\mathcal{M}_{6,0}^{\dagger})u = (2k+b+c+1)\frac{\partial u}{\partial x}$ when $d=0$. The relationship (6) is equivalent to the relation given by $\mathcal{M}_{0,1}$ in Theorem 1 when $d=0$. Finally, (7) follows from the fact that $(\mathcal{M}_{1,0}\mathcal{M}_{0,3} - \mathcal{M}_{0,5}^{\dagger}\mathcal{M}_{6,0}^{\dagger})u = -(2k+b+c+1)\frac{\partial u}{\partial z}$.

The derivatives of weighted versions of $P_{n,k}^{(a,b,c)}$ also satisfy sparse recurrence relations, which are analogous to an expression for the derivative of a weighted Jacobi polynomial [7, 18.9.16].

Corollary 2 *The following recurrence relations hold:*

$$-(2k+b+c+1)\tfrac{\partial}{\partial x}\left(x^a y^b z^c P_{n,k}^{(a,b,c)}\right) = x^{a-1}y^b z^{c-1}\Big((k+c)(n-k+1)P_{n+1,k}^{(a-1,b,c-1)} + (k+1)(n-k+a)P_{n+1,k+1}^{(a-1,b,c-1)}\Big),$$

$$\tfrac{\partial}{\partial y}\left(x^a y^b z^c P_{n,k}^{(a,b,c)}\right) = -(k+1)x^a y^{b-1} z^{c-1} P_{n+1,k+1}^{(a,b-1,c-1)},$$

$$(2k+b+c+1)\tfrac{\partial}{\partial z}\left(x^a y^b z^c P_{n,k}^{(a,b,c)}\right) = x^{a-1}y^{b-1}z^c\Big((k+b)(n-k+1)P_{n+1,k}^{(a-1,b-1,c)} - (k+1)(n-k+a)P_{n+1,k+1}^{(a-1,b-1,c)}\Big).$$

Proof The first recurrence follows from

$$(\mathcal{M}_{0,2}^{\dagger}\mathcal{M}_{1,0}^{\dagger} + \mathcal{M}_{0,4}\mathcal{M}_{6,0})u = (2k+b+c+1)(cx - az - xz\tfrac{\partial}{\partial x})u = -(2k+b+c+1)x^{1-a}z^{1-c}\tfrac{\partial}{\partial x}(x^a z^c u).$$

The second recurrence holds since

$$\mathcal{M}_{0,1}^{\dagger}u = (cy - bz - yz\tfrac{\partial}{\partial y})u = -y^{1-b}z^{1-c}\tfrac{\partial}{\partial x}(y^b z^c u).$$

The third recurrence is derived from the fact that

$$(\mathcal{M}_{0,3}^{\dagger}\mathcal{M}_{1,0}^{\dagger} - \mathcal{M}_{0,5}\mathcal{M}_{6,0})u = (2k+b+c+1)(bx - ay + xy(\tfrac{\partial}{\partial y} - \tfrac{\partial}{\partial x}))u$$
$$= (2k+b+c+1)x^{1-a}y^{1-b}z^{-c}\tfrac{\partial}{\partial z}(x^a y^b z^c u).$$

Conversion

Recurrence relations for conversion allow us to express $P_{n,k}^{(a,b,c)}$ in terms of Jacobi polynomials on the triangle with different parameters. Here, we give the recurrence relations that increment the parameters, which are analogues of [7, 18.9.3]. Similar relations can be found in [14, Prop. 4.4].

Corollary 3 *The following recurrence relations hold:*

$$(2n+a+b+c+2)P_{n,k}^{(a,b,c)} = (n+k+a+b+c+2)P_{n,k}^{(a+1,b,c)}$$
$$+ (n+k+b+c+1)P_{n-1,k}^{(a+1,b,c)}, \tag{8}$$
$$(2n+a+b+c+2)(2k+b+c+1)P_{n,k}^{(a,b,c)}$$
$$= (n+k+a+b+c+2)(k+b+c+1)P_{n,k}^{(a,b+1,c)}$$
$$- (n-k+a)(k+b+c+1)P_{n-1,k}^{(a,b+1,c)} + (k+c)(n+k+b+c+1)P_{n-1,k-1}^{(a,b+1,c)}$$
$$- (k+c)(n-k+1)P_{n,k-1}^{(a,b+1,c)}, \tag{9}$$
$$(2n+a+b+c+2)(2k+b+c+1)P_{n,k}^{(a,b,c)}$$
$$= (n+k+a+b+c+2)(k+b+c+1)P_{n,k}^{(a,b,c+1)}$$
$$- (n-k+a)(k+b+c+1)P_{n-1,k}^{(a,b,c+1)} - (k+b)(n+k+b+c+1)P_{n-1,k-1}^{(a,b,c+1)}$$
$$+ (k+b)(n-k+1)P_{n,k-1}^{(a,b,c+1)}. \tag{10}$$

Proof The recurrence relation in (8) follows from the fact that $(\mathcal{M}_{30} + \mathcal{M}_{50})u = (2n+a+b+c+2)u$ when $d = 0$. Since $(\mathcal{M}_{2,0} - \mathcal{M}_{4,0}^{\dagger})u = (2n+a+b+c+d+2)u$ and $(\mathcal{M}_{2,0}^{\dagger} - \mathcal{M}_{4,0})u = (2n+a+b+c+d+2)(1-x)u$, we obtain

$$(\mathcal{M}_{0,2}\mathcal{M}_{2,0} - \mathcal{M}_{0,2}\mathcal{M}_{4,0}^{\dagger} - \mathcal{M}_{0,4}^{\dagger}\mathcal{M}_{2,0}^{\dagger} + \mathcal{M}_{0,4}^{\dagger}\mathcal{M}_{4,0})u = (2k+b+c+1)(2n+a+b+c+d+2)u,$$

The recurrence relation (10) immediately follows. Similarly, (9) holds since

$$(\mathcal{M}_{0,3}\mathcal{M}_{2,0} - \mathcal{M}_{0,3}\mathcal{M}_{4,0}^{\dagger} + \mathcal{M}_{0,5}^{\dagger}\mathcal{M}_{2,0}^{\dagger} - \mathcal{M}_{0,5}^{\dagger}\mathcal{M}_{4,0})u = (2k+b+c+1)(2n+a+b+c+d+2)u.$$

Multiplication

Recurrence relations for multiplication allow one to express $xP_{n,k}^{(a,b,c)}$, $yP_{n,k}^{(a,b,c)}$, and $zP_{n,k}^{(a,b,c)}$ in terms of a sum of Jacobi polynomials on the triangle with potentially different parameters. The recurrences in Corollary 4 are analogous to the recurrence relations for $P_n^{(a,b)}$ found in [7, 18.9.6].

Corollary 4 *The following recurrence relations hold:*

$$(2n+a+b+c+2)xP_{n,k}^{(a,b,c)} = (n-k+a)P_{n,k}^{(a-1,b,c)} + (n-k+1)P_{n+1,k}^{(a-1,b,c)}, \tag{11}$$

$$\begin{aligned}(2k+b+c+1)(2n+a+b+c+2)yP_{n,k}^{(a,b,c)} &= (k+b)(n+k+b+c+1)P_{n,k}^{(a,b-1,c)} \\ &- (k+1)(n-k+a)P_{n,k+1}^{(a,b-1,c)} - (k+b)(n-k+1)P_{n+1,k}^{(a,b-1,c)} \\ &+ (k+1)(n+k+a+b+c+2)P_{n+1,k+1}^{(a,b-1,c)},\end{aligned} \tag{12}$$

$$\begin{aligned}(2k+b+c+1)(2n+a+b+c+2)zP_{n,k}^{(a,b,c)} &= (k+c)(n+k+b+c+1)P_{n,k}^{(a,b,c-1)} \\ &+ (k+1)(n-k+a)P_{n,k+1}^{(a,b,c-1)} - (k+c)(n-k+1)P_{n+1,k}^{(a,b,c-1)} \\ &- (k+1)(n+k+a+b+c+2)P_{n+1,k+1}^{(a,b,c-1)}.\end{aligned} \tag{13}$$

Proof The recurrence relation in (11) follows from the fact that $(\mathcal{M}_{3,0}^{\dagger}+\mathcal{M}_{5,0}^{\dagger})u = (2n+a+b+c+d+2)xu$. Since

$$(\mathcal{M}_{03}^{\dagger}\mathcal{M}_{20}^{\dagger}-\mathcal{M}_{03}^{\dagger}\mathcal{M}_{40}+\mathcal{M}_{05}\mathcal{M}_{20}-\mathcal{M}_{05}\mathcal{M}_{40}^{\dagger})u = (2k+b+c+1)(2n+a+b+c+d+2)yu$$

holds, we find that (12) is satisfied. Finally, (13) follows from

$$(\mathcal{M}_{02}^{\dagger}\mathcal{M}_{20}^{\dagger}-\mathcal{M}_{02}^{\dagger}\mathcal{M}_{40}+\mathcal{M}_{04}\mathcal{M}_{40}^{\dagger}-\mathcal{M}_{04}\mathcal{M}_{20})u = (2k+b+c+1)(2n+a+b+c+d+2)zu.$$

Combining the recurrence relations in Corollaries 3 and 4, we can derive expressions for $xP_{n,k}^{(a,b,c)}$, $yP_{n,k}^{(a,b,c)}$, and $zP_{n,k}^{(a,b,c)}$ in terms of a sum of Jacobi polynomials on the triangle with parameters (a, b, c). These are analogous to the three-term recurrence relation for Jacobi polynomials [7, 18.9.2]. Since these recurrence relations are long, we refer the reader to [2, pp. 80–81].

Differential Eigenvalue Problems

The polynomials $P_{n,k}^{(a,b,c)}$ are eigenfunctions for second-order differential operators (see [2, (5.3.4)] and [14, Prop. 4.11]), and the ladder operators.

Theorem 2 *The polynomial $P_{n,k}^{(a,b,c)}$ satisfies two second-order differential eigenproblems:*

$$zy\frac{\partial^2}{\partial y^2}P_{n,k}^{(a,b,c)} + ((1+b)(1-x)-(2+b+c)y)\frac{\partial}{\partial y}P_{n,k}^{(a,b,c)} = -k(k+b+c+1)P_{n,k}^{(a,b,c)}$$

and

$$\begin{aligned} &x(1-x)\frac{\partial^2}{\partial x^2}P_{n,k}^{(a,b,c)} - 2xy\frac{\partial^2}{\partial x\partial y}P_{n,k}^{(a,b,c)} + y(1-y)\frac{\partial^2}{\partial y^2}P_{n,k}^{(a,b,c)} \\ &\quad + (a+1-(a+b+c+3)x)\frac{\partial}{\partial x}P_{n,k}^{(a,b,c)} + (b+1-(a+b+c+3)y)\frac{\partial}{\partial y}P_{n,k}^{(a,b,c)} \\ &= -n(n+a+b+c+d+2)P_{n,k}^{(a,b,c,d)}(x,y). \end{aligned}$$

Proof The first equation follows from

$$\mathcal{M}_{0,1}\mathcal{M}_{0,1}^{\dagger}P_{n,k}^{(a,b,c,d)}(x,y) = k(k+b+c+1)P_{n,k}^{(a,b,c,d)}(x,y)$$

and the second from

$$\left[\frac{\mathcal{M}_{0,1}\mathcal{M}_{0,1}^{\dagger} - k(k+b+c+1)}{1-x} + \mathcal{M}_{3,0}\mathcal{M}_{3,0}^{\dagger} + \mathcal{M}_{4,0}\mathcal{M}_{4,0}^{\dagger}\right] P_{n,k}^{(a,b,c,d)}(x,y)$$
$$= (1+a-k+n)(3+t+2n)P_{n,k}^{(a,b,c,d)}(x,y).$$

Application: Calculating Laplacians

We can use the recurrence relationships in this paper to calculate partial derivatives too. Slevinsky's fast triangle transform [11] (which builds on his fast spherical harmonic transform [10]) as implemented in the FastTransform multithreaded C code [12] gives an efficient and stable routine for calculating the expansion coefficients on the triangle in $O(d^2 \log^2 d)$ operations, where d is the polynomial degree. The partial derivative recurrences (see Corollary 1) show us how to calculate the expansion coefficients of $\frac{\partial p}{\partial x}$ and $\frac{\partial p}{\partial y}$ with coefficients associated to the parameters $(a+1, b, c+1)$ and $(a, b+1, c+1)$, respectively. Moreover, Corollary 3 informs us how to convert from expansions with parameters (a, b, c) to $(a+1, b, c)$, $(a, b+1, c)$ and $(a, b, c+1)$. We can combine these various recurrences relations to compute

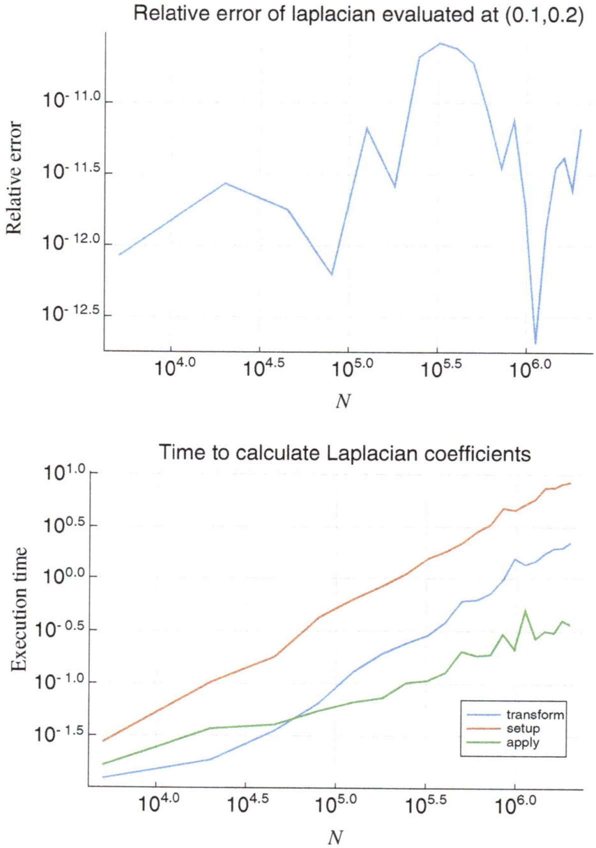

Fig. Top: Error when evaluating the Laplacian of $f(x, y) = \cos(nxy/40)$ at $(x, y) = (0.1, 0.2)$ by expanding in degree $N = (n + 1)(n + 2)/2$ Jacobi polynomials on the triangle with $(a, b, c) = (0, 0, 0)$ and using the recurrences. Bottom: Execution times of (1) the fast transform, (2) constructing the recurrences as 8 banded-block-banded matrices, and (3) applying the matrices

the coefficients of the Laplacian in the basis $P_{n,k}^{(2,2,2)}$ in an optimal complexity of $O(N) = O(d^2)$ operations, where d is the polynomial degree and $N = d(d+1)/2$ is the total number of degrees of freedom.

For example, the Laplacian of $f(x, y) = \cos(nxy/40)$ can be computed by first approximating f on the unit right-angled triangle to within machine precision by a polynomial, and then employing various recurrence relationships to calculate its Laplacian (Fig.). To do this efficiently, we store the recurrence relations as banded-block-banded matrices to take advantage of fast banded matrix-vector multiplication.

Conclusion

We introduce ladder operators for systematically deriving sparse recurrence relations for differentiation, conversion, and multiplication of Jacobi and orthogonal polynomials on the triangle. We use these recurrences to efficiently apply partial differential operators, in particular for calculating Laplacians. The importance of these relationships is that they allow general linear partial differential operators with polynomial coefficients to be represented as sparse operators acting on orthogonal polynomial expansions. This application will be the topic of a subsequent paper.

References

1. Derezinski, J., Majewski, P.: From conformal group to symmetries of hypergeometric type equations. Symmetry Integrability Geometry: Methods Appl. **12**, 108 (2016)
2. Dunkl, C.F., Xu, Y.: Orthogonal Polynomials of Several Variables, 2nd edn. Cambridge University Press, Cambridge (2014)
3. Karniadakis, G., Sherwin, S.: Spectral/hp Element Methods for Computational Fluid Dynamics. Oxford University Press, Oxford (2013)
4. Koornwinder, T.: Two-variable analogues of the classical orthogonal polynomials. In: Theory and Application of Special Functions (Proceedings of an Advanced Seminar Sponsored by the Mathematics Research Center, the University of Wisconsin, Madison, 1975), pp. 435–495
5. Miller, Jr., W.: Lie theory and generalizations of the hypergeometric functions. SIAM J. Appl. Math. **25**, 226–235 (1973)
6. Miller, Jr., W.: Symmetries of differential equations. The hypergeometric and Euler–Darboux equations. SIAM J. Math. Anal. **4**, 314–328 (1973)
7. Olver, F.W.J., Lozier, D.W., Boisvert, R.F, Clark, C.W. (eds): NIST Handbook of Mathematical Functions. Cambridge University Press, Cambridge (2010)
8. Olver, S., Townsend, A.: A fast and well-conditioned spectral method. SIAM Rev. **55**, 462–489 (2013)
9. Proriol, J.: Sur une famille de polynômes á deux variables orthogonaux dans un triangle. C. R. Acad. Sci. Paris **257**, 2459–2461 (1957)
10. Slevinsky, R.M.: Fast and backward stable transforms between spherical harmonic expansions and bivariate Fourier series. Appl. Comp. Harm. Anal. **47**(3), 585–606 (2019)
11. Slevinsky, R.M.: Conquering the pre-computation in two-dimensional harmonic polynomial transforms. Preprint arXiv:1711.07866 (2017)
12. Slevinsky, R.M.: FastTransforms. v0.1. https://github.com/MikaelSlevinsky/FastTransforms
13. Vasil, G.M., Burns, K.J., Lecoanet, D., Olver, S., Brown, B.P., Oishi, J.S.: Tensor calculus in polar coordinates using Jacobi polynomials. J. Comp. Phys. **325**, 53–73 (2016)
14. Xu, Y.: Approximation and orthogonality in Sobolev spaces on a triangle. Constr. Approx. **46**, 349–434 (2017)

Part II
Contributed Papers

4

On the Order Reduction of Entropy Stable DGSEM for the Compressible Euler Equations

Florian J. Hindenlang and Gregor J. Gassner

Introduction

Discontinuous Galerkin spectral element collocation method (DGSEM) with either Legendre-Gauss or Legendre-Gauss-Lobatto (LGL) nodes (see e.g. [14]) are among the most efficient variants in the class of element based high order methods, such as e.g. discontinuous Galerkin, flux reconstruction, or summation-by-parts (SBP) finite differences. In particular, the LGL variant, starting in [9], turned out to be similar to a SBP finite difference approximation with simultaneous-approximate-term technique (SAT). This relationship allowed to construct conservative skew-symmetric approximations, e.g. [9, 10, 21], and later enabled DGSEM-LGL approximations that are discretely entropy stable, e.g. [1, 3, 6, 8, 13, 17, 19, 20], and/or kinetic energy preserving [12]. These novel variants of nodal split form DG methods feature drastically increased non-linear robustness towards aliasing induced instabilities and favourable properties regarding the simulation of unresolved turbulence, e.g. [7, 23].

In addition to the very robust dissipative entropy stable versions, it is also possible to construct virtually dissipation free variants by choosing appropriate element interface numerical fluxes. These entropy conserving variants all show an odd-even behavior when experimentally testing the order of convergence, e.g. [9, 21], where the observed convergence order for even polynomial degrees

F. J. Hindenlang
Max Planck Institute for Plasma Physics, Garching, Germany
e-mail: florian.hindenlang@ipp.mpg.de

G. J. Gassner (✉)
Department for Mathematics and Computer Science, Center for Data and Simulation Science, University of Cologne, Cologne, Germany
e-mail: ggassner@math.uni-koeln.de

N is N and for odd N is $N + 1$. Lately, a discussion emerged in the community, with interesting debates during the recent ICOSAHOM conference in London, where researchers reported non-optimal convergence behavior of the entropy stable DGSEM-LGL even with dissipative numerical surface fluxes, e.g. [6].

This paper contributes to this discussion and presents results of an experimental convergence order study for the compressible Euler equations with (1) the standard DGSEM with either Gauss and LGL nodes, (2) the entropy stable DGSEM with LGL nodes. For these nodal schemes, we test the convergence order with different numerical surface fluxes and report the results depending on the Mach number of the test case. The remainder of the paper is organized as follows: in the next section we describe the numerical model for our numerical experiments, in Sect. we present our observed experimental convergence orders for different configurations and draw our conclusion in Sect.

Numerical Model

We consider the compressible Euler equations defined in the domain $\Omega \subset \mathbb{R}^3$

$$\mathbf{u}_t + \sum_{i=1}^{3} \frac{\partial \mathbf{f}_i}{\partial x_i} = \mathbf{0}. \tag{1}$$

The state vector contains the conservative variables and the advective flux components are

$$\mathbf{u} = \begin{bmatrix} \varrho \\ \varrho\vec{v} \\ E \end{bmatrix} = \begin{bmatrix} \varrho \\ \varrho v_1 \\ \varrho v_2 \\ \varrho v_3 \\ E \end{bmatrix}, \quad \mathbf{f}_1 = \begin{bmatrix} \varrho v_1 \\ \varrho v_1^2 + p \\ \varrho v_1 v_2 \\ \varrho v_1 v_3 \\ (E+p)v_1 \end{bmatrix}, \quad \mathbf{f}_2 = \begin{bmatrix} \varrho v_2 \\ \varrho v_2 v_1 \\ \varrho v_2^2 + p \\ \varrho v_2 v_3 \\ (E+p)v_2 \end{bmatrix}, \quad \mathbf{f}_3 = \begin{bmatrix} \varrho v_3 \\ \varrho v_3 v_1 \\ \varrho v_3 v_2 \\ \varrho v_3^2 + p \\ (E+p)v_3 \end{bmatrix}. \tag{2}$$

Here, ϱ, $\vec{v} = (v_1, v_2, v_3)^T$, p, E are the mass density, fluid velocities, pressure and total energy. We close the system with the ideal gas assumption, which relates the total energy and pressure

$$p = (\gamma - 1)\left(E - \frac{1}{2}\varrho \left\|\vec{v}\right\|^2\right), \tag{3}$$

where γ denotes the adiabatic coefficient.

For our discretization, we subdivide the domain into non-overlapping hexahedral elements. For each element, we define a transfinite mapping to a unit reference space and use this mapping to transform the Eq. (1) from physical to reference space. A weak form is created by taking the inner product of the transformed equation with a test function. We use integration-by-parts for the flux term and approximate the resulting weak form as follows: the conservative variables are approximated by a polynomial in reference space with degree N, interpolated at the Gauss or LGL nodes. The volume fluxes are replaced by a standard interpolation of the non-linear flux function at the same Gauss/LGL nodes (standard DGSEM-Gauss or DGSEM-LGL), see e.g. [14]. For the LGL variant, we are also able to introduce the split form volume integral based on entropy conserving and kinetic energy preserving numerical volume fluxes (Split-DGSEM), e.g. [12] and [22], resulting in either the entropy conserving or entropy stable DGSEM variants, depending on the choice of numerical surface flux.

Convergence Results

In this section, we compare the convergence of the standard DGSEM and the entropy conservative and entropy stable discretization for different choices of the numerical flux and polynomial degrees $N = 2, 3, 4, 5$.

We choose the test case of a two-dimensional density wave, with a constant pressure and transported with a constant velocity, which was proposed for one-dimensional convergence tests in [4]. The density evolves as

$$\varrho(x_1, x_2, t) = 1 + 0.1 \sin\Big(\pi \left((x_1 - v_1 t) + (x_2 - v_2 t)\right)\Big) \tag{4}$$

with a prescribed velocity (v_1, v_2). The pressure is chosen as $p = 1/\gamma$ with $\gamma = 1.4$, so that the sound speed ranges between $c = 0.95 \ldots 1.05$. Thus, by changing the velocity, we change the Mach number of the flow $\mathrm{Ma} = |\vec{v}|/c$. Three Mach numbers are chosen: $\mathrm{Ma} \approx 0.2$ with $(v_1, v_2) = (0.1, 0.15)$, $\mathrm{Ma} \approx 1.0$ with $(v_1, v_2) = (0.7, 0.65)$ and $\mathrm{Ma} \approx 3.5$ with $(v_1, v_2) = (2.5, 2.4)$. The experimental order of convergence (EOC) is computed with the L_2 error of the density at $t = 1$.

The convergence study is performed with the open source, three-dimensional curvilinear split-form DG framework FLUXO (www.github.com/project-fluxo). As the test case is two-dimensional, we use fully periodic cartesian meshes of the domain $[-1, 1]^3$ with an equal number of elements in x- and y-directions and always one element in z-direction. Note that h_0 in the convergence tables refers to the coarsest mesh level, which is 4^2 elements for $N = 2, 3$ ($h_0 = 1/2$) and 2^2 elements for $N = 4, 5$ ($h_0 = 1$).

All simulation results are obtained with an explicit five stage, fourth order accurate low storage Runge–Kutta scheme [2], where a stable time step is computed according to the adjustable coefficient $CFL \in (0, 1]$ the local maximum wave speed, and the relative grid size, e.g. [11]. We made sure that the time integrator did not influence the spatial convergence order, by adjusting the CFL number accordingly.

Standard DGSEM

The convergence of the standard DGSEM with Gauss-Legendre nodes (DGSEM-Gauss) and with Legendre-Gauss-Lobatto (DGSEM-LGL) is shown in Tables for the three Mach numbers and two choices of the numerical flux, namely the HLL (Harten, Lax, van Leer) flux and the Roe flux. The results of the LLF (local Lax-Friedrichs) flux and the HLLC flux (HLL variant with three waves, C for 'contact' wave) are reported in the Appendix, as the HLL results are similar to LLF, and HLLC behaves exactly the same as Roe, see Tables . Details on the properties and the implementation of the LLF, HLL, HLLC, and Roe fluxes are found in the book of Toro [18] and the references therein.

For the HLL flux and the low Mach number Ma $= 0.2$, we observe an odd-even behavior with an *order reduction* for even polynomial degrees $N = 2, 4$. Also for Ma $= 1.0$, the convergence for even degrees is slightly affected, whereas for the high Mach number, all fluxes converge with full order. Comparing the L_2 errors of the finest mesh for HLL and Roe for the low Mach number, HLL is less accurate for $N = 2, 4$ and more accurate for $N = 3, 5$.

All numerical fluxes are approximate Riemann solvers, but the LLF and HLL only use the maximum wave speeds, whereas the HLLC and Roe also take the contact wave into account, and therefore keep the full order of the scheme for all Mach numbers for this test case.

Entropy Conservative and Entropy Stable DGSEM

Now, we investigate the order reduction of the entropy conservative and entropy stable discretizations. Here, the standard DGSEM volume integral is replaced by split-form formulation (Split-DGSEM) using a two-point entropy conservative and kinetic energy preserving flux (ECKEP). If we choose the ECKEP flux at the surface, we get an entropy-conserving scheme. For entropy stability, we can use the LLF or HLL flux directly at the surface, or use the ECKEP flux and add a dissipation term, which must still satisfy the entropy inequality condition. In Winters et al. [22], such dissipation terms are carefully derived, using either only the maximum wave

Table Experimental order of convergence of L_2 error to the exact density (4), using the standard DGSEM-Gauss with HLL and Roe fluxes

Volume disc. + Surface flux	Mach	Mesh level	$N = 2$		$N = 3$		$N = 4$		$N = 5$	
			$L_2(\varrho)$	EOC	$L_2(\varrho)$	EOC	$L_2(\varrho)$	EOC	$L_2(\varrho)$	EOC
DGSEM-Gauss + HLL	3.5	$h_0/2$	1.87e−04	3.34	8.57e−06	4.02	1.03e−05	5.02	6.76e−07	6.07
		$h_0/4$	2.27e−05	3.04	5.35e−07	4.00	3.30e−07	4.96	1.07e−08	5.99
		$h_0/8$	2.82e−06	3.00	3.34e−08	4.00	1.02e−08	5.01	1.66e−10	6.01
		$h_0/16$	3.53e−07	3.00 ■	2.09e−09	4.00 ■	3.22e−10	4.99 ■	2.60e−12	5.99 ■
	1.0	$h_0/2$	2.60e−04	2.55	5.92e−06	4.55	1.15e−05	4.39	6.74e−07	6.84
		$h_0/4$	3.74e−05	2.80	3.27e−07	4.18	4.94e−07	4.54	7.23e−09	6.54
		$h_0/8$	4.88e−06	2.94	1.95e−08	4.07	1.73e−08	4.84	9.98e−11	6.18
		$h_0/16$	6.18e−07	2.98 ■	1.20e−09	4.03 ■	5.60e−10	4.95 ■	1.52e−12	6.04 ■
	0.2	$h_0/2$	4.87e−04	1.70	4.36e−06	5.05	1.57e−05	2.56	9.80e−07	6.66
		$h_0/4$	1.08e−04	2.18	1.10e−07	5.31	9.90e−07	3.99	4.47e−09	7.77
		$h_0/8$	1.95e−05	2.46	5.84e−09	4.23	5.06e−08	4.29	3.96e−11	6.82
		$h_0/16$	3.06e−06	2.67 ☒	2.34e−10	4.64 ■	2.26e−09	4.49 ☒	4.57e−13	6.44 ■
DGSEM-Gauss + Roe	3.5	$h_0/2$	1.87e−04	3.34	8.57e−06	4.02	1.03e−05	5.02	6.76e−07	6.07
		$h_0/4$	2.27e−05	3.04	5.35e−07	4.00	3.30e−07	4.96	1.07e−08	5.99
		$h_0/8$	2.82e−06	3.00	3.34e−08	4.00	1.02e−08	5.01	1.66e−10	6.01
		$h_0/16$	3.53e−07	3.00 ■	2.09e−09	4.00 ■	3.22e−10	4.99 ■	2.60e−12	5.99 ■
	1.0	$h_0/2$	1.82e−04	3.07	8.76e−06	4.04	1.11e−05	5.07	6.95e−07	6.08
		$h_0/4$	2.26e−05	3.00	5.40e−07	4.02	3.44e−07	5.00	1.07e−08	6.01
		$h_0/8$	2.82e−06	3.00	3.35e−08	4.01	1.05e−08	5.04	1.67e−10	6.01
		$h_0/16$	3.53e−07	3.00 ■	2.09e−09	4.00 ■	3.25e−10	5.01 ■	2.61e−12	6.00 ■

(continued)

Table (continued)

Volume disc. + Surface flux	Mach	Mesh level	$N = 2$		$N = 3$		$N = 4$		$N = 5$	
			$L_2(\varrho)$	EOC	$L_2(\varrho)$	EOC	$L_2(\varrho)$	EOC	$L_2(\varrho)$	EOC
	0.2	$h_0/2$	2.14e−04	2.65	1.04e−05	3.78	1.16e−05	4.53	7.76e−07	5.78
		$h_0/4$	2.22e−05	3.26	5.49e−07	4.25	3.49e−07	5.05	1.08e−08	6.17
		$h_0/8$	2.82e−06	2.98	3.76e−08	3.87	1.04e−08	5.07	1.70e−10	5.99
		$h_0/16$	3.53e−07	3.00 ■	2.07e−09	4.19 ■	3.38e−10	4.94 ■	2.64e−12	6.01 ■

Full order is marked with ■ ($\gtrsim N + 1$) and an order reduction with ⊠

Table Experimental order of convergence of L_2 error to the exact density (4), using standard DGSEM-LGL with HLL and Roe fluxes

Volume disc. + Surface flux	Mach	Mesh level	$N = 2$		$N = 3$		$N = 4$		$N = 5$	
			$L_2(\varrho)$	EOC	$L_2(\varrho)$	EOC	$L_2(\varrho)$	EOC	$L_2(\varrho)$	EOC
DGSEM-LGL + HLL	3.5	$h_0/2$	1.22e−03	3.38	3.85e−05	4.03	4.34e−05	4.93	2.70e−06	5.93
		$h_0/4$	1.26e−04	3.27	2.41e−06	4.00	1.39e−06	4.96	4.33e−08	5.96
		$h_0/8$	1.48e−05	3.10	1.51e−07	4.00	4.34e−08	5.01	6.67e−10	6.02
		$h_0/16$	1.81e−06	3.03 ■	9.42e−09	4.00 ■	1.36e−09	4.99 ■	1.05e−11	5.98 ■
	1.0	$h_0/2$	1.04e−03	2.30	3.44e−05	4.44	4.11e−05	4.89	2.89e−06	6.33
		$h_0/4$	1.58e−04	2.72	1.88e−06	4.20	1.77e−06	4.54	3.85e−08	6.23
		$h_0/8$	2.11e−05	2.91	1.16e−07	4.02	6.25e−08	4.82	5.37e−10	6.16
		$h_0/16$	2.69e−06	2.97 ■	7.18e−09	4.01 ■	2.05e−09	4.93 ■	8.36e−12	6.00 ■
	0.2	$h_0/2$	1.20e−03	1.96	4.86e−05	3.51	5.16e−05	3.91	3.85e−06	5.31
		$h_0/4$	2.72e−04	2.14	1.85e−06	4.71	2.99e−06	4.11	4.25e−08	6.50
		$h_0/8$	5.57e−05	2.29	1.21e−07	3.94	1.56e−07	4.26	5.45e−10	6.28
		$h_0/16$	1.01e−05	2.47 ⊠	5.95e−09	4.34 ■	7.19e−09	4.44 ⊠	7.11e−12	6.26 ■
DGSEM-LGL + Roe	3.5	$h_0/2$	1.22e−03	3.38	3.85e−05	4.03	4.34e−05	4.93	2.70e−06	5.93
		$h_0/4$	1.26e−04	3.27	2.41e−06	4.00	1.39e−06	4.96	4.33e−08	5.96
		$h_0/8$	1.48e−05	3.10	1.51e−07	4.00	4.34e−08	5.01	6.67e−10	6.02
		$h_0/16$	1.81e−06	3.03 ■	9.42e−09	4.00 ■	1.36e−09	4.99 ■	1.05e−11	5.98 ■
	1.0	$h_0/2$	9.17e−04	2.86	3.96e−05	3.94	4.41e−05	4.94	2.76e−06	6.02
		$h_0/4$	1.15e−04	2.99	2.41e−06	4.04	1.47e−06	4.90	4.44e−08	5.96
		$h_0/8$	1.44e−05	3.00	1.51e−07	4.00	4.38e−08	5.07	6.84e−10	6.02
		$h_0/16$	1.80e−06	3.00 ■	9.42e−09	4.00 ■	1.35e−09	5.02 ■	1.08e−11	5.99 ■

(continued)

Table (continued)

Volume disc. + Surface flux	Mach	Mesh level	$N = 2$		$N = 3$		$N = 4$		$N = 5$	
			$L_2(\varrho)$	EOC	$L_2(\varrho)$	EOC	$L_2(\varrho)$	EOC	$L_2(\varrho)$	EOC
	0.2	$h_0/2$	9.26e−04	2.35	4.63e−05	3.45	4.26e−05	4.27	2.97e−06	5.37
		$h_0/4$	1.19e−04	2.96	2.40e−06	4.27	1.59e−06	4.74	4.48e−08	6.05
		$h_0/8$	1.43e−05	3.06	1.57e−07	3.93	4.34e−08	5.19	6.83e−10	6.04
		$h_0/16$	1.80e−06	2.99 ■	9.37e−09	4.07 ■	1.49e−09	4.86 ■	1.09e−11	5.97 ■

Full order is marked with ■ ($\gtrsim N + 1$) and an order reduction with ⊠

Table Experimental order ofconvergence of L_2 error tothe exact density (4), usingentropy conservative ECKEP flux and entropy stable HLL and ECKEP-Roe fluxes

Volume disc. + Surface flux	Mach	Mesh level	$N=2$		$N=3$		$N=4$		$N=5$	
			$L_2(\varrho)$	EOC	$L_2(\varrho)$	EOC	$L_2(\varrho)$	EOC	$L_2(\varrho)$	EOC
Split-DGSEM + ECKEP	3.5	$h_0/2$	1.62e−03	4.23	8.45e−05	2.77	5.80e−05	5.88	4.63e−06	5.63
		$h_0/4$	1.30e−04	3.64	7.14e−06	3.56	1.60e−06	5.18	8.38e−08	5.79
		$h_0/8$	1.05e−05	3.62 ■	7.90e−07	3.18	4.56e−08	5.13	3.59e−09	4.54
		$h_0/16$	1.69e−06	2.64 ⊠	9.58e−08	3.04 ⊠	1.23e−09	5.21 ■	1.18e−10	4.93 ⊠
	1.0	$h_0/2$	1.41e−03	3.89	9.45e−05	2.34	7.71e−05	5.86	3.49e−06	6.00
		$h_0/4$	1.25e−04	3.49	1.26e−05	2.90	1.98e−06	5.28	5.00e−08	6.12
		$h_0/8$	1.48e−05	3.09	1.60e−06	2.98	4.06e−08	5.61	1.22e−09	5.36
		$h_0/16$	1.32e−06	3.48 ■	2.01e−07	3.00 ⊠	1.13e−09	5.17 ■	3.85e−11	4.98 ⊠
	0.2	$h_0/2$	1.13e−03	2.26	8.03e−05	2.98	5.95e−05	4.14	4.11e−06	5.14
		$h_0/4$	1.13e−04	3.32	1.02e−05	2.97	1.85e−06	5.01	1.33e−07	4.95
		$h_0/8$	1.12e−05	3.34 ■	1.29e−06	2.99	4.02e−08	5.52	4.21e−09	4.98
		$h_0/16$	1.98e−06	2.50 ⊠	1.61e−07	3.00 ⊠	1.43e−09	4.81 ■	1.32e−10	5.00 ⊠
Split-DGSEM + HLL	3.5	$h_0/2$	1.23e−03	3.36	3.88e−05	4.06	4.49e−05	4.94	3.01e−06	5.97
		$h_0/4$	1.27e−04	3.27	2.42e−06	4.00	1.43e−06	4.97	4.84e−08	5.96
		$h_0/8$	1.48e−05	3.10	1.51e−07	4.00	4.44e−08	5.01	7.48e−10	6.02
		$h_0/16$	1.81e−06	3.03 ■	9.46e−09	4.00 ■	1.40e−09	4.99 ■	1.18e−11	5.98 ■
	1.0	$h_0/2$	1.04e−03	2.30	3.48e−05	4.44	4.16e−05	4.90	3.48e−06	6.10
		$h_0/4$	1.58e−04	2.72	1.88e−06	4.21	1.79e−06	4.54	4.66e−08	6.23
		$h_0/8$	2.11e−05	2.90	1.16e−07	4.02	6.39e−08	4.81	6.08e−10	6.26
		$h_0/16$	2.69e−06	2.97 ■	7.20e−09	4.01 ■	2.11e−09	4.92 ■	9.28e−12	6.03 ■

(continued)

Table (continued)

Volume disc. + Surface flux	Mach	Mesh level	$N=2$		$N=3$		$N=4$		$N=5$	
			$L_2(\varrho)$	EOC	$L_2(\varrho)$	EOC	$L_2(\varrho)$	EOC	$L_2(\varrho)$	EOC
	0.2	$h_0/2$	1.20e−03	1.96	4.90e−05	3.51	5.23e−05	3.92	4.47e−06	5.16
		$h_0/4$	2.72e−04	2.14	1.87e−06	4.71	3.02e−06	4.12	6.14e−08	6.19
		$h_0/8$	5.58e−05	2.29	1.21e−07	3.95	1.57e−07	4.26	6.18e−10	6.64
		$h_0/16$	1.01e−05	2.47 ☒	5.98e−09	4.34 ■	7.28e−09	4.43 ☒	8.20e−12	6.23 ■
Split-DGSEM + ECKEP-Roe	3.5	$h_0/2$	1.23e−03	3.36	3.88e−05	4.06	4.49e−05	4.94	3.01e−06	5.97
		$h_0/4$	1.27e−04	3.27	2.42e−06	4.00	1.43e−06	4.97	4.84e−08	5.96
		$h_0/8$	1.48e−05	3.10	1.51e−07	4.00	4.44e−08	5.01	7.48e−10	6.02
		$h_0/16$	1.81e−06	3.03 ■	9.46e−09	4.00 ■	1.40e−09	4.99 ■	1.18e−11	5.98 ■
	1.0	$h_0/2$	9.18e−04	2.86	3.98e−05	3.94	4.51e−05	4.92	3.08e−06	5.95
		$h_0/4$	1.15e−04	2.99	2.42e−06	4.04	1.51e−06	4.90	4.97e−08	5.95
		$h_0/8$	1.44e−05	3.00	1.51e−07	4.00	4.49e−08	5.07	7.66e−10	6.02
		$h_0/16$	1.80e−06	3.00 ■	9.46e−09	4.00 ■	1.38e−09	5.02 ■	1.21e−11	5.99 ■
	0.2	$h_0/2$	9.26e−04	2.35	4.65e−05	3.45	4.37e−05	4.26	3.26e−06	5.34
		$h_0/4$	1.19e−04	2.96	2.40e−06	4.27	1.63e−06	4.74	5.03e−08	6.02
		$h_0/8$	1.43e−05	3.06	1.58e−07	3.93	4.45e−08	5.19	7.66e−10	6.04
		$h_0/16$	1.80e−06	2.99 ■	9.40e−09	4.07 ■	1.53e−09	4.86 ■	1.22e−11	5.97 ■

Full order is marked with ■ ($\gtrsim N+1$) and an order reduction with ☒

speed (LLF-type) or incorporating all waves (Roe-type), which we will refer to as ECKEP-LLF and ECKEP-Roe fluxes.

In Table , we summarize the convergence of the dissipation-free ECKEP flux, the HLL and ECKEP-Roe flux. The results for LLF and ECKEP-LLF fluxes are found in the Appendix in Table , as they have the same convergence and error levels as the HLL flux. As expected, the dissipation-free surface flux (ECKEP) produces an order reduction for all Mach numbers for $N = 3, 5$, and for $N = 2$ full order is not kept in the last refinement step.

If we simply use the HLL flux, we have an entropy stable scheme, but an order reduction for $N = 2, 4$ can be observed for the low Mach number flow, analogously to the standard DGSEM-LGL scheme. Interestingly, the odd-even behavior switches between entropy conserving and entropy stable fluxes.

The ECKEP-Roe entropy stable flux accounts for all waves of the Riemann problem and adjusts the dissipation for each wave accordingly, which gives full order convergence for all Mach numbers.

Conclusions

In this work, we report the convergence of standard DGSEM Gauss and Gauss-Lobatto schemes to entropy conservative (EC) and entropy stable (ES) DGSEM schemes for the Euler equations, as there have been findings of order reduction for EC and ES schemes. We choose a simple density transport test case on a periodic domain and investigate the influence of the Mach number of the transport velocity.

The EC scheme is dissipation free and an order reduction is observed by the convergence study presented here, confirming many similar observations found in literature. We also confirm that the ES scheme can have an order reduction for low Mach numbers, but only if the entropy stable numerical flux relies on simple approximate Riemann solvers such as local Lax-Friedrichs or HLL. If all waves are accounted for in the dissipation term of the entropy stable flux as presented in [22], the full order is observed for all Mach numbers. In addition, we reproduce the same behavior for the standard DGSEM Gauss and Gauss-Lobatto schemes, where the LLF and HLL fluxes suffer from order reduction at low Mach number, and HLLC and Roe fluxes have full order for all Mach numbers.

We want to emphasize that the present convergence study should be seen merely as an observation, confirming that the numerical flux can have strong influence on the convergence order for both the standard DGSEM and the entropy stable DGSEM. Also, we stress that in our tests the order reduction is related to the form of the dissipation term in the numerical surface flux and is not related to the insufficient integration precision of the LGL-quadrature.

Based on the observations presented in this work, a possible explanation for the loss of convergence for the density transport at low Mach numbers when using LLF and HLL fluxes is the form of dissipation from the approximate Riemann solver. In the case of the density transport, the exact solution follows the characteristic with velocity $\vec{v}$. However, the approximate Riemann solver LLF and HLL consider only two waves with maximum velocity $\sim (|\vec{v}| + c)$ and do not consider the contact wave with velocity $\vec{v}$. Thus, the contact wave is dissipated proportional to $\sim (|\vec{v}| + c)$ and not to $|\vec{v}|$. For low Mach numbers, where $c > |\vec{v}|$, this causes over-upwinding. Over-upwinding was discussed in [5, 15]. It is not intuitive at first, but over-upwinding (over-penalization) can lead to a reduction of the in-built dissipation of the DG scheme, getting wave-propagation characteristics similar to a continuous Galerkin method [16]. This loss of in-built dissipation could be an explanation for the even-odd behavior we observed. However, it is still unclear why numerical surface fluxes with no in-built dissipation that are symmetric, e.g. EC flux, lead to an odd-even behavior in the convergence order and why numerical surface fluxes with over-upwinding, i.e. reduced dissipation due to over-penalization, cause an opposite even-odd behavior. What supports the explanation is the recovery of full convergence order for LLF and HLL when the difference in wave speed becomes smaller for higher Mach numbers, i.e. no over-upwinding. In contrast to LLF and HLL, the HLLC and Roe solvers take specifically the contact wave into account and adjust the dissipation accordingly and thus avoid strong over-upwinding by construction. In our tests, we always observe full convergence order for all Mach numbers for HLLC and Roe.

Lastly we note that a convergence study using a manufactured solution technique can be misleading, as full convergence order is found independent of the choice of numerical flux. Hence, the introduction of a source term to balance the prescribed solution overcomes possible deficiencies of the surface fluxes, showing the limit of the manufactured solution technique in this context. In the Appendix, the convergence results of a manufactured solution are reported.

Acknowledgements Gregor Gassner thanks the European Research Council for funding through the ERC Starting Grant "An Exascale aware and Un-crashable Space-Time-Adaptive Discontinuous Spectral Element Solver for Non-Linear Conservation Laws" (`Extreme`), ERC grant agreement no. 714487. Florian Hindenlang thanks Eric Sonnendrücker and the Max-Planck Institute for Plasma Physics in Garching for their constant support. We would also like to thank all participants of the ICOSAHOM 2018 for the valuable discussions on the topic of entropy stable schemes, which motivated this work.

Appendix

Additional Convergence Results

In this section, we present additional convergence results of the density wave test case for the DGSEM-Gauss and DGSEM-LGL with LLF and HLLC fluxes in Tables and also the entropy stable schemes with LLF and ECKEP-The results for LLF-type fluxes behave like the HLL flux, and for the HLLC flux like the Roe-type fluxes presented in other Table.

Table Experimental order of convergence of L_2 error to the exact density (4), using DGSEM-Gauss with LLF and HLLC fluxes

Volume disc. + Surface flux	Mach	Mesh level	$N=2$		$N=3$		$N=4$		$N=5$	
			$L_2(\varrho)$	EOC	$L_2(\varrho)$	EOC	$L_2(\varrho)$	EOC	$L_2(\varrho)$	EOC
DGSEM-Gauss + LLF	3.5	$h_0/2$	2.42e−04	2.97	6.43e−06	4.41	1.05e−05	4.51	6.68e−07	6.60
		$h_0/4$	3.24e−05	2.90	3.71e−07	4.11	4.32e−07	4.60	8.04e−09	6.38
		$h_0/8$	4.15e−06	2.96	2.27e−08	4.03	1.47e−08	4.87	1.15e−10	6.13
		$h_0/16$	5.22e−07	2.99 ■	1.41e−09	4.01 ■	4.73e−10	4.96 ■	1.77e−12	6.02 ■
	1.0	$h_0/2$	3.13e−04	2.29	4.59e−06	4.84	1.18e−05	3.90	6.69e−07	7.40
		$h_0/4$	5.30e−05	2.56	2.25e−07	4.35	6.08e−07	4.28	5.56e−09	6.91
		$h_0/8$	7.43e−06	2.83	1.29e−08	4.12	2.47e−08	4.62	6.79e−11	6.35
		$h_0/16$	9.61e−07	2.95 ■	7.65e−10	4.07 ■	8.53e−10	4.85 ■	9.97e−13	6.09 ■
	0.2	$h_0/2$	4.95e−04	1.69	4.33e−06	5.07	1.58e−05	2.53	9.88e−07	6.68
		$h_0/4$	1.12e−04	2.15	1.06e−07	5.35	1.01e−06	3.97	4.46e−09	7.79
		$h_0/8$	2.06e−05	2.44	5.47e−09	4.28	5.23e−08	4.27	3.86e−11	6.85
		$h_0/16$	3.29e−06	2.65 ⊠	2.15e−10	4.67 ■	2.38e−09	4.46 ⊠	4.35e−13	6.47 ■
DGSEM-Gauss + HLLC	3.5	$h_0/2$	1.87e−04	3.34	8.57e−06	4.02	1.03e−05	5.02	6.76e−07	6.07
		$h_0/4$	2.27e−05	3.04	5.35e−07	4.00	3.30e−07	4.96	1.07e−08	5.99
		$h_0/8$	2.82e−06	3.00	3.34e−08	4.00	1.02e−08	5.01	1.66e−10	6.01
		$h_0/16$	3.53e−07	3.00 ■	2.09e−09	4.00 ■	3.22e−10	4.99 ■	2.60e−12	5.99 ■
	1.0	$h_0/2$	1.82e−04	3.07	8.76e−06	4.04	1.11e−05	5.07	6.95e−07	6.08
		$h_0/4$	2.26e−05	3.00	5.40e−07	4.02	3.44e−07	5.00	1.07e−08	6.01
		$h_0/8$	2.82e−06	3.00	3.35e−08	4.01	1.05e−08	5.04	1.67e−10	6.01
		$h_0/16$	3.53e−07	3.00 ■	2.09e−09	4.00 ■	3.25e−10	5.01 ■	2.61e−12	6.00 ■

(continued)

Table (continued)

Volume disc. + Surface flux	Mach	Mesh level	$N = 2$		$N = 3$		$N = 4$		$N = 5$	
			$L_2(\varrho)$	EOC	$L_2(\varrho)$	EOC	$L_2(\varrho)$	EOC	$L_2(\varrho)$	EOC
	0.2	$h_0/2$	2.14e−04	2.65	1.04e−05	3.78	1.16e−05	4.53	7.76e−07	5.78
		$h_0/4$	2.22e−05	3.26	5.49e−07	4.25	3.49e−07	5.05	1.08e−08	6.17
		$h_0/8$	2.82e−06	2.98	3.76e−08	3.87	1.04e−08	5.07	1.70e−10	5.99
		$h_0/16$	3.53e−07	3.00 ■	2.07e−09	4.19 ■	3.38e−10	4.94 ■	2.64e−12	6.01 ■

Full order is marked with ■ ($\gtrsim N+1$) and an order reduction with ⊠

Table Experimental order of convergence of L_2 error to the exact density (4), using DGSEM-GL with LLF and HLLC fluxes

Volume disc. + Surface flux	Mach	Mesh level	$N = 2$		$N = 3$		$N = 4$		$N = 5$	
			$L_2(\varrho)$	EOC	$L_2(\varrho)$	EOC	$L_2(\varrho)$	EOC	$L_2(\varrho)$	EOC
DGSEM-LGL + LLF	3.5	$h_0/2$	1.23e−03	3.11	3.44e−05	4.36	4.02e−05	4.82	2.91e−06	6.14
		$h_0/4$	1.51e−04	3.02	2.00e−06	4.10	1.61e−06	4.64	3.80e−08	6.26
		$h_0/8$	1.89e−05	3.00	1.23e−07	4.02	5.55e−08	4.86	5.46e−10	6.12
		$h_0/16$	2.36e−06	3.00 ■	7.66e−09	4.01 ■	1.79e−09	4.95 ■	8.54e−12	6.00 ■
	1.0	$h_0/2$	1.10e−03	1.97	3.15e−05	4.83	4.02e−05	4.93	2.85e−06	6.73
		$h_0/4$	2.03e−04	2.43	1.68e−06	4.23	1.98e−06	4.34	3.68e−08	6.28
		$h_0/8$	2.99e−05	2.76	1.03e−07	4.02	8.35e−08	4.57	4.86e−10	6.24
		$h_0/16$	3.94e−06	2.93 ■	6.39e−09	4.01 ■	2.96e−09	4.82 ■	7.53e−12	6.01 ■
	0.2	$h_0/2$	1.21e−03	1.95	4.88e−05	3.51	5.20e−05	3.90	3.89e−06	5.31
		$h_0/4$	2.77e−04	2.12	1.86e−06	4.72	3.05e−06	4.09	4.27e−08	6.51
		$h_0/8$	5.78e−05	2.26	1.21e−07	3.94	1.61e−07	4.24	5.47e−10	6.29
		$h_0/16$	1.06e−05	2.44 ⊠	5.98e−09	4.34 ■	7.57e−09	4.41 ⊠	7.16e−12	6.26 ■
DGSEM-LGL + HLLC	3.5	$h_0/2$	1.22e−03	3.38	3.85e−05	4.03	4.34e−05	4.93	2.70e−06	5.93
		$h_0/4$	1.26e−04	3.27	2.41e−06	4.00	1.39e−06	4.96	4.33e−08	5.96
		$h_0/8$	1.48e−05	3.10	1.51e−07	4.00	4.34e−08	5.01	6.67e−10	6.02
		$h_0/16$	1.81e−06	3.03 ■	9.42e−09	4.00 ■	1.36e−09	4.99 ■	1.05e−11	5.98 ■
	1.0	$h_0/2$	9.17e−04	2.86	3.96e−05	3.94	4.41e−05	4.94	2.76e−06	6.02
		$h_0/4$	1.15e−04	2.99	2.41e−06	4.04	1.47e−06	4.90	4.44e−08	5.96
		$h_0/8$	1.44e−05	3.00	1.51e−07	4.00	4.38e−08	5.07	6.84e−10	6.02
		$h_0/16$	1.80e−06	3.00 ■	9.42e−09	4.00 ■	1.35e−09	5.02 ■	1.08e−11	5.99 ■

(continued)

Table 5 continued)

Volume disc. + Surface flux	Mach	Mesh level	$N = 2$		$N = 3$		$N = 4$		$N = 5$	
			$L_2(\varrho)$	EOC	$L_2(\varrho)$	EOC	$L_2(\varrho)$	EOC	$L_2(\varrho)$	EOC
	0.2	$h_0/2$	9.26e−04	2.35	4.63e−05	3.45	4.26e−05	4.27	2.97e−06	5.37
		$h_0/4$	1.19e−04	2.96	2.40e−06	4.27	1.59e−06	4.74	4.48e−08	6.05
		$h_0/8$	1.43e−05	3.06	1.57e−07	3.93	4.34e−08	5.19	6.83e−10	6.04
		$h_0/16$	1.80e−06	2.99 ■	9.37e−09	4.07 ■	1.49e−09	4.86 ■	1.09e−11	5.97 ■

Full order is marked with ■ ($\gtrsim N + 1$) and an order reduction with ⊠

Table Experimental order of convergence of L_2 error to the exact density (4), using entropy stable LLF and ECKEP-LLF flux

Volume disc. + Surface flux	Mach	Mesh level	$N = 2$		$N = 3$		$N = 4$		$N = 5$	
			$L_2(\varrho)$	EOC	$L_2(\varrho)$	EOC	$L_2(\varrho)$	EOC	$L_2(\varrho)$	EOC
Split-DGSEM + LLF	3.5	$h_0/2$	1.24e−03	3.11	3.49e−05	4.37	4.16e−05	4.85	3.37e−06	6.05
		$h_0/4$	1.52e−04	3.03	2.01e−06	4.12	1.64e−06	4.67	4.49e−08	6.23
		$h_0/8$	1.89e−05	3.00	1.23e−07	4.03	5.67e−08	4.85	6.21e−10	6.18
		$h_0/16$	2.36e−06	3.00 ■	7.68e−09	4.01 ■	1.84e−09	4.95 ■	9.58e−12	6.02 ■
	1.0	$h_0/2$	1.09e−03	1.97	3.21e−05	4.82	4.07e−05	4.94	3.77e−06	6.34
		$h_0/4$	2.03e−04	2.43	1.69e−06	4.25	1.99e−06	4.35	4.69e−08	6.33
		$h_0/8$	2.99e−05	2.76	1.04e−07	4.03	8.46e−08	4.56	5.57e−10	6.39
		$h_0/16$	3.94e−06	2.93 ■	6.41e−09	4.02 ■	3.03e−09	4.80 ■	8.38e−12	6.05 ■
	0.2	$h_0/2$	1.21e−03	1.95	4.92e−05	3.50	5.26e−05	3.91	4.52e−06	5.15
		$h_0/4$	2.77e−04	2.12	1.88e−06	4.71	3.07e−06	4.10	6.21e−08	6.19
		$h_0/8$	5.79e−05	2.26	1.22e−07	3.95	1.63e−07	4.24	6.21e−10	6.64
		$h_0/16$	1.06e−05	2.44 ⊠	6.01e−09	4.34 ■	7.66e−09	4.41 ⊠	8.26e−12	6.23 ■
Split-DGSEM + ECKEP-LLF	3.5	$h_0/2$	1.24e−03	3.10	3.49e−05	4.37	4.16e−05	4.84	3.38e−06	6.06
		$h_0/4$	1.52e−04	3.03	2.01e−06	4.12	1.64e−06	4.67	4.49e−08	6.23
		$h_0/8$	1.89e−05	3.01	1.23e−07	4.03	5.68e−08	4.85	6.21e−10	6.18
		$h_0/16$	2.36e−06	3.00 ■	7.68e−09	4.01 ■	1.84e−09	4.95 ■	9.58e−12	6.02 ■
	1.0	$h_0/2$	1.10e−03	1.97	3.20e−05	4.83	4.08e−05	4.94	3.77e−06	6.36
		$h_0/4$	2.04e−04	2.43	1.69e−06	4.25	2.00e−06	4.35	4.69e−08	6.33
		$h_0/8$	3.00e−05	2.76	1.04e−07	4.03	8.49e−08	4.56	5.57e−10	6.40
		$h_0/16$	3.95e−06	2.93 ■	6.41e−09	4.02 ■	3.04e−09	4.81 ■	8.38e−12	6.05 ■

(continued)

Table (continued)

Volume disc. + Surface flux	Mach	Mesh level	$N = 2$		$N = 3$		$N = 4$		$N = 5$	
			$L_2(\varrho)$	EOC	$L_2(\varrho)$	EOC	$L_2(\varrho)$	EOC	$L_2(\varrho)$	EOC
	0.2	$h_0/2$	1.21e−03	1.95	4.93e−05	3.51	5.27e−05	3.91	4.53e−06	5.15
		$h_0/4$	2.78e−04	2.12	1.88e−06	4.71	3.08e−06	4.10	6.21e−08	6.19
		$h_0/8$	5.79e−05	2.26	1.22e−07	3.95	1.63e−07	4.24	6.21e−10	6.65
		$h_0/16$	1.07e−05	2.44 ☒	6.01e−09	4.34 ■	7.67e−09	4.41 ☒	8.26e−12	6.23 ■

Full order is marked with ■ ($\gtrsim N + 1$) and an order reduction with ☒

Manufactured Solution with Source Term

Here, we run a convergence test with the method of manufactured solutions. To do so, we assume a two-dimensional solution of the form

$$\mathbf{u} = \left[\varrho\,,\varrho v_1\,,\varrho v_2\,,\varrho v_3\,,E\right]^T = \left[g\,,g\,,g\,,0\,,g^2\right]^T \tag{5}$$

$$\text{with}\;\; g = g(x_1, x_2, t) = 0.5\sin(2\pi(x_1 + x_2 - t)) + 2.$$

Note that the average Mach number in the domain is Ma $= 0.8$. Inserting (5) into the Euler equations, and using the fact that spatial and time derivatives are $g' = \partial_{x_1} g = \partial_{x_2} g = -\partial_t g$, we get an additional residual

$$\mathbf{u}_t + \sum_{i=1}^{3} \frac{\partial \mathbf{f}_i}{\partial x_i} = \begin{pmatrix} g' \\ (3\gamma - 2)g' + 2(\gamma - 1)gg' \\ (3\gamma - 2)g' + 2(\gamma - 1)gg' \\ 0 \\ (6\gamma - 2)g' + 2(2\gamma - 1)gg' \end{pmatrix} \tag{6}$$

To solve the inhomogeneous problem, we subtract the residual from the approximate solution in each Runge–Kutta step. Moreover, we run the test case up to the final time $t = 1.0$.

In the convergence results for the standard DGSEM Gauss and Gauss-Lobatto, we see that the LLF flux still leads to an order reduction for $N = 2, 4$, whereas full order is found for the HLL, HLLC and Roe fluxes, see Tables.

In Table the entropy conservative scheme shows again an order reduction for $N = 3, 5$, and the LLF-Type dissipation too, for $N = 2, 4$, and for this test case, all entropy stable schemes exhibit full order.

Table Experimental order of convergence of L_2 error of density for the manufactured solution (5), using DGSEM-Gauss with LLF, HLL, HLLC and Roe fluxes

Volume disc. + Surface flux	Mesh level	$N=2$		$N=3$		$N=4$		$N=5$	
		$L_2(\varrho)$	EOC	$L_2(\varrho)$	EOC	$L_2(\varrho)$	EOC	$L_2(\varrho)$	EOC
DGSEM-Gauss + LLF	$h_0/2$	2.30e−03	2.20	4.54e−05	5.34	1.13e−04	6.35	4.52e−05	4.38
	$h_0/4$	4.81e−04	2.25	1.99e−06	4.52	4.78e−06	4.56	2.37e−07	7.58
	$h_0/8$	9.48e−05	2.34	1.02e−07	4.28	2.88e−07	4.05	1.53e−09	7.28
	$h_0/16$	1.57e−05	2.60 ⊠	6.25e−09	4.03 ■	1.73e−08	4.05 ⊠	1.57e−11	6.61 ■
DGSEM-Gauss + HLL	$h_0/2$	1.24e−03	2.84	5.46e−05	4.35	1.32e−04	5.22	1.47e−05	6.30
	$h_0/4$	1.17e−04	3.41	3.36e−06	4.02	2.89e−06	5.51	1.44e−07	6.67
	$h_0/8$	1.41e−05	3.04	1.85e−07	4.18	7.14e−08	5.34	1.74e−09	6.37
	$h_0/16$	1.76e−06	3.00 ■	1.07e−08	4.11 ■	2.15e−09	5.05 ■	2.24e−11	6.28 ■
DGSEM-Gauss + HLLC	$h_0/2$	1.24e−03	2.84	5.46e−05	4.35	1.32e−04	5.22	1.47e−05	6.30
	$h_0/4$	1.17e−04	3.41	3.36e−06	4.02	2.89e−06	5.51	1.44e−07	6.67
	$h_0/8$	1.41e−05	3.04	1.85e−07	4.18	7.14e−08	5.34	1.74e−09	6.37
	$h_0/16$	1.76e−06	3.00 ■	1.07e−08	4.11 ■	2.15e−09	5.05 ■	2.24e−11	6.28 ■
DGSEM-Gauss + Roe	$h_0/2$	1.24e−03	2.84	5.46e−05	4.35	1.32e−04	5.22	1.47e−05	6.30
	$h_0/4$	1.17e−04	3.41	3.36e−06	4.02	2.89e−06	5.51	1.44e−07	6.67
	$h_0/8$	1.41e−05	3.04	1.85e−07	4.18	7.14e−08	5.34	1.74e−09	6.37
	$h_0/16$	1.76e−06	3.00 ■	1.07e−08	4.11 ■	2.15e−09	5.05 ■	2.24e−11	6.28 ■

Full order is marked with ■ ($\gtrsim N+1$) and an order reduction with ⊠

Table Experimental order of convergence of L_2 error of density for the manufactured solution (5), using DGSEM-LGL with LLF, HLL, HLLC and Roe fluxes

Volume disc. + Surface flux	Mesh level	$N = 2$		$N = 3$		$N = 4$		$N = 5$	
		$L_2(\varrho)$	EOC	$L_2(\varrho)$	EOC	$L_2(\varrho)$	EOC	$L_2(\varrho)$	EOC
DGSEM-LGL + LLF	$h_0/2$	7.36e−03	2.85	3.15e−04	4.38	5.69e−04	5.78	9.33e−05	5.34
	$h_0/4$	1.33e−03	2.47	1.43e−05	4.46	2.04e−05	4.80	9.27e−07	6.65
	$h_0/8$	2.79e−04	2.25	7.99e−07	4.16	8.81e−07	4.54	9.34e−09	6.63
	$h_0/16$	5.31e−05	2.39 ⊠	4.72e−08	4.08 ■	5.94e−08	3.89 ⊠	1.37e−10	6.09 ■
DGSEM-LGL + HLL	$h_0/2$	5.32e−03	3.16	2.52e−04	4.00	3.84e−04	5.36	4.33e−05	6.24
	$h_0/4$	5.99e−04	3.15	1.38e−05	4.19	1.43e−05	4.75	4.58e−07	6.56
	$h_0/8$	7.25e−05	3.05	7.69e−07	4.17	2.92e−07	5.61	7.08e−09	6.02
	$h_0/16$	9.02e−06	3.01 ■	4.74e−08	4.02 ■	7.77e−09	5.23 ■	1.10e−10	6.01 ■
DGSEM-LGL + HLLC	$h_0/2$	5.32e−03	3.16	2.52e−04	4.00	3.84e−04	5.36	4.33e−05	6.24
	$h_0/4$	5.99e−04	3.15	1.38e−05	4.19	1.43e−05	4.75	4.58e−07	6.56
	$h_0/8$	7.25e−05	3.05	7.69e−07	4.17	2.92e−07	5.61	7.08e−09	6.02
	$h_0/16$	9.02e−06	3.01 ■	4.74e−08	4.02 ■	7.77e−09	5.23 ■	1.10e−10	6.01 ■
DGSEM-LGL + Roe	$h_0/2$	5.32e−03	3.16	2.52e−04	4.00	3.84e−04	5.36	4.33e−05	6.24
	$h_0/4$	5.99e−04	3.15	1.38e−05	4.19	1.43e−05	4.75	4.58e−07	6.56
	$h_0/8$	7.25e−05	3.05	7.69e−07	4.17	2.92e−07	5.61	7.08e−09	6.02
	$h_0/16$	9.02e−06	3.01 ■	4.74e−08	4.02 ■	7.77e−09	5.23 ■	1.10e−10	6.01 ■

Full order is marked with ■ ($\gtrsim N + 1$) and an order reduction with ⊠

Table Experimental order of convergence of L_2 error of density for the manufactured solution (5), using entropy conservative and entropy stable schemes

Volume disc. + Surface flux	Mesh level	$N = 2$		$N = 3$		$N = 4$		$N = 5$	
		$L_2(\varrho)$	EOC	$L_2(\varrho)$	EOC	$L_2(\varrho)$	EOC	$L_2(\varrho)$	EOC
Split-DGSEM + ECKEP	$h_0/2$	1.31e−02	3.53	1.28e−03	2.88	5.62e−03	2.49	1.12e−03	4.05
	$h_0/4$	1.30e−03	3.34	1.13e−04	3.49	3.12e−04	4.17	5.97e−06	7.56
	$h_0/8$	1.23e−04	3.40	1.24e−05	3.20	2.00e−06	7.29	9.45e−08	5.98
	$h_0/16$	1.76e−05	2.80 ■	1.67e−06	2.89 ⊠	3.41e−08	5.87 ■	3.17e−09	4.90 ⊠
Split-DGSEM + LLF	$h_0/2$	7.60e−03	3.13	3.70e−04	3.91	6.56e−04	4.44	1.28e−04	4.75
	$h_0/4$	1.63e−03	2.22	1.90e−05	4.29	3.09e−05	4.41	2.01e−06	6.00
	$h_0/8$	3.40e−04	2.26	9.89e−07	4.26	1.83e−06	4.08	1.80e−08	6.80
	$h_0/16$	6.14e−05	2.47 ⊠	6.41e−08	3.95 ■	9.15e−08	4.32 ⊠	2.17e−10	6.37 ■
Split-DGSEM + ECKEP-LLF	$h_0/2$	7.66e−03	3.21	3.74e−04	3.98	6.79e−04	4.45	1.33e−04	4.75
	$h_0/4$	1.64e−03	2.23	1.90e−05	4.30	3.08e−05	4.46	2.06e−06	6.01
	$h_0/8$	3.41e−04	2.26	9.88e−07	4.26	1.83e−06	4.07	1.79e−08	6.85
	$h_0/16$	6.16e−05	2.47 ⊠	6.40e−08	3.95 ■	9.17e−08	4.32 ⊠	2.17e−10	6.37 ■
Split-DGSEM + HLL	$h_0/2$	5.82e−03	3.10	3.01e−04	3.85	5.11e−04	4.35	7.06e−05	5.27
	$h_0/4$	7.06e−04	3.04	2.04e−05	3.88	1.67e−05	4.94	1.08e−06	6.03
	$h_0/8$	8.63e−05	3.03	1.16e−06	4.14	5.08e−07	5.04	1.67e−08	6.02
	$h_0/16$	1.08e−05	3.00 ■	7.20e−08	4.01 ■	1.62e−08	4.97 ■	2.64e−10	5.98 ■
Split-DGSEM + ECKEP-Roe	$h_0/2$	5.81e−03	3.11	3.01e−04	3.85	5.11e−04	4.35	7.06e−05	5.27
	$h_0/4$	7.06e−04	3.04	2.04e−05	3.88	1.67e−05	4.94	1.08e−06	6.03
	$h_0/8$	8.63e−05	3.03	1.16e−06	4.14	5.08e−07	5.04	1.67e−08	6.02
	$h_0/16$	1.08e−05	3.00 ■	7.20e−08	4.01 ■	1.62e−08	4.97 ■	2.64e−10	5.98 ■

Full order is marked with ■ ($\gtrsim N + 1$) and an order reduction with ⊠

References

1. Bohm, M., Winters, A.R., Gassner, G.J., Derigs, D., Hindenlang, F., Saur, J.: An entropy stable nodal discontinuous Galerkin method for the resistive MHD equations. Part I: Theory and numerical verification. J. Comput. Phys. 108076 (2018). https://doi.org/10.1016/j.jcp.2018.06.027
2. Carpenter, M., Kennedy, C.: Fourth-order 2N-storage Runge-Kutta schemes. Technical Report NASA TM 109111 (1994)
3. Carpenter, M., Fisher, T., Nielsen, E., Frankel, S.: Entropy stable spectral collocation schemes for the Navier–Stokes equations: discontinuous interfaces. SIAM J. Sci. Comput. **36**(5), B835–B867 (2014)
4. Chan, J.: On discretely entropy conservative and entropy stable discontinuous Galerkin methods. J. Comput. Phys. **362**, 346–374 (2018)
5. Chan, J., Warburton, T.: On the penalty stabilization mechanism for upwind discontinuous Galerkin formulations of first order hyperbolic systems. Comput. Math. Appl. **74**(12), 3099–3110 (2017)
6. Chen, T., Shu, C.-W.: Entropy stable high order discontinuous Galerkin methods with suitable quadrature rules for hyperbolic conservation laws. J. Comput. Phys. **345**, 427–461 (2017)
7. Flad, D., Gassner, G.: On the use of kinetic energy preserving DG-schemes for large eddy simulation. J. Comput. Phys. **350**, 782–795 (2017)
8. Friedrich, L., Winters, A.R., Del Rey Fernández, D.C., Gassner, G.J., Parsani, M., Carpenter, M.H.: An entropy stable h/p non-conforming discontinuous Galerkin method with the summation-by-parts property. J. Sci. Comput. **77**, 689–725 (2018)
9. Gassner, G.: A skew-symmetric discontinuous Galerkin spectral element discretization and its relation to SBP-SAT finite difference methods. SIAM J. Sci. Comput. **35**(3), A1233–A1253 (2013)
10. Gassner, G.J.: A kinetic energy preserving nodal discontinuous Galerkin spectral element method. Int. J. Numer. Methods Fluids **76**(1), 28–50 (2014)
11. Gassner, G., Kopriva, D.A.: A comparison of the dispersion and dissipation errors of Gauss and Gauss–Lobatto discontinuous Galerkin spectral element methods. SIAM J. Sci. Comput. **33**, 2560–2579 (2011)
12. Gassner, G.J., Winters, A.R., Kopriva, D.A.: Split form nodal discontinuous Galerkin schemes with summation-by-parts property for the compressible Euler equations. J. Comput. Phys. **327**, 39–66 (2016)
13. Gassner, G.J., Winters, A.R., Hindenlang, F.J., Kopriva, D.A.: The BR1 scheme is stable for the compressible Navier–Stokes equations. J. Sci. Comput. **77**, 154–200 (2018)
14. Kopriva, D.A.: Implementing Spectral Methods for Partial Differential Equations. Scientific Computation. Springer, Berlin (2009)
15. Manzanero, J., Ferrer, E., Rubio, G., Valero, E.: On the role of numerical dissipation in stabilising under-resolved turbulent simulations using discontinuous Galerkin methods. Preprint arXiv:1805.10519 (2018)
16. Moura, R., Sherwin, S., Peirò, J.: Eigensolution analysis of spectral/hp continuous Galerkin approximations to advection-diffusion problems: insights into spectral vanishing viscosity. J. Comput. Phys. **307**, 401–422 (2016)
17. Parsani, M., Carpenter, M.H., Nielsen, E.J.: Entropy stable wall boundary conditions for the three-dimensional compressible Navier–Stokes equations. J. Comput. Phys. **292**, 88–113 (2015)
18. Toro, E.: Riemann Solvers and Numerical Methods for Fluid Dynamics. Springer, Berlin (2009)
19. Wintermeyer, N., Winters, A.R., Gassner, G.J., Kopriva, D.A.: An entropy stable nodal discontinuous Galerkin method for the two dimensional shallow water equations on unstructured curvilinear meshes with discontinuous bathymetry. J. Comput. Phys. (2016, submitted). arXiv:1509.07096 [math.NA]

20. Wintermeyer, N., Winters, A.R., Gassner, G.J., Warburton, T.: An entropy stable discontinuous Galerkin method for the shallow water equations on curvilinear meshes with wet/dry fronts accelerated by GPUs. J. Comput. Phys. **375**, 447–480 (2018)
21. Winters, A.R., Gassner, G.J.: A comparison of two entropy stable discontinuous Galerkin spectral element approximations for the shallow water equations with non-constant topography. J. Comput. Phys. **301**, 357–376 (2015)
22. Winters, A.R., Derigs, D., Gassner, G.J., Walch, S.: A uniquely defined entropy stable matrix dissipation operator for high Mach number ideal MHD and compressible Euler simulations. J. Comput. Phys. **332**, 274–289 (2017)
23. Winters, A.R., Moura, R.C., Mengaldo, G., Gassner, G.J., Walch, S., Peiro, J., Sherwin, S.J.: A comparative study on polynomial dealiasing and split form discontinuous Galerkin schemes for under-resolved turbulence computations. J. Comput. Phys. **372**, 1–21 (2018)

5

HPS Accelerated Spectral Solvers for Time Dependent Problems: Part I, Algorithms

Tracy Babb, Per-Gunnar Martinsson, and Daniel Appelö

Introduction

In this chapter, part two in a two part series, describes a sequence of numerical experiments demonstrating the performance of a highly computationally efficient solver for equations of the form

$$\kappa \frac{\partial u}{\partial t} = \mathscr{L}u(\boldsymbol{x}, t) + g(u, \boldsymbol{x}, t), \quad \boldsymbol{x} \in \Omega, t > 0, \tag{1}$$

with initial data $u(\boldsymbol{x}, 0) = u_0(\boldsymbol{x})$. Here $\mathscr{L}$ is an elliptic operator acting on a fixed domain Ω and f is lower order, possibly nonlinear terms. We take κ to be real or imaginary, allowing for parabolic and Schrödinger type equations.

The "Hierarchial Poincaré–Steklov (HPS)" solver has already been demonstrated to be a highly competitive spectrally accurate solver for elliptic problems [1, 4, 7] and has also been used together with a class of exponential integrators [5], to evolve solutions to hyperbolic differential equations. As just mentioned, the focus here is on differential equations in the form (1) whose discretization leads to stiff system of ODE that can beneficially be advanced in time using Explicit, Singly Diagonally Implicit Runge–Kutta (ESDIRK) methods. ESDIRK methods offer the advantages of stiff accuracy and L-stability and are well suited for the HPS algorithm as they only require a single matrix factorization. They are also easily combined with

T. Babb · D. Appelö (✉)
University of Colorado, Boulder, CO, USA
e-mail: tracy.babb@colorado.edu; daniel.appelo@colorado.edu

P.-G. Martinsson
University of Texas, Austin, TX, USA
e-mail: pgm@ices.utexas.edu

explicit Runge–Kutta method leading to so called Additive Runge–Kutta (ARK) methods [6].

To this end we investigate the stability and accuracy that is obtained when combining high-order time-stepping schemes with the HPS method for solving elliptic equations. We restrict attention to relatively simple geometries (rectangles) but note that the method can without difficulty be generalized to domains that can be expressed as a union of rectangles, possibly mapped via curvilinear smooth parameter maps.

The rest of this chapter is organized as follows. In other Sect. we present results illustrating that the order reduction phenomena for DIRK methods observed in [8] can be circumvented when formulating the time stepping in terms of slopes (with boundary conditions differentiated in time) rather than formulating it in terms of stage solutions. we present numerical results for Schrödingers equation in two dimensions. We present numerical results for a nonlinear problem, viscous Burgers' equation in two dimensions. Finally, in Sect. we summarize and conclude. For a longer description of the method we refer to thee first part of this paper and to [2].

Time Dependent Boundary Conditions

This section discusses time-dependent boundary conditions within the two different Runge–Kutta formulations. In particular, we investigate the order reduction that has been documented in [8] for implicit Runge–Kutta methods and earlier in [3] for explicit Runge–Kutta methods.

In this first experiment, introduced in [8], we solve the heat equation in one dimension

$$u_t = u_{xx} + f(t), \qquad x \in [0, 2], \quad t > 0. \tag{2}$$

We set the initial data, Dirichlet boundary conditions and the forcing $f(t)$ so that exact solution is $u(x, t) = \cos(t)$. This example is designed to eliminate the effect of the spatial discretization, with the solution being constant in space and allows for the study of possible order reduction near the boundaries.

We use the HPS scheme in space and use 32 leafs with $p = 32$ Chebyshev nodes per leaf. We apply the third, fourth, and fifth order ESDIRK methods from [6]. We consider solving for the intermediate solutions, or as we refer to it below "the u_i formulation" with the boundary condition enforced as $u_i^n = \cos(t_n + c_i \Delta t)$. We also consider solving for the stages, which we refer to as "the k_i formulation" with boundary conditions imposed as $k_i^n = -\sin(x, t_n + c_i \Delta t)$.

Error reduction for time dependent boundary conditions has been studied both in the context of explicit Runge–Kutta methods in e.g. [3] and more recently for implicit Runge–Kutta methods in [8]. In [8] the authors report observed orders of accuracy equal to two (for the solution u) for DIRK methods of order 2, 3, and 4 for

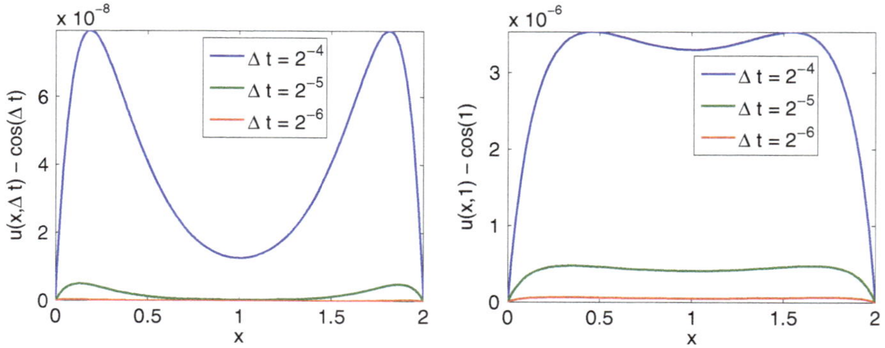

(a) (b)

Fig. The error in solving (2). Results are for a third order ESDIRK. (**a**) Displays the single step error which converges with fourth order of accuracy. (**b**) Displays the global error at $t = 1$ converging at third order. Both errors converge at one order higher than what is expected from the analysis in [8]

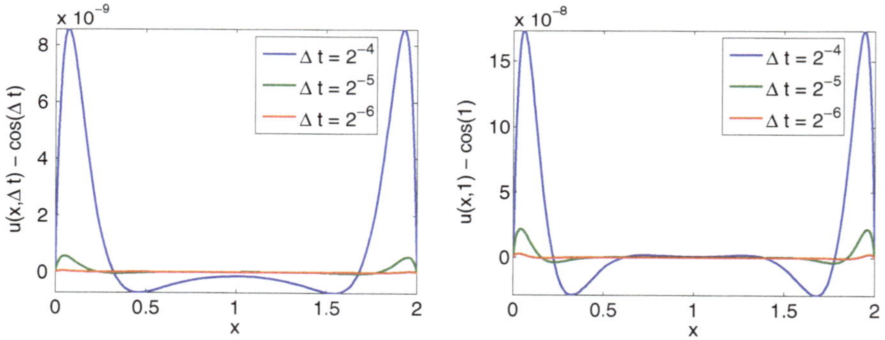

(a) (b)

Fig. The error in solving (2). Results are for a fifth order ESDIRK. (**a**) Displays the single step error which converges with fourth order of accuracy. (**b**) Displays the global error at $t = 1$ converging at third order. Both errors converge at one order higher than what is expected from the analysis in [8] but still lower than expected

the problem (2) discretized with a finite difference method on a fine grid (the spatial errors are zero) using the u_i formulation.

Figures in chap. show the error for the third and fifth order ESDIRK methods, respectively, as a function of x for a single step and at the final time $t = 1$. Figure shows the maximum error for the third, fourth, and fifth order methods as a function of time step Δt after a single step and at the final time $t = 1$.

In general, for a method of order p we expect that the single step error decreases as Δt^{p+1} while the global error decreases as Δt^p. However, with time dependent

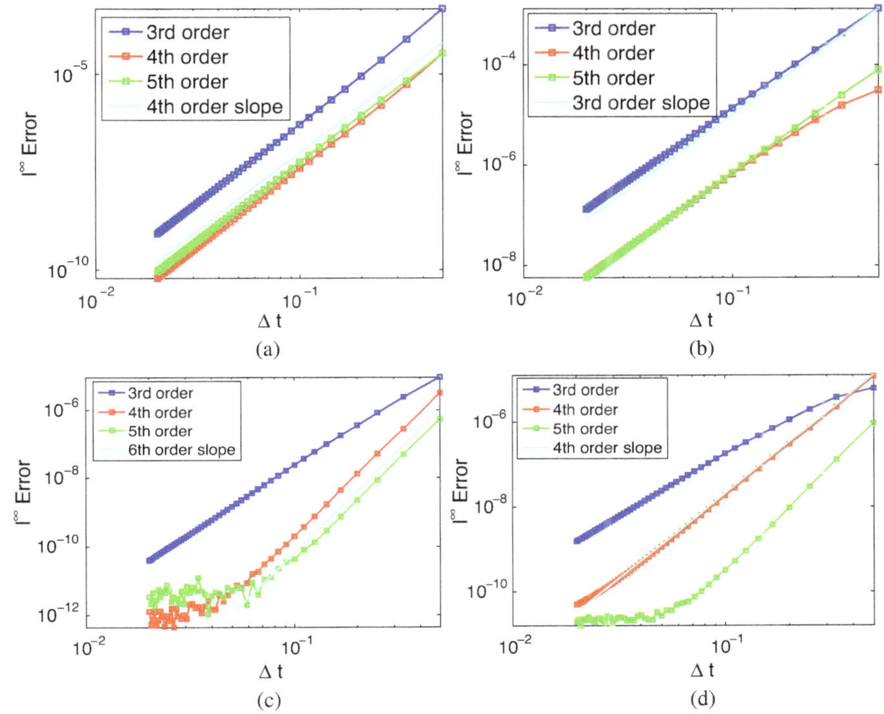

Fig. The maximum error (here denoted l^∞) in solving (2) for the third, fourth, and fifth order ESDIRK methods for a sequence of decreasing time steps. (**a**, **c**) are errors after one time step and (**b**, **d**) are the errors at time $t = 1$. The top row are for the u_i formulation and the bottom row is for the k_i formulation. Note that the k_i formulation is free of order reduction

boundary conditions implemented as $u_i^n = \cos(t_n + c_i \Delta t)$ the results in [8] indicate that the rate of convergence will not exceed two for the single step or global error.

The results for the third order method ($p = 3$) displayed in Fig. show that the single step error decreases as Δt^{p+1} while the global error decreases as Δt^p, which is better than the results documented in [8]. However, we still see that a boundary layer appears to be forming, but it is of the same order as the error away from the boundary. The results for the fifth order method ($p = 5$) displayed in Fig. show that the single step error decreases as Δt^4 while the global error decreases as Δt^3, which is still better than the results documented in [8]. However, the boundary layer is giving order reduction from Δt^{p+1} for the single step error and Δt^p for the global error. We note that our observations differ from those in [8] but that this possibly can be attributed to the use of a ESDIRK method rather than a DIRK method.

We repeat the experiment but now we use the k_i formulation for Runge–Kutta methods and for the boundary condition we enforce $k_i^n = -\sin(t_n + c_i \Delta t)$. The intuition here is that k_i^n is an approximation to u_t at time $t_n + c_i \Delta t$ and we use the value of u_t for the boundary condition of k_i^n. Intuitively we expect that the fact that we reduce the index of the system of differential algebraic equation in the u_i formulation by differentiating the boundary conditions can restore the design order of accuracy.

In the previous examples the Runge–Kutta method introduced an error on the interior while the solution on the boundary was exact. If the error on the boundary is on the same order of magnitude as the error on the interior then the error in u_{xx} is of the correct order, but when the value of u is exact on the boundary it introduces a larger error in u_{xx}. In the k_i formulation, for each intermediate stage we find $u_{xx} = 0$ and then $k_i^n = -\sin(t_n + c_i \Delta t)$ on the interior and on the boundary. So at a fixed time the solution is constant in x and a boundary layer does not form. Additionally, the error is constant in x at any fixed time and for a method of order p we obtain the expected behavior where the single step error decreases as Δt^{p+1} and the global error decreases as Δt^p.

Figure shows the maximum error for the third, fourth, and fifth order methods as a function of time step Δt after a single step and at the final time $t = 1$. The results show that the methods behave exactly as we expect. The single step error behaves as Δt^{p+1} for the third and fifth order methods and Δt^{p+2} for the fourth order method. The fourth order method gives sixth order error in a single step because the exact solution is $u(x, t) = \cos(t)$, which has every other derivative equal to zero at $t = 0$ and for a single step we start at $t = 0$. The global error behaves as Δt^p for each method.

Schrödinger Equation

Next we consider the Schrödinger equation for $u = u(x, y, t)$

$$\begin{aligned} i\hbar u_t &= -\frac{\hbar^2}{2M}\Delta u + V(x, y)u, \quad t > 0, \quad (x, y) \in [x_l, x_r] \times [y_b, y_t], \\ u(x, y, 0) &= u_0(x, y). \end{aligned} \tag{3}$$

Here we nondimensionalize in a way equivalent to setting $M = 1, \hbar = 1$ in the above equation. We choose the potential to be the harmonic potential

$$V(x, y) = \frac{1}{2}\left(x^2 + y^2\right).$$

This leads to an exact solution

$$u(x, y, t) = Ae^{-it}e^{-\frac{(x^2+y^2)}{2}}, \tag{4}$$

where we set $A = 1/\sqrt{\sqrt{\pi}}$ and solve until $t = 2\pi$ on the domain $(x, y) \in [-8, 8]^2$.

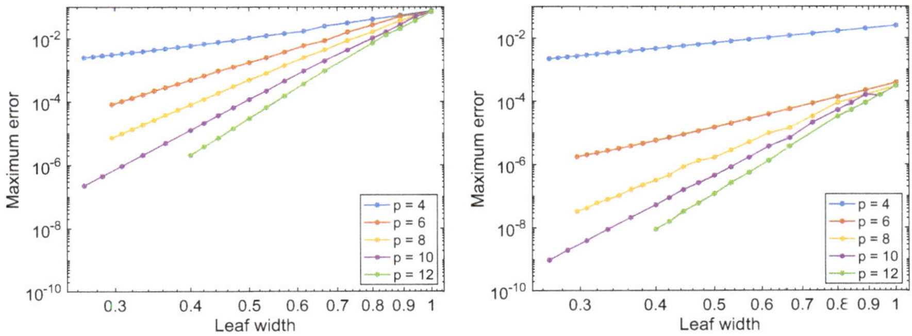

Fig. Error in the Schrödinger equation as a function of leaf size. The exact solution is given in Eq. (4)

Table: Estimated rates of convergence for different Runge–Kutta methods and different orders of approximation

p	4	6	8	10	12
ESDIRK3	2.59	5.73	7.72	9.69	11.47
ESDIRK4	1.89	6.47	7.82	9.76	11.69
ESDIRK5	1.84	4.42	7.69	9.71	11.48

The computational domain is subdivided into $n_x \times n_y$ panels with $p \times p$ points on each panel. To begin, we study the order of accuracy with respect to leaf size. To eliminate the effect of time-stepping errors we scale $\Delta t = h^{p/q_{\rm RK}}$, where $q_{\rm RK}$ is the order of the Runge–Kutta method. In Fig. we display the errors as a function of the leaf size for $p = 4, 6, 8, 10, 12, 16$ and for the third and fifth order Runge–Kutta methods ($q_{RK} = 3, 5$). The rates of convergence are found for all three Runge–Kutta methods and summarized in Table. As can be seen from the table, $p = 4$ appears to converge at second order, while for higher p we generally observe a rate of convergence approaching to p.

In this problem the efficiency of the method is limited by the order of the Runge–Kutta methods. However, as our methods are unconditionally stable we may enhance the efficiency by using Richardson extrapolation to achieve a highly accurate solution in time. We solve the same problem, but now we fix $p = 12$ and take $5 \cdot 2^n$ time steps, with $n = 0, 1, \ldots, 5$. For the third order ESDIRK method we use 60×60 leaf boxes. For the fourth order ESDIRK method we use 90×90 leaf boxes. For the fifth order ESDIRK method we use 120×120 leaf boxes. Table shows that we can easily achieve much higher accuracy by using Richardson extrapolation.

Finally, we solve a problem without an analytic solution. In this problem the initial data

$$u(x, y, t) = 3\sin(x)\sin(y)e^{-(x^2+y^2)},$$

Table Estimated errors at the final time after Richardson extrapolation

q_{RK}/ extrapolations	0	1	2	3	4	5	6
3	1.32 (−1)	1.01 (−2)	1.27 (−4)	1.17 (−5)	6.98 (−8)	8.62 (−10)	7.40 (−6)
4	2.70 (−4)	6.46 (−6)	1.23 (−7)	2.95 (−10)	1.59 (−11)	3.70 (−14)	1.20 (−11)
5	1.28 (−3)	9.67 (−6)	6.30 (−8)	1.86 (−10)	4.11 (−13)	9.27 (−14)	5.08 (−11)

The notation $d(-p)$ means $d \cdot 10^{-p}$

Table Errors computed against a p and h refined solution

p/panels	2	4	8	16	32
8	1.11 (0)	1.39 (−1)	8.74 (−3)	1.50 (−4)	2.45 (−6)
Rate	*	3.00	3.99	5.87	5.92
10	5.87 (−1)	3.16 (−2)	4.62 (−4)	6.17 (−6)	5.21 (−8)
Rate	*	4.21	6.10	6.22	6.89

The errors are maximum errors at the final time $t = 4$. The notation $d(-p)$ means $d \cdot 10^{-p}$

interacts with the weak and slightly non-symmetric potential

$$V(x, y) = 1 - e^{-(x+0.9y)^4},$$

allowing the solution to reach the boundary where we impose homogenous Dirichlet conditions.

We evolve the solution until time $t = 4$ using $p = 8$ and 10 and 2, 4, 8, 16 and 32 leaf boxes in each direction of a domain of size 12×12. The errors computed against a reference solution with $p = 12$ and with 32 leaf boxes can be found in Table.

In Fig. we display snapshots of the magnitude of the solution at the initial time $t = 0$, the intermediate times $t \approx 1.07$, $t \approx 1.68$ and at the final time $t = 4.0$.

Burgers' Equation in Two Dimensions

As a first step towards a full blown flow solver we solve Burgers' equation in two dimensions using the additive Runge–Kutta methods described in the first part of this paper. Precisely, we solve the system

$$\mathbf{u}_t + \mathbf{u} \cdot \nabla \mathbf{u} = \varepsilon \Delta \mathbf{u}, \quad \mathbf{x} \in [-\pi, \pi]^2, \quad t > 0, \tag{5}$$

where $\mathbf{u} = [u(x, y, t), v(x, y, t)]^T$ is the vector containing the velocities in the x and y directions.

The first problem we solve uses the initial condition $\mathbf{u} = 5[-y, x]^T \exp(-3r^2)$ and the boundary conditions are taken to be no-slip boundary conditions on all sides.

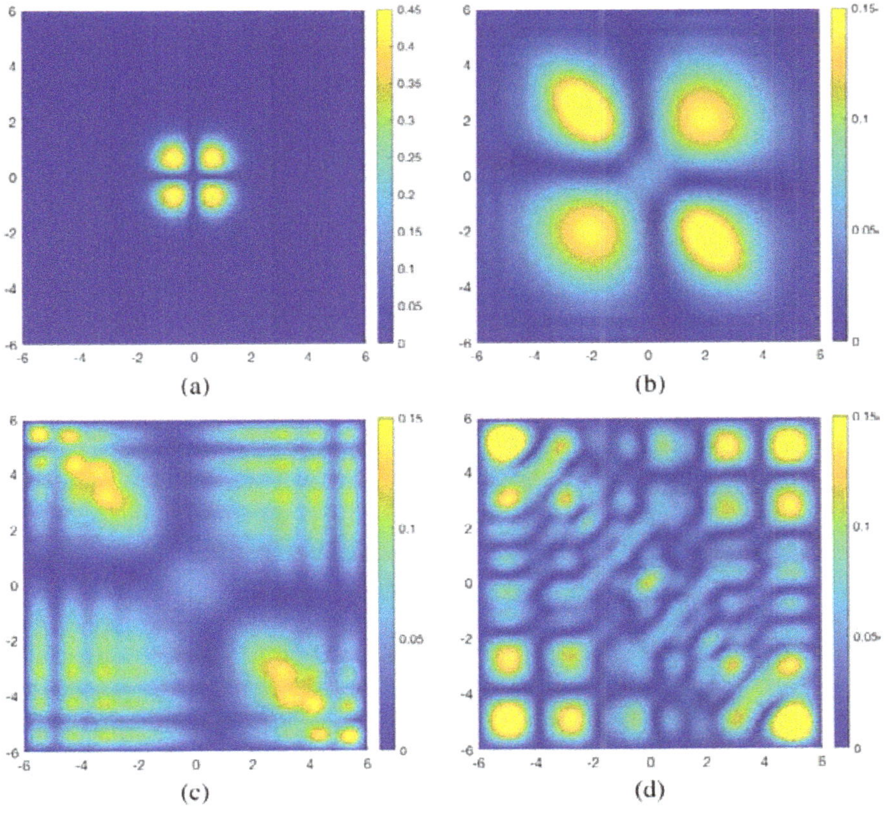

Fig. Snapshots of the magnitude of the solution at the initial time (**a**) $t = 0$, the intermediate times (**b**) $t \approx 1.07$, (**c**) $t \approx 1.68$ and at the final time (**d**) $t = 4.0$

We solve the problem using 24×24 leafs, $p = 24$, $\varepsilon = 0.005$, and the fifth order ARK method found in [6]. We use a time step of $k = 1/80$ and solve until time $t_{\max} = 5$. The low viscosity combined with the initial condition produces a rotating flow resembling a vortex that steepens up over time.

In Fig. we can see the velocities at times $t = 0.5$ and $t = 1$. The fluid rotates and expands out and eventually forms a shock like transition. This creates a sharp flow region with large gradients resulting in a flow that may be difficult to resolve with a low order accurate method. These sharp gradients can be seen in the two vorticity plots in Fig. along with the speed and vorticity plots in Fig.

In our second experiment we consider a cross stream of orthogonal flows. We use an initial condition of

$$\mathbf{u} = [8y\,e^{-36\left(\frac{y}{2}\right)^8}, -8xe^{-36\left(\frac{x}{2}\right)^8}]^T, \tag{6}$$

and time independent boundary conditions that are compatible with the initial data.

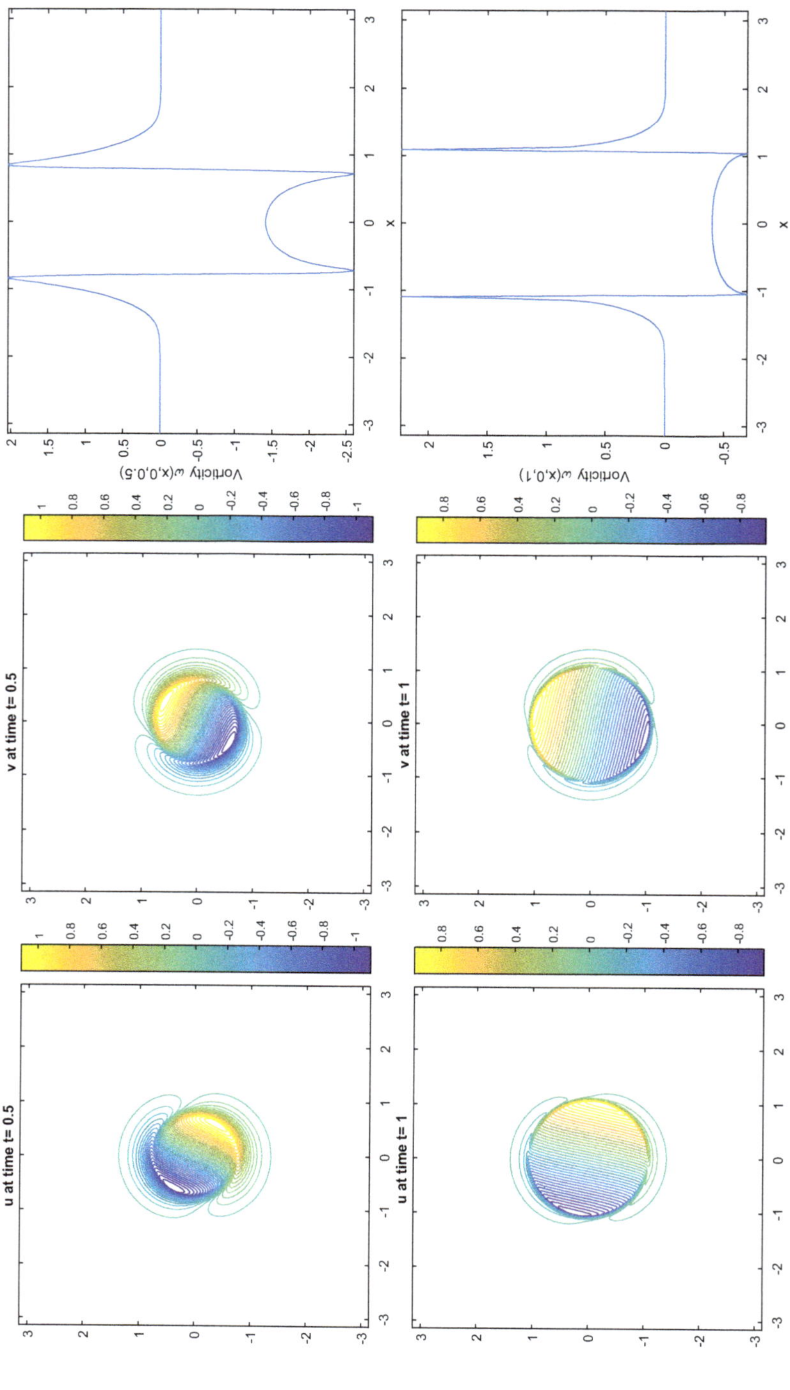

Fig. The four plots on the left show the velocities in the x and y directions at times t =0.5 a n d t = 1. We can see the fluid rotating and expanding. The two plots in the third column show the vorticity at these times. We see sharp gradients near the edge of the rotating fluid

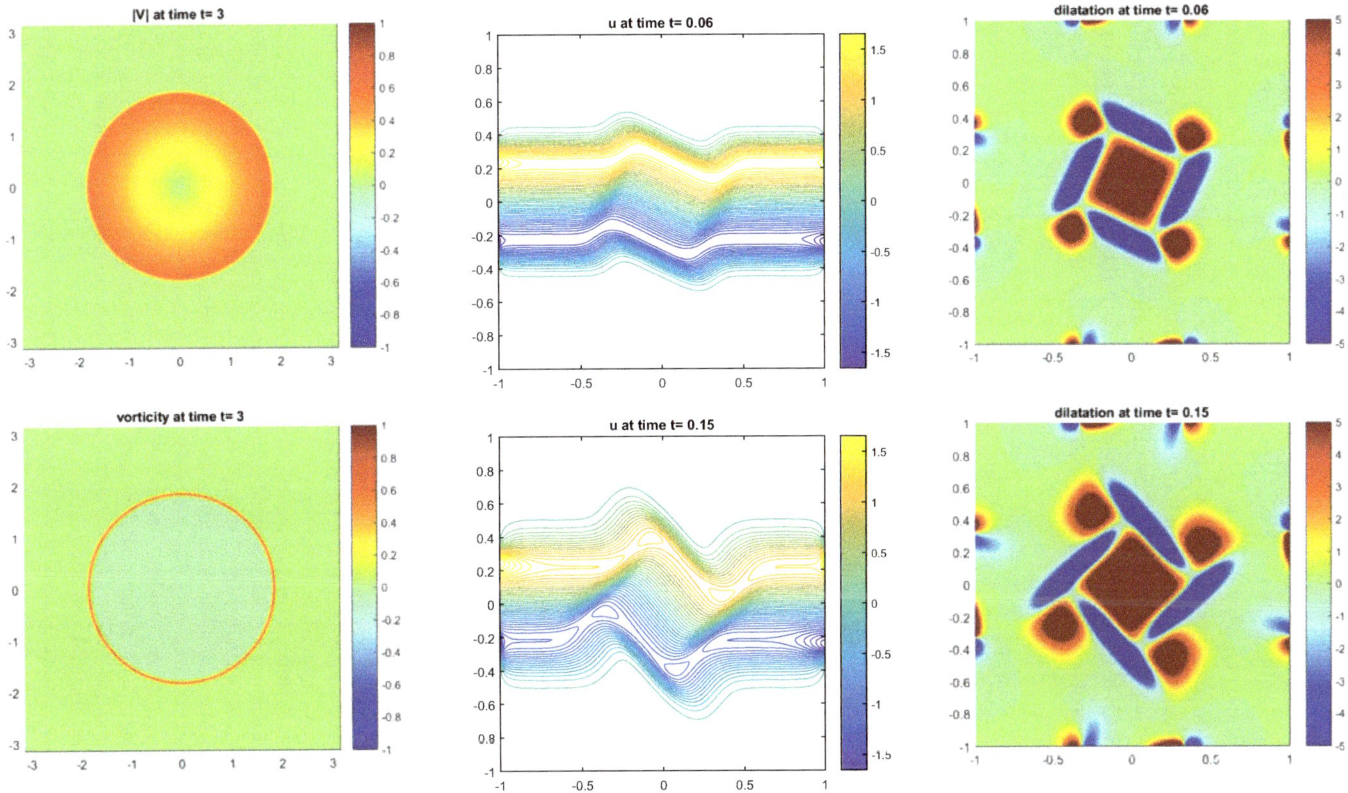

Fig. The plots in the first column correspond to the rotating shock problem. The second and third columns show the velocity in the x direction and the dilatation at times $t = 0.06$ and $t = 0.15$ for the cross flow problem

This initial horizontal velocity drops to zero quickly as we approach $|y| = 0.5$. For $|y| < 0.5$ the exponential term approaches exp(0) and the velocity behaves like $u = 8y$. The flow has changed slightly by $t = 0.06$, but we can see in Fig. the flow is moving to the right for $y > 0$ and the flow is moving the left for $y < 0$ and all significant behavior is in $|y| < 0.5$. A plot of the velocity v would show similar behavior. We also use 24×24 leafs, $p = 24$, $\epsilon = 0.025$, $k = 1/200$, and $t_{\max} = 0.75$. We show plots of the horizontal velocity u and the dilatation at time $t = 0.06$ and $t = 0.15$. We only show plots before time $t = 0.15$ when the fluid is hardest to resolve and we observe that after $t = 0.15$ the cross streams begin to dissipate. This problem contains sharp interfaces inside $\mathbf{x} \in [-0.5, 0.5]^2$.

In this two part series we have demonstrated that the spectrally accurate Hierarchial Poincaré–Steklov solver can be easily extended to handle time dependent PDE problems with a parabolic principal part by using ESDIRK methods. We have outlined the advantages of the two possible ways to formulate implicit Runge–Kutta methods within the HPS scheme and demonstrated the capabilities on both linear and non-linear examples.

There are many avenues for future work, for example:

- Extension of the solvers to compressible and incompressible flows.
- Application of the current solvers to inverse and optimal design problems, in particular for problems where changes in parameters do not require new factorizations.

References

1. Babb, T., Gillman, A., Hao, S., Martinsson, P.: An accelerated Poisson solver based on multidomain spectral discretization. BIT Numer. Math. **58**, 851–879 (2018)
2. Babb, T., Martinsson, P.-G., Appelö, D.: HPS accelerated spectral solvers for time dependent problems (2018). arXiv:1811.04555
3. Carpenter, M., Gottlieb, D., Abarbanel, S., Don, W.-S.: The theoretical accuracy of Runge–Kutta time discretizations for the initial boundary value problem: a study of the boundary error. SIAM J. Sci. Comput. **16**, 1241–1252 (1995)
4. Gillman, A., Martinsson, P.: A direct solver with $\mathcal{O}(N)$ complexity for variable coefficient elliptic PDEs discretized via a high-order composite spectral collocation method. SIAM J. Sci. Comput. **36**, A2023–A2046 (2014). arXiv:1307.2665
5. Haut, T., Babb, T., Martinsson, P., Wingate, B.: A high-order scheme for solving wave propagation problems via the direct construction of an approximate time-evolution operator. IMA J. Numer. Anal. **36**, 688–716 (2016)
6. Kennedy, C., Carpenter, M.: Additive Runge–Kutta schemes for convection-diffusion-reaction equations. Appl. Numer. Math. **44**, 139–181 (2003)

7. Martinsson, P.: A direct solver for variable coefficient elliptic PDEs discretized via a composite spectral collocation method. J. Comp. Phys. **242**, 460–479 (2013)
8. Rosales, R., Seibold, B., Shirokoff, D., Zhou, D.: Order reduction in high-order Runge–Kutta methods for initial boundary value problems (2017). arXiv:1712.00897

6

HPS Accelerated Spectral Solvers for Time Dependent Problems: Part II, Numerical Experiments

Tracy Babb, Per-Gunnar Martinsson, and Daniel Appelö

Introduction

In this chapter describes a highly computationally efficient solver for equations of the form

$$\kappa \frac{\partial u}{\partial t} = \mathscr{L}u(\boldsymbol{x}, t) + h(u, \boldsymbol{x}, t), \quad \boldsymbol{x} \in \Omega, t > 0, \tag{1}$$

with initial data $u(\boldsymbol{x}, 0) = u_0(\boldsymbol{x})$. Here $\mathscr{L}$ is an elliptic operator acting on a fixed domain Ω and h is lower order, possibly nonlinear terms. We take κ to be real or imaginary, allowing for parabolic and Schrödinger type equations. We desire the benefits that can be gained from an implicit solver, such as L-stability and stiff accuracy, which means that the computational bottleneck will be the solution of a sequence of elliptic equations set on Ω. In situations where the elliptic equation to be solved is the same in each time-step, it is highly advantageous to use a *direct* (as opposed to *iterative*) solver. In a direct solver, an approximate solution operator to the elliptic equation is built once. The cost to build it is typically higher than the cost required for a single elliptic solve using an iterative method such as multigrid, but the upside is that after it has been built, each subsequent solve is very fast. In this chapter, we argue that a particularly efficient direct solver to use in this context is a method obtained by combining a multidomain spectral collocation discretization (a

T. Babb · D. Appelö (✉)
University of Colorado, Boulder, CO, USA
e-mail: tracy.babb@colorado.edu; daniel.appelo@colorado.edu

P.-G. Martinsson
University of Texas, Austin, TX, USA
e-mail: pgm@ices.utexas.edu

so-called "patching method", see e.g. Ch. 5.13 in [3]) with a nested dissection type solver. It has recently been demonstrated [1, 7, 12] that this combined scheme, which we refer to as a "Hierarchial Poincaré–Steklov (HPS)" solver, can be used with very high local discretization orders (up to $p = 20$ or higher) without jeopardizing either speed or stability, as compared to lower order methods.

In this chapter, we investigate the stability and accuracy that is obtained when combining high-order time-stepping schemes with the HPS method for solving elliptic equations. We restrict attention to relatively simple geometries (mostly rectangles). The method can without substantial difficulty be generalized to domains that can naturally be expressed as a union of rectangles, possibly mapped via curvilinear smooth parameter maps.

A longer version of this chapter with additional details is available at [2]. Also note that the conclusions are deferred to Part II of this paper (same issue).

The Hierarchical Poincaré–Steklov Method

In this section, we describe a computationally efficient and highly accurate technique for solving an elliptic PDE of the form

$$\begin{aligned} [Au](\boldsymbol{x}) &= g(\boldsymbol{x}), \qquad & \boldsymbol{x} \in \Omega, \\ u(\boldsymbol{x}) &= f(\boldsymbol{x}), \qquad & \boldsymbol{x} \in \Gamma, \end{aligned} \tag{2}$$

where Ω is a domain with boundary Γ, and where A is a variable coefficient elliptic differential operator

$$\begin{aligned} [Au](\boldsymbol{x}) = -c_{11}(\boldsymbol{x})[\partial_1^2 u](\boldsymbol{x}) - 2c_{12}(\boldsymbol{x})[\partial_1\partial_2 u](\boldsymbol{x}) - c_{22}(\boldsymbol{x})[\partial_2^2 u](\boldsymbol{x}) \\ + c_1(\boldsymbol{x})[\partial_1 u](\boldsymbol{x}) + c_2(\boldsymbol{x})[\partial_2 u](\boldsymbol{x}) + c(\boldsymbol{x})\, u(\boldsymbol{x}) \end{aligned}$$

with smooth coefficients. In the present context, (2) represents an elliptic solve that is required in an implicit time-descretization technique of a parabolic PDE, as discussed in Sect. . For simplicity, let us temporarily suppose that the domain Ω is rectangular; the extension to more general domains is discussed in Remark 1.

Our ambition here is merely to provide a high level description of the method; for implementation details, we refer to [1, 2, 7–9, 12, 13].

Discretization

We split the domain Ω into $n_1 \times n_2$ boxes, each of size $h \times h$. Then on each box, we place a $p \times p$ tensor product grid of Chebyshev nodes, as shown in Fig. . W e u s e collocation to discretize the PDE (2). With $\{\boldsymbol{x}_i\}_{i=1}^N$ denoting the collocation points,

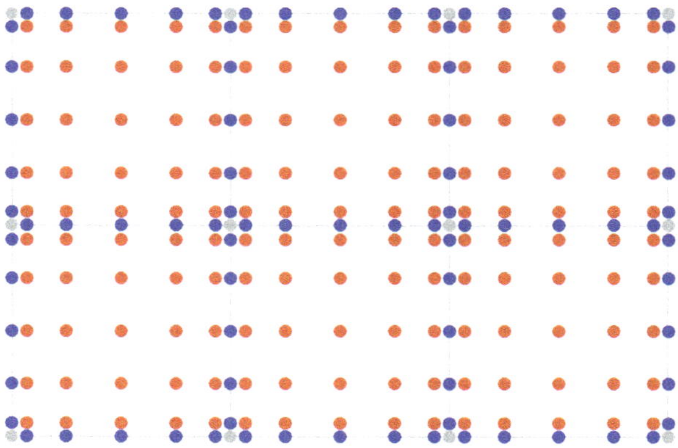

Fig. The domain Ω is split into $n_1 \times n_2$ squares, each of size $h \times h$. In the figure, $n_1 = 3$ and $n_2 = 2$. Then on each box, a $p \times p$ tensor product grid of Chebyshev nodes is placed, shown for $p = 7$. At red nodes, the PDE (2) is enforced via collocation of the spectral differentiation matrix. At the blue nodes, we enforce continuity of the normal fluxes. Observe that the corner nodes (gray) are excluded from consideration

the vector u that represents our approximation to the solution u of (2) is given simply by $\mathsf{u}(i) \approx u(\boldsymbol{x}_i)$. We then discretize (2) as follows:

1. For each collocation node that is *internal* to a box (red nodes in Fig.), we enforce (2) by directly collocating the spectral differential operator supported on the box, as described in, e.g., Trefethen [15].
2. For each collocation node on an *edge* between two boxes (blue nodes in Fig.), we enforce that the normal fluxes across the edge be continuous. For instance, for a node $\boldsymbol{x}_i$ on a vertical line, we enforce that $\partial u/\partial x_1$ is continuous across the edge by equating the values for $\partial u/\partial x_1$ obtained by spectral differentiation of the boxes to the left and to the right of the edge. For an edge node that lies on the external boundary Γ, simply evaluate the normal derivative at the node, as obtained by spectral differentiation in the box that holds the node.
3. All corner nodes (gray in Fig.) are dropped from consideration. For an elliptic operator of the form (2) with $c_{12} = 0$, it turns out that these values do not contribute to any of the spectral derivatives on the interior nodes, which means that the method without corner nodes is mathematically equivalent to the method with corner nodes, see [5, in chap.] for details. When $q_2 \neq 0$, one must in order to drop the corner nodes include an extrapolation operator when evaluating the terms involving the spectral representation of the mixed derivative $\partial^2 u/\partial x_1 \partial x_2$. This may lead to a slight drop in the order of convergence, but the difference is hardly noticeable in practice, and the exclusion of corner nodes greatly simplifies the implementation of the method.

Since we exclude the corner nodes from consideration, the total number of nodes in the grid equals $N = (p-2)\big(p\,n_1\,n_2 + n_1 + n_2\big) \approx p^2\,n_1\,n_2$. The discretization procedure described then results in an $N \times N$ matrix A. For a node i, the value of $\mathsf{A}(i,:)\mathsf{u}$ depends on what type of node i is:

$$\mathsf{A}(i,:)\mathsf{u} \approx \begin{cases} [Au](\boldsymbol{x}_i) \text{ for any interior (red) node,} \\ 0 \text{ for any edge node (blue) not on } \Gamma, \\ \partial u/\partial n \text{ for any edge node (blue) on } \Gamma. \end{cases}$$

This matrix A can be used to solve BVPs with a variety of different boundary conditions, including Dirichlet, Neumann, Robin, and periodic [12].

In many situations, a simple uniform mesh of the type shown in Fig. is not optimal, since the regularity in the solution may vary greatly, due to corner singularities, localized loads, etc. The HPS method can easily be adapted to handle local refinement. The essential difficulty that arises is that when boxes of different sizes are joined, the collocation nodes along the joint boundary will not align. It is demonstrated in [1, 5] that this difficulty can stably and efficiently be handled by incorporating local interpolation operators.

A Hierarchical Direct Solver

A key observation in previous work on the HPS method is that the sparse linear system that results from the discretization technique described in Section is particularly well suited for direct solvers, such as the well-known multifrontal solvers that compute an LU-factorization of a sparse matrix. The key is to minimize fill-in by using a so called nested dissection ordering [4, 6]. Such direct solvers are very powerful in a situation where a sequence of linear systems with the same coefficient matrix needs to be solved, since each solve is very fast once the coefficient matrix has been factorized. This is precisely the environment under consideration here. The particular advantage of combining the multidomain spectral collocation discretization described in Section, say that the time required for factorizing the matrix is *independent* of the local discretization order. As we will see in the numerical experiments, this enables us to attain both very high accuracy, and very high computational efficiency.

Remark 1 (General Domains) For simplicity we restrict attention to rectangular domains in this chapter. The extension to domains that can be mapped to a union of rectangles via smooth coordinate maps is relatively straight-forward, since the method can handle variable coefficient operators [12,]. Some care must be exercised since singularities may arise at intersections of parameter maps, which may require local refinement to maintain high accuracy.

The direct solver described exactly mimics the classical nested dissection method, and has the same asymptotic complexity of $O(N^{1.5})$ for the "build" (or "factorization") stage, and then $O(N \log N)$ cost for solving a system once the coefficient matrix has been factorized. Storage requirements are also $O(N \log N)$. A more precise analysis of the complexity that takes into account the dependence on the order p of the local discretization shows [1] that $T_{\text{build}} \sim N\,p^4 + N^{1.5}$, and $T_{\text{solve}} \sim N\,p^2 + N \log N$.

Time-Stepping Methods

For high-order time-stepping of (1), we use the so called Explicit, Singly Diagonally Implicit Runge–Kutta (ESDIRK) methods. These methods have a Butcher diagram with a constant diagonal γ and are of the form

$$
\begin{array}{c|ccccccc}
0 & 0 & & & & & & \\
2\gamma & \gamma & \gamma & & & & & \\
c_3 & a_{3,1} & a_{3,2} & \gamma & & & & \\
\vdots & \vdots & \vdots & \ddots & \ddots & & & \\
c_{s-1} & a_{s-1,1} & a_{s-1,2} & a_{s-1,3} & \cdots & \gamma & & \\
1 & b_1 & b_2 & b_3 & \cdots & b_{s-1} & \gamma \\
\hline
 & b_1 & b_2 & b_3 & \cdots & b_{s-1} & \gamma
\end{array}
$$

ESDIRK methods offer the advantages of stiff accuracy and L-stability. They are particularly attractive when used in conjunction with direct solvers since the elliptic solve required in each stage involves the same coefficient matrix $(I - h\gamma\mathscr{L})$, where h is the time-step.

In general we split the right hand side of (1) into a stiff part, $F^{[1]}$, that will be treated implicitly using ESDIRK methods, and a part, $F^{[2]}$, that will be treated explicitly (with a Butcher table denoted $\hat{c}$, $\hat{A}$, and $\hat{b}$). Precisely we will use the Additive Runge–Kutta (ARK) methods by Carpenter and Kennedy [11], of order 3, 4 and 5.

We may choose to formulate the Runge–Kutta method in terms of either solving for slopes or solving for stage solutions. We denote these the k_i formulation and the u_i formulation, respectively. When solving for slopes the stage computation is

$$k_i^n = F^{[1]}(t_n + c_i \Delta t, u^n + \Delta t \sum_{j=1}^{s} a_{ij} k_j^n + \Delta t \sum_{j=1}^{s} \hat{a}_{ij} l_j^n), \quad i = 1, \ldots, s, \tag{3}$$

$$l_i^n = F^{[2]}(t_n + c_i \Delta t, u^n + \Delta t \sum_{j=1}^{s} a_{ij} k_j^n + \Delta t \sum_{j=1}^{s} \hat{a}_{ij} l_j^n), \quad i = 1, \ldots, s. \tag{4}$$

Note that the explicit nature of (4) is encoded in the fact that the elements on the diagonal and above in $\hat{A}$ are zero. Once the slopes have been computed the solution at the next time-step is assembled as

$$u^{n+1} = u^n + \Delta t \sum_{j=1}^{s} b_j k_j^n + \Delta t \sum_{j=1}^{s} \hat{b}_j l_j^n. \tag{5}$$

If the method is instead formulated in terms of solving for the stage solutions the implicit solves take the form

$$u_i^n = u^n + \Delta t \sum_{j=1}^{s} \left(a_{ij} F^{[1]}(t_n + c_j \Delta t, u_j^n) + \hat{a}_{ij} F^{[2]}(t_n + c_j \Delta t, u_j^n) \right),$$

and the explicit update for u^{n+1} is given by

$$u^{n+1} = u^n + \Delta t \sum_{j=1}^{s} b_j (F^{[1]}(t_n + c_j \Delta t, u_j^n) + F^{[2]}(t_n + c_j \Delta t, u_j^n)).$$

The two formulations are algebraically equivalent but offer different advantages. For example, when working with the slopes we do not observe (see experiments presented in the second part of this paper) any order reduction due to time-dependent boundary conditions (see e.g. the analysis by Rosales et al. [14]). On the other hand and as discussed in some detail below, in solving for the slopes the HPS framework requires an additional step to enforce continuity.

We note that it is generally preferred to solve for the slopes when implementing implicit Runge–Kutta methods, particularly when solving very stiff problems where the influence of roundoff (or solver tolerance) errors can be magnified by the Lipschitz constant when solving for the stages directly.

Remark 2 The HPS method for elliptic solves was previously used in [10], which considered a linear hyperbolic equation

$$\frac{\partial u}{\partial t} = \mathscr{L} u(\boldsymbol{x}, t), \quad \boldsymbol{x} \in \Omega, t > 0,$$

where $\mathscr{L}$ is a skew-Hermitian operator. The evolution of the numerical solution can be performed by approximating the propagator $\exp(\tau \mathscr{L}) : L^2(\Omega) \to L^2(\Omega)$ via a rational approximation

$$\exp(\tau \mathscr{L}) \approx \sum_{m=-M}^{M} b_m (\tau \mathscr{L} - \alpha_m)^{-1}.$$

If application of $(\tau \mathscr{L} - \alpha_m)^{-1}$ to the current solution can be reduced to the solution of an elliptic-type PDE it is straightforward to apply the HPS scheme to each term in the approximation. A drawback with this approach is that multiple operators must be formed and it is also slightly more convenient to time step non-linear equations using the Runge–Kutta methods we use here.

There are two modifications to the HPS algorithm that are necessitated by the use of ARK time integrators, we discuss these in the next two subsections.

Neumann Data Correction in the Slope Formulation

In the HPS algorithm the PDE is enforced on interior nodes and continuity of the normal derivative is enforced on the leaf boundary. Now, due to the structure of the update formula (5), if at some time u^n has an error component in the null space of the operator that is used to solve for a slope k_i, then this will remain throughout the solution process. Although this does not affect the stability of the method it may result in loss of relative accuracy as the solution evolves. As a concrete example consider the heat equation

$$u_t = u_{xx}, \quad x \in [0,2], t > 0, \tag{6}$$

with the initial data $u(x,0) = 1 - |x-1|$, and with homogenous Dirichlet boundary conditions. We discretize this on two leaves which we denote by α and β.

Now in the k_i formulation, we solve several PDEs for the k_i values and update the solution as

$$u^{n+1} = u^n + \Delta t \sum_{j=1}^{s} b_j k_j^n.$$

Here, even though the individual slopes have continuous derivatives the kink in u^n will be propagated to u^{n+1}. In this particular example we would end up with the incorrect steady state solution $u(x,t) = 1 - |x-1|$.

Fortunately, this can easily be mitigated by adding a consistent penalization of the jump in the derivative of the solution during the merging of two leaves (for details see Section in [1]). That is, if we denote the jump by $[[\cdot]]$ we replace the condition $0 = [[Tk + h^k]]$ where Tk is the derivative from the homogenous part and h^k is the derivative for the particular solution (of the slope) by the condition $[[Tk + h^k - \Delta t^{-1} h^u]] = 0$. In comparison to [1] we get the slightly modified merge formula

$$\mathsf{k}_{i,3} = \left(\mathsf{T}^{\alpha}_{3,3} - \mathsf{T}^{\beta}_{3,3}\right)^{-1} \left(\mathsf{T}^{\beta}_{3,2}\mathsf{k}_{i,2} - \mathsf{T}^{\alpha}_{3,1}\mathsf{k}_{i,1} + \mathsf{h}^{k,\beta}_3 - \mathsf{h}^{k,\alpha}_3 - \frac{1}{\Delta t}(\mathsf{h}^{u,\alpha}_3 - \mathsf{h}^{u,\beta}_3)\right),$$

along with the modified equation for the fluxes of the particular solution on the parent box

$$\begin{bmatrix} \mathsf{v}_1 \\ \mathsf{v}_2 \end{bmatrix} = \left(\begin{bmatrix} \mathsf{T}^{\alpha}_{1,1} & 0 \\ 0 & \mathsf{T}^{\beta}_{2,2} \end{bmatrix} + \begin{bmatrix} \mathsf{T}^{\alpha}_{1,3} \\ \mathsf{T}^{\beta}_{2,3} \end{bmatrix} (\mathsf{T}^{\alpha}_{3,3} - \mathsf{T}^{\beta}_{3,3})^{-1} [- \mathsf{T}^{\alpha}_{3,1} \mid \mathsf{T}^{\beta}_{3,2}] \right) \begin{bmatrix} \mathsf{k}_{i,1} \\ \mathsf{k}_{i,2} \end{bmatrix} +$$
$$\begin{bmatrix} \mathsf{h}^{k,\alpha}_1 \\ \mathsf{h}^{k,\beta}_2 \end{bmatrix} + \begin{bmatrix} \mathsf{T}^{\alpha}_{1,3} \\ \mathsf{T}^{\beta}_{2,3} \end{bmatrix} (\mathsf{T}^{\alpha}_{3,3} - \mathsf{T}^{\beta}_{3,3})^{-1} (\mathsf{h}^{\beta}_3 - \mathsf{h}^{\alpha}_3 - \frac{1}{\Delta t}(\mathsf{h}^{u,\alpha}_3 - \mathsf{h}^{u,\beta}_3)).$$

Due to space we must refer to [1] for a detailed discussion of these equations. Briefly, $h^{k,\alpha}$ and $h^{k,\beta}$ above denote the spectral derivative on each child's boundary for the particular solution to the PDE for k_i and are already present in [1]. However, $h^{u,\alpha}$ and $h^{u,\beta}$, which denote the spectral derivative of u^n on the boundary from each child box, are new additions.

The above initial data is of course extreme but we note that the problem persists for any non-polynomial initial data with the size of the (stationary) error depending on resolution of the simulation. We further note that the described penalization removes this problem without affecting the accuracy or performance of the overall algorithm.

Remark 3 Although for linear constant coefficient PDE it may be possible to project the initial data in a way so that interior affine functions do not cause the difficulty above, for greater generality, we have chosen to enforce the extra penalization throughout the time evolution.

Remark 4 When utilizing the u_i formulation in a purely implicit problem we do not encounter the difficulty described above. This is because we enforce continuity of the derivative in u^n_s when solving

$$(I - \Delta t \gamma \mathscr{L}) u^n_s = u^n + \Delta t \mathscr{L}\Big(\sum_{j=1}^{s-1} a_{sj} u^n_j \Big) + \Delta t \sum_{j=1}^{s-1} a_{sj} g(x, t_n + c_j \Delta t),$$

followed by the update $u^{n+1} = u^n_s$.

Enforcing Continuity in the Explicit Stage

The second modification is to the first explicit stage in the k_i formulation. Solving a problem with no forcing this stage is simply

$$k^n_1 = \mathscr{L}(u_n).$$

When, for example, $\mathscr{L}$ is the Laplacian, we must evaluate it on all nodes on the interior of the physical domain. This includes the nodes on the boundary between two leafs where the spectral approximation to the Laplacian can be different if we use values from different leaves. The seemingly obvious choice, replacing the Laplacian on the leaf boundary by the average, leads to instability. However, stability can be restored if we enforce $k_1^n = \mathscr{L}(u_n)$ on the interior of each leaf and continuity of the derivative across each leaf boundary. Algorithmically, this is straightforward as these are the same conditions that are enforced in the regular HPS algorithm, except in this case we simply have an identity equation for k_1 on the interior nodes instead of a full PDE.

Although it is convenient to enforce continuity of the derivative using the regular HPS algorithm it can be done in a more efficient fashion by forming a separate system of equations involving only data on the leaf boundary nodes. In a single dimension on a discretization with n leafs this reduces the work associated with enforcing continuity of the derivative across leaf boundary nodes from solving $n \times (p-1)-1$ equations for $n \times (p-1)-1$ unknowns to solving a tridiagonal system of equations $n-1$ equations for $n-1$ unknowns.

In two dimensions the system is slightly different, but if we have $n \times n$ leafs with $p \times p$ Chebyshev nodes on each leaf then eliminating the explicit equations for the interior nodes reduces the system to $(p-2) \times 2n$ independent tridiagonal systems of $n-1$ equations with $n-1$ unknowns for a total of $(p-2) \times 2n \times (n-1)$ equations with $(p-2) \times 2n \times (n-1)$ unknowns.

When the u_i formulation is used for a fully implicit problem the intermediate stage values still requires us to evaluate $\mathscr{L}u^n$, but this quantity only enters through the body load in the intermediate stage PDEs. The explicit first stage in this formulation is simply $u_1^n = u^n$. Furthermore, while we must calculate

$$u^{n+1} = u^n + \Delta t \mathscr{L}\Big(\sum_{j=1}^{s} a_{sj} u_j^n\Big),$$

this is equivalent to u_s^n since $b_j = a_{sj}$ and we simply take $u^{n+1} = u_s^n$.

When both explicit and implicit terms are present, we proceed differently. Now, the values of u_i^n look almost identical to the implicit case and we still avoid the problem of an explicit "solve" in u_1^n, but we also have

$$u^{n+1} = u^n + \Delta t \sum_{j=1}^{s} b_j (F^{[1]}(t_n + c_j \Delta t, u_j^n) + F^{[2]}(t_n + c_j \Delta t, u_j^n))$$

The ESDIRK method has the property that $b_j = a_{sj}$, but for the explicit Runge–Kutta method we have $b_j \neq \hat{a}_{sj}$. When the explicit operator $F^{[2]}$ does not contain partial derivatives we need not enforce continuity of the derivative and can simply reformulate the method as

$$u^{n+1} = u_s^n + \Delta t \sum_{j=1}^{s} (a_{sj} - \hat{a}_{sj}) F^{[2]}(t_n + c_j \Delta t, u_j^n)$$

Boundary Conditions

The above description for Runge–Kutta methods does not address how to impose boundary conditions for a system of ODEs resulting from a discretization of a PDE. In particular, the different formulations incorporate boundary conditions in slightly different ways.

In this work we consider Dirichlet, Neumann, and periodic boundary conditions. For periodic boundary conditions the intermediate stage boundary conditions are enforced to be periodic for both formulations. As the k_i stage values are approximations to the time derivative of u, the imposed Dirichlet boundary conditions for $x \in \Gamma$ are $k_i^n = u_t(x, t_n + c_i \Delta t)$. When solving for u_i one may attempt to enforce boundary conditions using $u_i = u(x, t + c_i \Delta t), x \in \Gamma$. However, as demonstrated in part two of this series and discussed in detail in [14], this results in order reduction for time dependent boundary conditions.

In the HPS algorithm, Neumann or Robin boundary conditions are mapped to Dirichlet boundary conditions using the linear Dirichlet to Neumann operator as discussed for example in [1].

References

1. Babb, T., Gillman, A., Hao, S., Martinsson, P.: An accelerated Poisson solver based on multidomain spectral discretization. BIT Numer. Math. **58**, 851–879 (2018)
2. Babb, T., Martinsson, P.-G., Appelö, D.: HPS accelerated spectral solvers for time dependent problems (2018). arXiv:1811.04555
3. Canuto, C., Hussaini, M.Y., Quarteroni, A., Zang, T.A.: Spectral Methods: Evolution to Complex Geometries and Applications to Fluid Dynamics. Springer, Berlin (2007)
4. Duff, I., Erisman, A., Reid, J.: Direct Methods for Sparse Matrices. Oxford University Press, Oxford (1989)
5. Geldermans, P., Gillman, A.: An adaptive high order direct solution technique for elliptic boundary value problems. SIAM J. Sci. Comput. **41**(1), A292–A315 (2019). arXiv:1710.08787
6. George, A.: Nested dissection of a regular finite element mesh. SIAM J. Numer. Anal. **10**, 345–363 (1973)
7. Gillman, A., Martinsson, P.: A direct solver with $\mathcal{O}(N)$ complexity for variable coefficient elliptic PDEs discretized via a high-order composite spectral collocation method. SIAM J. Sci. Comput. **36**, A2023–A2046 (2014). arXiv:1307.2665
8. Gillman, A., Barnett, A., Martinsson, P.-G.: A spectrally accurate direct solution technique for frequency-domain scattering problems with variable media. BIT Numer. Math. **55**, 141–170 (2015)

9. Hao, S., Martinsson, P.: A direct solver for elliptic PDEs in three dimensions based on hierarchical merging of Poincaré-Steklov operators. J. Comput. Appl. Math. **308**, 419–434 (2016)
10. Haut, T., Babb, T., Martinsson, P., Wingate, B.: A high-order scheme for solving wave propagation problems via the direct construction of an approximate time-evolution operator. IMA J. Numer. Anal. **36**, 688–716 (2016)
11. Kennedy, C., Carpenter, M.: Additive Runge–Kutta schemes for convection-diffusion-reaction equations. Appl. Numer. Math. 44, 139–181 (2003)
12. Martinsson, P.: A direct solver for variable coefficient elliptic PDEs discretized via a composite spectral collocation method. J. Comput. Phys. **242**, 460–479 (2013)
13. Martinsson, P.: The hierarchical Poincaré-Steklov (HPS) solver for elliptic PDEs: a tutorial (2015). arXiv:1506.01308
14. Rosales, R., Seibold, B., Shirokoff, D., Zhou, D.: Order reduction in high-order Runge–Kutta methods for initial boundary value problems (2017). arXiv:1712.00897
15. Trefethen, L.: Spectral Methods in Matlab. SIAM, Philadelphia (2000)

7

Stabilised Hybrid Discontinuous Galerkin Methods for the Stokes Problem with Non-Standard Boundary Conditions

Gabriel R. Barrenechea, Michał Bosy, and Victorita Dolean

Introduction

The interest of this paper is to discretise the Stokes problem with non-standard boundary conditions. In [1], a hybrid discontinuous Galerkin (hdG) method was proposed and analysed for this problem. The finite element method used was the combination of BDM elements of order k for the velocity, and discontinuous elements of order $k-1$ for the pressure. In this paper we increase the order of the pressure space to k, while keeping the order for the velocity space fixed as k. Since this pair does not satisfy the inf-sup condition, a stabilisation term needs to be added.

The stabilisation term referred to above can be built using a diversity of approaches, but, roughly speaking, the stabilisation can be residual or non-residual. In [8] the authors added a mesh-dependent term penalising the gradient of the pressure to the formulation. Later, in [14] this method was restricted and reinterpreted

G. R. Barrenechea
Department of Mathematics and Statistics, University of Strathclyde, Glasgow, UK
e-mail: gabriel.barrenechea@strath.ac.uk

M. Bosy
Department of Mathematics and Statistics, University of Strathclyde, Glasgow, UK

Dipartimento di Matematica "F. Casorati", Universitá degli Studi di Pavia, Pavia, Italy
e-mail: michal.bosy@unipv.it

V. Dolean (✉)
Department of Mathematics and Statistics, University of Strathclyde, Glasgow, UK

University Côte d'Azur, CNRS, LJAD, Nice Cedex, France
e-mail: work@victoritadolean.com

as a Petrov–Galerkin scheme leading to the first consistent stabilised method, and further developments were presented in the works [7] and [13]. For a review of different residual stabilised finite element methods for the Stokes problem, see the review paper [2].

Now, due to their nature, residual methods include unphysical couplings to the formulation, and modify all the entries of the stiffness matrix. Hence, non-residual methods where only a positive semi-definite term penalising the pressure is added have also being proposed. Examples of this type of methods are the pressure gradient projection [9] and local pressure gradient stabilisation [3]. The methods just mentioned typically use two nested meshes in order to build the method. Thus, to avoid this complication, the local pressure gradient stabilisation has been also presented on the same mesh in [12]. Additionally, methods that use fluctuations of the pressure gradient are not effective when the finite element space for pressure is the piecewise constant space. The usual way to overcome this is to add pressure jumps to the formulation, as it has been done, e.g., in [16]. These have been shown to be very effective, but they do somehow temper with the data structure of the code. To avoid this, the authors in [10] present an approach that is based on polynomial-pressure-projection. This method works for low order of polynomials as was shown in [4], and preserves symmetry of the original equation.

In the light of the discussion of the previous paragraphs, in this work we propose a stabilised hdG method for the Stokes problem with non-standard boundary conditions. The method is reminiscent of the Dorhmann–Bochev method (from [10]), but uses the same velocity space used in the hdG method from [1].

Notations and Model Problem

Let Ω be an open polygonal domain in $\mathbb{R}^2$ with Lipschitz boundary $\Gamma := \partial\Omega$. We use boldface font for tensor or vector variables e.g. $\boldsymbol{u}$ is a velocity vector field. The scalar variables will be italic e.g. p denotes pressure scalar value. We define the stress tensor $\boldsymbol{\sigma} := \nu\nabla\boldsymbol{u} - p\boldsymbol{I}$ (where $\nu > 0$ is the fluid viscosity and $\boldsymbol{I}$ is the identity matrix) and the flux as $\boldsymbol{\sigma_n} := \boldsymbol{\sigma}\ \boldsymbol{n}$. In addition, we denote normal and tangential components as follows $u_n := \boldsymbol{u}\cdot\boldsymbol{n}$, $u_t := \boldsymbol{u}\cdot\boldsymbol{t}$, $\sigma_{nn} := \boldsymbol{\sigma_n}\cdot\boldsymbol{n}$, where $\boldsymbol{n}$ is the outward unit normal vector to the boundary Γ and $\boldsymbol{t}$ is a vector tangential to Γ such that $\boldsymbol{n}\cdot\boldsymbol{t} = 0$.

For $D \subset \Omega$, we use the standard $L^2(D)$ space with the following norm

$$\|f\|_D^2 := \int_D f^2\,d\boldsymbol{x} \text{ for all } f \in L^2(D).$$

Let us define, for $m \in \mathbb{N}$, the following Sobolev spaces

$$H^m(D) := \left\{v \in L^2(D) : \forall\, |\boldsymbol{\alpha}| \leq m\ \partial^{\boldsymbol{\alpha}} v \in L^2(D)\right\},$$
$$H(div, D) := \left\{\boldsymbol{v} \in [L^2(D)]^2 : \nabla \cdot \boldsymbol{v} \in L^2(D)\right\},$$

where, for $\boldsymbol{\alpha} = (\alpha_1, \alpha_2) \in \mathbb{N}^2$, $|\boldsymbol{\alpha}| = \alpha_1 + \alpha_2$, and $\partial^{\boldsymbol{\alpha}} = \frac{\partial^{|\boldsymbol{\alpha}|}}{\partial x_1^{\alpha_1} \partial x_2^{\alpha_2}}$. In addition, we will use the standard semi-norm and norm for the Sobolev space $H^m(D)$

$$|f|^2_{H^m(D)} := \sum_{|\boldsymbol{\alpha}|=m} \|\partial^{\boldsymbol{\alpha}} f\|^2_D, \quad \|f\|^2_{H^m(D)} := \sum_{k=0}^{m} |f|^2_{H^k(D)}\ \forall\, f \in H^m(D).$$

In this work, we consider the two dimensional Stokes problem with tangential-velocity and normal-flux (TVNF) boundary conditions

$$\begin{cases} -\nu \Delta \boldsymbol{u} + \nabla p &= \boldsymbol{f} \ \text{ in } \Omega, \\ \nabla \cdot \boldsymbol{u} &= 0 \ \text{ in } \Omega, \\ \sigma_{nn} &= g \ \text{ on } \Gamma, \\ u_t &= 0 \ \text{ on } \Gamma, \end{cases} \tag{1}$$

where $\boldsymbol{u} : \bar{\Omega} \to \mathbb{R}^2$ is the unknown velocity field, $p : \bar{\Omega} \to \mathbb{R}$ the pressure, $\nu > 0$ the viscosity, which is considered to be constant, and $\boldsymbol{f} \in [L^2(\Omega)]^2$, $g \in L^2(\Gamma)$ are given functions. The restriction to homogeneous Dirichlet conditions on u_t is made only to simplify the presentation.

Let $\{\mathcal{T}_h\}_{h>0}$ be a regular family of triangulations of $\bar{\Omega}$ made of triangles. For each triangulation $\mathcal{T}_h$, $\mathcal{E}_h$ denotes the set of its edges. In addition, for each of element $K \in \mathcal{T}_h$, $h_K := \text{diam}(K)$, and we denote $h := \max_{K \in \mathcal{T}_h} h_K$. We define following Sobolev spaces on the triangulation $\mathcal{T}_h$ and the set of all edges in $\mathcal{E}_h$

$$L^2(\mathcal{E}_h) := \left\{v : v|_E \in L^2(E)\ \forall\, E \in \mathcal{E}_h\right\},$$
$$H^m(\mathcal{T}_h) := \left\{v \in L^2(\Omega) : v|_K \in H^m(K)\ \forall\, K \in \mathcal{T}_h\right\} \text{ for } m \in \mathbb{N},$$

with the corresponding broken norms.

Now we will introduce the finite element spaces that discretise the above spaces. Let $k \geq 1$. We start by introducing the velocity and pressure spaces. To discretise the velocity $\boldsymbol{u}$ we use the Brezzi–Douglas–Marini space (see [5, Section 2.3.1]) of order $k \geq 1$ defined by

$$\boldsymbol{BDM}_h^k := \left\{\boldsymbol{v_h} \in H(div, \Omega) : \boldsymbol{v_h}|_K \in \left[\mathbb{P}_k(K)\right]^2\ \forall\, K \in \mathcal{T}_h\right\}.$$

Associated to this space, we introduce the BDM projection $\Pi^k : [H^1(\Omega)]^2 \to \boldsymbol{BDM}_h^k$ defined in [5, Section]. The pressure is discretised using the following space

$$Q_h^k := \left\{ q_h \in L^2(\Omega) : q_h|_K \in \mathbb{P}_k(K) \ \forall\, K \in \mathcal{T}_h \right\}.$$

Associated to this space we define the local $L^2(K)$-projection $\Psi_K^k : L^2(K) \to \mathbb{P}_k(K)$ for each $K \in \mathcal{T}_h$ defined as follows. For every $w \in L^2(K)$, $\Psi_K^k(w)$ is the unique element of $\mathbb{P}_k(K)$ satisfying $\int_K \Psi_K^k(w) v_h dx = \int_K w v_h dx \quad \forall\, v_h \in \mathbb{P}_k(K)$, and we define the continuous projection $\Psi^k|_K = \Psi_K^k$ for all $K \in \mathcal{T}_h$.

The last ingredient needed in the method described below is a finite element space associated to a family of Lagrange multipliers associated to the edges of the triangulation. These multipliers will be denoted by $\tilde{u}$ and are meant to approximate the tangential trace of the velocity $\boldsymbol{u}$ on the edges of the triangulation. For this, and in order to propose a discretisation with fewer degrees of freedom, we discretise the Lagrange multiplier $\tilde{u}$ using the space

$$M_{h,0}^{k-1} := \left\{ \tilde{v}_h \in L^2(\mathcal{E}_h) : \tilde{v}_h|_E \in \mathbb{P}_{k-1}(E) \ \forall\, E \in \mathcal{E}_h,\ \tilde{v}_h = 0 \text{ on } \Gamma \right\}.$$

Furthermore, we introduce for all $E \in \mathcal{E}_h$ the $L^2(E)$-projection $\Phi_E^{k-1} : L^2(E) \to \mathbb{P}_{k-1}(E)$ defined as follows. For every $\tilde{w} \in L^2(E)$, $\Phi_E^{k-1}(\tilde{w})$ is the unique element of $\mathbb{P}_{k-1}(E)$ satisfying $\int_E \Phi_E^{k-1}(\tilde{w})\tilde{v}_h\, ds = \int_E \tilde{w}\tilde{v}_h\, ds \quad \forall\, \tilde{v}_h \in \mathbb{P}_{k-1}(E)$, and we denote $\Phi^{k-1} : L^2(\mathcal{E}_h) \to M_h^{k-1}$ defined as $\Phi^{k-1}|_E := \Phi_E^{k-1}$ for all $E \in \mathcal{E}_h$.

The Stabilised Method

Our approach is to write the discrete problem with the same degree of polynomials for velocity and pressure spaces. In other words, denoting $\boldsymbol{V_h} := \boldsymbol{BDM}_h^k \times M_{h,0}^{k-1}$, we want to use the space $\boldsymbol{V_h} \times Q_h^k$, instead of $\boldsymbol{V_h} \times Q_h^{k-1}$ as it was done in [1]. To do this, we need the proper stabilisation term, because this choice of spaces does not guarantee inf-sup stability.

The first ingredient in the definition of the stabilised method for (1) we use the same bilinear forms as in [1], this is

$$a\Big((\boldsymbol{w_h}, \tilde{w}_h), (\boldsymbol{v_h}, \tilde{v}_h)\Big) := \sum_{K \in \mathcal{T}_h} \Bigg(\int_K \nu \nabla \boldsymbol{w_h} : \nabla \boldsymbol{v_h}\, d\boldsymbol{x}$$

$$- \int_{\partial K} \nu \left(\boldsymbol{\partial_n w_h}\right)_t \left((\boldsymbol{v_h})_t - \tilde{v}_h \right) ds + \varepsilon \int_{\partial K} \nu\big((\boldsymbol{w_h})_t - \tilde{w}_h \big) \left(\boldsymbol{\partial_n v_h}\right)_t\, ds$$

$$+\nu\frac{\tau}{h_K}\int_{\partial K}\Phi^{k-1}\big((\boldsymbol{w_h})_t-\tilde{w}_h\big)\Phi^{k-1}\big((\boldsymbol{v_h})_t-\tilde{v}_h\big)\,ds\Big)$$

$$b\left(\left(\boldsymbol{v_h},\tilde{v}_h\right),q_h\right):=-\sum_{K\in\mathcal{T}_h}\int_K q_h\nabla\cdot\boldsymbol{v_h}\,d\boldsymbol{x},$$

where $\varepsilon\in\{-1,1\}$ and $\tau>0$ is a stabilisation parameter. In addition, to compensate for the non-inf-sup stability of the finite element spaces we have chosen, we introduce the bilinear form

$$s\left(p_h,q_h\right):=\frac{1}{\nu}\int_\Omega\left(p_h-\Psi^{k-1}p_h\right)\left(q_h-\Psi^{k-1}q_h\right)\,d\boldsymbol{x}.$$

With these ingredients we can now present the finite element method analysed in this work: *Find* $\left(\boldsymbol{u_h},\tilde{u}_h,p_h\right)\in\boldsymbol{V_h}\times Q_h^k$ *such that for all* $\left(\boldsymbol{v_h},\tilde{v}_h,q_h\right)\in\boldsymbol{V_h}\times Q_h^k$

$$A\left(\left(\boldsymbol{u_h},\tilde{u}_h,p_h\right),\left(\boldsymbol{v_h},\tilde{v}_h,q_h\right)\right)=\int_\Omega \boldsymbol{f}\boldsymbol{v_h}\,d\boldsymbol{x}+\int_\Gamma g(\boldsymbol{v_h})_n\,ds, \tag{2}$$

where

$$\begin{aligned}A\left(\left(\boldsymbol{u_h},\tilde{u}_h,p_h\right),\left(\boldsymbol{v_h},\tilde{v}_h,q_h\right)\right):=&a\left(\left(\boldsymbol{u_h},\tilde{u}_h\right),\left(\boldsymbol{v_h},\tilde{v}_h\right)\right)+b\left(\left(\boldsymbol{v_h},\tilde{v}_h\right),p_h\right)\\&+b\left(\left(\boldsymbol{u_h},\tilde{u}_h\right),q_h\right)-s\left(p_h,q_h\right).\end{aligned}$$

Well-Posedness of the Discrete Problem

Let us consider the following norm on $\boldsymbol{V_h}$ (see [1, Lemma 3.2] for a proof that this is actually a norm in $\boldsymbol{V_h}$)

$$|||\left(\boldsymbol{w_h},\tilde{w}_h\right)|||^2:=\nu\sum_{K\in\mathcal{T}_h}\left(|\boldsymbol{w_h}|^2_{H^1(K)}+h_K\left\|\boldsymbol{\partial_n w_h}\right\|^2_{\partial K}+\frac{\tau}{h_K}\left\|\Phi^{k-1}\big((\boldsymbol{w_h})_t-\tilde{w}_h\big)\right\|^2_{\partial K}\right).$$

The first step towards proving the stability of Method (2) is the following weak inf-sup condition for b.

Lemma 1 *There exist constants* $C_1,C_2>0$*, independent of* h_K *and* ν*, such that*

$$\sup_{(\boldsymbol{v_h},\tilde{v}_h)\in\boldsymbol{V_h}}\frac{b\left(\left(\boldsymbol{v_h},\tilde{v}_h\right),q_h\right)}{|||\left(\boldsymbol{v_h},\tilde{v}_h\right)|||}\geq C_1\left\|q_h\right\|_\Omega-C_2\left\|q_h-\Psi^{k-1}q_h\right\|_\Omega\quad\forall q_h\in Q_h^k. \tag{3}$$

Proof We consider an arbitrary $q_h \in Q_h^k$. Let $\tilde{\Omega}$ be a convex, open, Lipschitz set such that $\Omega \subset \tilde{\Omega}$, and let us consider following extension

$$\hat{q}_h := \begin{cases} q_h \text{ in } & \Omega \\ 0 \text{ in } \tilde{\Omega} \setminus \Omega \end{cases}.$$

Let now $\boldsymbol{\phi}$ be the unique weak solution of the problem

$$\begin{cases} -\Delta \boldsymbol{\phi} = \hat{q}_h \text{ in } \tilde{\Omega} \\ \quad \boldsymbol{\phi} = 0 \quad \text{on } \partial\Omega \end{cases}.$$

Since $\tilde{\Omega}$ is convex, then $\boldsymbol{\phi} \in H^2(\tilde{\Omega})$. Then $\boldsymbol{w} := \nabla\boldsymbol{\phi}|_\Omega$ belongs to $[H^1(\Omega)]^2$, and for $\tilde{w} := w_t$,

$$b\left(\left(\boldsymbol{w}, \tilde{w}\right), q_h\right) = \|q_h\|_\Omega^2 \quad \forall q_h \in Q_h^k. \tag{4}$$

In addition, applying standard regularity results, see [5, Section , we get

$$\|\boldsymbol{w}\|_{H^1(\Omega)} \le \|\nabla\boldsymbol{\phi}\|_{H^1(\tilde{\Omega})} \le c_1\|q_h\|_\Omega. \tag{5}$$

In [1, Lemma 3.5] it is shown that there exists a Fortin operator $\boldsymbol{\Pi} : \left[H^1(\Omega)\right]^2 \to \boldsymbol{V_h}$ satisfying the following condition: for all $\boldsymbol{v} \in [H^1(\Omega)]^2$ the following holds

$$b\left(\left(\boldsymbol{v}, \tilde{v}\right), q_h\right) = b\left(\boldsymbol{\Pi}\left(\boldsymbol{v}\right), q_h\right) \quad \forall\, q_h \in Q_h^{k-1}, \tag{6}$$

$$|||\boldsymbol{\Pi}\left(\boldsymbol{v}\right)||| \le C\sqrt{\nu}\|\boldsymbol{v}\|_{H^1(\Omega)}. \tag{7}$$

Let $\left(\boldsymbol{w_h}, \tilde{w}_h\right) := \boldsymbol{\Pi}\left(\boldsymbol{w}\right)$, then thanks to (6), (4) and the continuity of b (see [1, Lemma 3.3])

$$\begin{aligned} b\left(\left(\boldsymbol{w_h}, \tilde{w}_h\right), q_h\right) &= b\left(\left(\boldsymbol{w}, \tilde{w}\right), q_h\right) - b\left(\left(\boldsymbol{w} - \boldsymbol{w}_h, \tilde{w} - \tilde{w}_h\right), q_h - \Psi^{k-1}q_h\right) \\ &\ge \|q_h\|_\Omega^2 - c_2\sqrt{\sum_{K\in\mathcal{T}_h} |\boldsymbol{w_h} - \boldsymbol{w}|_{H^1(K)}^2}\left\|q_h - \Psi^{k-1}q_h\right\|_\Omega. \end{aligned}$$

Using the approximation properties of the BDM interpolation operator (see [5, Preposition 2.5.1]) and (5)

$$\begin{aligned} b\left(\left(\boldsymbol{w_h}, \tilde{w}_h\right), q_h\right) &\ge \left(\frac{1}{c_1}\|q_h\|_\Omega - c_2c_3\left\|q_h - \Psi^{k-1}q_h\right\|_\Omega\right)|\boldsymbol{w}|_{H^1(\Omega)} \\ &\ge \left(C_1\|q_h\|_\Omega - C_2\left\|q_h - \Psi^{k-1}q_h\right\|_\Omega\right)|||\left(\boldsymbol{w_h}, \tilde{w}_h\right)|||, \end{aligned}$$

where, in the last estimate we have used the stability of the Fortin operator $\boldsymbol{\Pi}$ in the $|||\cdot|||$ norm (7). This proves the result with $C_1 = \frac{1}{C\sqrt{\nu c_1}}$ and $C_2 = \frac{c_2 c_3}{C\sqrt{\nu}}$. □

Before showing an inf-sup condition, we prove the continuity of bilinear form A.

Lemma 2 *There exists a constant $C > 0$ such that, for all $\left(\boldsymbol{w_h}, \tilde{w}_h\right), \left(\boldsymbol{v_h}, \tilde{v}_h\right) \in \boldsymbol{V_h}$ and $r_h, q_h \in Q_h^k$, we have*

$$\left|A\left(\left(\boldsymbol{w_h}, \tilde{w}_h, r_h\right), \left(\boldsymbol{v_h}, \tilde{v}_h, q_h\right)\right)\right| \leq C|||\left(\boldsymbol{w_h}, \tilde{w}_h, r_h\right)|||_h|||\left(\boldsymbol{v_h}, \tilde{v}_h, q_h\right)|||_h. \tag{8}$$

Proof We use the continuity of the bilinear forms (see [1, Lemma 3.3]) and the fact that the projection is a bounded operator. □

The final step towards stability is proving the inf-sup condition for bilinear form A.

Lemma 3 *There exists $\beta > 0$ independent of h_K such that for all $\left(\boldsymbol{w_h}, \tilde{u}_h, r_h\right) \in \boldsymbol{V_h} \times Q_h^k$ the following holds*

$$\sup_{\left(\boldsymbol{v_h}, \tilde{v}_h, q_h\right) \in \boldsymbol{V_h} \times Q_h^k} \frac{A\left(\left(\boldsymbol{w_h}, \tilde{w}_h, r_h\right), \left(\boldsymbol{v_h}, \tilde{v}_h, q_h\right)\right)}{|||\left(\boldsymbol{v_h}, \tilde{v}_h, q_h\right)|||_h} \geq \beta|||\left(\boldsymbol{w_h}, \tilde{w}_h, r_h\right)|||_h. \tag{9}$$

As a consequence, Problem (2) *is well-posed.*

Proof Let $\left(\boldsymbol{w_h}, \tilde{w}_h, r_h\right) \in \boldsymbol{V_h} \times Q_h^k$. The idea of the proof is to construct an appropriate $\left(\boldsymbol{v_h}, \tilde{v}_h, q_h\right)$ such that

$$A\left(\left(\boldsymbol{w_h}, \tilde{w}_h, r_h\right), \left(\boldsymbol{v_h}, \tilde{v}_h, q_h\right)\right) \geq c|||\left(\boldsymbol{w_h}, \tilde{w}_h, r_h\right)|||_h\,|||\left(\boldsymbol{v_h}, \tilde{v}_h, q_h\right)|||_h.$$

To achieve that we use coercivity of a (see [1, Lemma 3.4]), continuity of c (see [1, Lemma 3.3]) and Lemma 2. For details see [6]. □

Error Analysis

In this section we present the error estimates for the method. The addition of the stabilising bilinear form $s(\cdot,\cdot)$ introduced a consistency error. However according to [4], this should not be viewed as a serious flaw, as this consistency error can be bounded in an optimal way. The following result is the first step towards that goal.

Lemma 4 *Let $\left(\boldsymbol{u}, p\right) \in \left[H^1(\Omega) \cap H^2\left(\mathcal{T}_h\right)\right]^2 \times L^2(\Omega)$ be the solution of the problem* (1) *and $\tilde{u} = u_t$ on all edges of $\mathcal{E}_h$. If $\left(\boldsymbol{u_h}, \tilde{u}_h, p_h\right) \in \boldsymbol{V_h} \times Q_h^k$ solves* (2), *then for all* $\left(\boldsymbol{v_h}, \tilde{v}_h, q_h\right) \in \boldsymbol{V_h} \times Q_h^k$ *the following holds*

$$A\left(\left(\boldsymbol{u}-\boldsymbol{u_h}, \tilde{u}-\tilde{u}_h, p-p_h\right), \left(\boldsymbol{v_h}, \tilde{v}_h, q_h\right)\right) = s\left(p, q_h\right). \tag{10}$$

Next, we introduce the following norm

$$|||(\boldsymbol{u}, \tilde{u}, p)|||_h := |||(\boldsymbol{u}, \tilde{u})||| + \frac{1}{\sqrt{\nu}}\|p\|_{\Omega}, \tag{11}$$

and prove the following variant of Cea's lemma [11, Lemma 2.28] for this stabilised Stokes problem.

Lemma 5 *Let* $(\boldsymbol{u}, p) \in \left[H^1(\Omega) \cap H^2(\mathcal{T}_h)\right]^2 \times L^2(\Omega)$ *be the solution of the problem* (1) *and* $\tilde{u} = u_t$ *on all edges of* $\mathcal{E}_h$. *If* $\left(\boldsymbol{u_h}, \tilde{u}_h, p_h\right) \in \boldsymbol{V_h} \times Q_h^k$ *solves* (2), *then there exists* $C > 0$, *independent of* h *and* ν, *such that*

$$\begin{aligned} |||\left(\boldsymbol{u}-\boldsymbol{u_h}, \tilde{u}-\tilde{u}_h, p-p_h\right)|||_h \leq & C \inf_{(\boldsymbol{v_h}, \tilde{v}_h, q_h) \in \boldsymbol{V_h} \times Q_h^k} |||\left(\boldsymbol{u}-\boldsymbol{v_h}, \tilde{u}-\tilde{v}_h, p-q_h\right)|||_h \\ & + \frac{C}{\sqrt{\nu}}\left\|p-\Psi^{k-1}p\right\|_{\Omega}. \end{aligned} \tag{12}$$

Proof It is a combination of Lemmas 1, 2 and 3. For details see [6]. □

Lemma 6 *Let* $(\boldsymbol{u}, p) \in \left[H^1(\Omega) \cap H^2(\mathcal{T}_h)\right]^2 \times L^2(\Omega)$ *be the solution of the problem* (1) *and* $\tilde{u} = u_t$ *on all edges of* $\mathcal{E}_h$. *If* $\left(\boldsymbol{u_h}, \tilde{u}_h, p_h\right) \in \boldsymbol{V_h} \times Q_h^k$ *solves* (2), *then there exists* $C > 0$, *independent of* h *and* ν, *such that*

$$|||\left(\boldsymbol{u}-\boldsymbol{u_h}, \tilde{u}-\tilde{u}_h, p-p_h\right)|||_h \leq C h^k \left(\sqrt{\nu}\|\boldsymbol{u}\|_{H^{k+1}(\mathcal{T}_h)} + \frac{1}{\sqrt{\nu}}\|p\|_{H^k(\mathcal{T}_h)}\right).$$

Proof It is a combination of [1, Lemmas 3.8] and Lemma 5 with the local L^2-projection approximation [11, Theorem 1.103]. □

Numerical Experiments

The computational domain is the unit square $\Omega = (0, 1)^2$. We present the results for $k = 1$, that is the discrete space is given by $\boldsymbol{BDM_h^1} \times M_{h,0}^0 \times Q_h^1$. We test both the symmetric method ($\varepsilon = -1$) and the non-symmetric method ($\varepsilon = 1$). We have followed the recommendation given in [15, Section] and taken $\tau = 6$.

We choose the right hand side f and the boundary condition g such that the exact solution is given by

$$\boldsymbol{u} = \operatorname{curl}\left[\left(1-\cos((1-x)^2)\right)\sin(x^2)\sin(y^2)\left(1-\cos((1-y)^2)\right)\right], \quad p = \tan(xy).$$

In Fig. a and b we depict the errors for both the symmetric and non-symmetric cases, respectively. We can see that they not only validate the theory from Sect, but also perform an optimal h^2 convergence rate for $\|\boldsymbol{u}-\boldsymbol{u}_h\|_\Omega$. Furthermore, we observe an increased order of convergence for $\|p-p_h\|_\Omega$. In fact, the error seems to decrease with $O(h^{3/2})$, rather than the $O(h)$ predicted by the theory.

To stress the last point made in the previous paragraph, in Table we compare the L^2 error of the pressure $(||p-p_h||_\Omega)$ for hdG method introduced in [1] and stabilised hdG method from Sect. Columns $p_h \in Q_h^0$ are associated with hdG method and $p_h \in Q_h^1$ with stabilised hdG ones. There, we confirm that the pressure

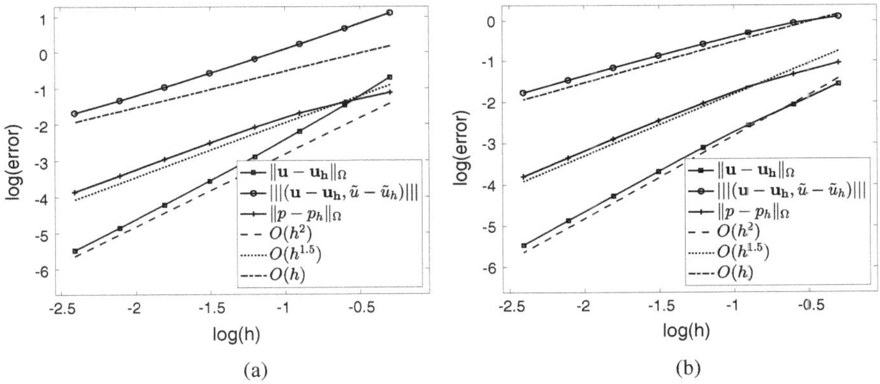

Fig. Convergence the stabilised method with $k = 1$. (**a**) Symmetric bilinear form ($\varepsilon = -1$).(**b**) Non-symmetric bilinear form ($\varepsilon = 1$)

Table Comparison of the error of the pressure $||p-p_h||_\Omega$

	Symmetric bilinear form ($\varepsilon=-1$)		Non-symmetric bilinear form ($\varepsilon=1$)	
h	$p_h \in Q_h^0$	$p_h \in Q_h^1$	$p_h \in Q_h^0$	$p_h \in Q_h^1$
2^{-1}	0.152296	0.077228	0.159019	0.090624
2^{-2}	0.082775	0.041790	0.084875	0.047488
2^{-3}	0.042620	0.020500	0.043313	0.009449
2^{-4}	0.021357	0.008338	0.021513	0.003516
2^{-5}	0.010676	0.003083	0.010707	0.001269
2^{-6}	0.005340	0.001105	0.005346	0.002171
2^{-7}	0.002671	0.000392	0.002672	0.000453
2^{-8}	0.001336	0.000139	0.001336	0.000161

error for the stabilised version is much smaller than the one for the inf-sup stable case, in addition to having an increased order of convergence.

Conclusion

In this work we have applied the idea introduced in [10] to stabilise the hdG method proposed in [1] for the Stokes problem with TVNF boundary conditions. The method adds a simple, symmetric, term to the formulation, and allowed us to use a higher order pressure space, which, in turn, improved the pressure convergence (although a proof of this fact is, in general, not available). This approach was also applied to NVTF boundary conditions (see [6]) and can be used for other discontinuous Galerkin methods that deal with Stokes or nearly incompressible elasticity problems.

Future testing using higher order discretisations is needed to assess whether this approach provides an increase of the convergence rate for the pressure. Thus, the numerical tests with higher order of polynomials for discontinuous finite methods is interest for further research to look for the improvement of the convergence.

References

1. Barrenechea, G.R., Bosy, M., Dolean, V., Nataf, F., Tournier, P.-H.: Hybrid discontinuous Galerkin discretisation and domain decomposition preconditioners for the Stokes problem. Comput. Methods Appl. Math. **19**(4), 703–722 (2019)
2. Barth, T., Bochev, P.B., Gunzburger, M., Shadid, J.: A taxonomy of consistently stabilized finite element methods for the Stokes problem. SIAM J. Sci. Comput. **25**(5), 1585–1607 (2004)
3. Becker, R., Braack, M.: A finite element pressure gradient stabilization for the Stokes equations based on local projections. Calcolo **38**(4), 173–199 (2001)
4. Bochev, P.B., Dohrmann, C.R., Gunzburger, M.D.: Stabilization of low-order mixed finite elements for the Stokes equations. SIAM J. Numer. Anal. **44**(1), 82–101 (2006)
5. Boffi, D., Brezzi, F., Fortin, M.: Mixed Finite Element Methods and Applications. Springer Series in Computational Mathematics, vol. 44. Springer, Heidelberg (2013)
6. Bosy, M.: Efficient discretisation and domain decomposition preconditioners for incompressible fluid mechanics. Ph.D. Thesis, University of Strathclyde (2017). https://doi.org/10.13140/RG.2.2.24947.17444
7. Brezzi, F., Douglas, J., Jr.: Stabilized mixed methods for the Stokes problem. Numer. Math. **53**(1–2), 225–235 (1988)
8. Brezzi, F., Pitkäranta, J.: On the stabilization of finite element approximations of the Stokes equations. In: Efficient Solutions of Elliptic Systems (Kiel, 1984). Notes on Numerical Fluid Mechanics, vol. 10, pp. 11–19. Friedr. Vieweg, Braunschweig (1984)
9. Codina, R., Blasco, J.: Analysis of a pressure-stabilized finite element approximation of the stationary Navier-Stokes equations. Numer. Math. **87**(1), 59–81 (2000)
10. Dohrmann, C.R., Bochev, P.B.: A stabilized finite element method for the Stokes problem based on polynomial pressure projections. Int. J. Numer. Methods Fluids **46**(2), 183–201 (2004)
11. Ern, A., Guermond, J.L.: Theory and Practice of Finite Elements. Applied Mathematical Sciences, vol. 159. Springer, New York (2004)

12. Ganesan, S., Matthies, G., Tobiska, L.: Local projection stabilization of equal order interpolation applied to the Stokes problem. Math. Comput. **77**(264), 2039–2060 (2008)
13. Hughes, T.J.R., Franca, L.P.: A new finite element formulation for computational fluid dynamics. VII. The Stokes problem with various well-posed boundary conditions: symmetric formulations that converge for all velocity/pressure spaces. Comput. Methods Appl. Mech. Eng. **65**(1), 85–96 (1987)
14. Hughes, T.J.R., Franca, L.P., Balestra, M.: A new finite element formulation for computational fluid dynamics. V. Circumventing the Babuška-Brezzi condition: a stable Petrov-Galerkin formulation of the Stokes problem accommodating equal-order interpolations. Comput. Methods Appl. Mech. Eng. **59**(1), 85–99 (1986)
15. Lehrenfeld, C.: Hybrid discontinuous Galerkin methods for solving incompressible flow problems. Dissertation, Rheinisch-Westfälischen Technischen Hochschule Aachen (2010)
16. Silvester, D.J., Kechkar, N.: Stabilised bilinear-constant velocity-pressure finite elements for the conjugate gradient solution of the Stokes problem. Comput. Methods Appl. Mech. Eng. **79**(1), 71–86 (1990)

8

RBF Based CWENO Method

Jan S. Hesthaven, Fabian Mönkeberg, and Sara Zaninelli

Introduction

A broad range of physical phenomena can be described by hyperbolic conservation laws of the form

$$\begin{aligned} u_t + f(u)_x &= 0, \qquad (x,t) \in \mathbb{R} \times \mathbb{R}_+, \\ u(0) &= u_0, \end{aligned} \tag{1}$$

with the conserved variables $u : \mathbb{R} \times \mathbb{R}_+ \to \mathbb{R}^N$ and the flux function $f : \mathbb{R}^N \to \mathbb{R}^N$. The nonlinear behavior of f can lead to complex solutions, most notably shocks. It is well-known that high-order methods give good results for smooth data, but for discontinuous ones spurious oscillations are introduced. A popular class of methods to solve (1) is the finite volume method, which is based on a discretization in space $\ldots < x_{i-1/2} < x_{i+1/2} < \ldots$ and the average values $\bar{u}_i$ of its cells $C_i = [x_{i-1/2}, x_{i+1/2}]$. It is defined by the semi-discrete scheme

$$\frac{\mathrm{d}\bar{u}_i}{\mathrm{d}t} = -\frac{F_{i+1/2} - F_{i-1/2}}{\Delta x}, \tag{2}$$

where the numerical flux term $F_{i+1/2}$ depends on the values $\{\bar{u}_{i-k}, \ldots, \bar{u}_{i+p-k}\}$ with $0 \leq k \leq p-1$. For more details we refer the reader to [15, 20, 22].

The class of essentially nonoscillatory (ENO) methods, introduced by Harten et al. [14], reduces spurious oscillations to a minimum. They are based on a monotone numerical flux function $F(u, v)$ and high-order accurate reconstruction $s_i(x)$ for

J. S. Hesthaven · F. Mönkeberg (✉) · S. Zaninelli
École Polytechnique Fédérale de Lausanne (EPFL), Lausanne, Switzerland
e-mail: jan.hesthaven@epfl.ch; fabian.monkeberg@epfl.ch; sara.zaninelli@epfl.ch

each cell i. The central idea is to choose the least oscillating interpolation function s_i and define the numerical flux $F_{i+1/2} = F(u^+_{i+1/2}, u^-_{i+1/2})$ with $u^\pm_{i+1/2}$ being the evaluation of s_{i+1} and s_i at the interface $x_{i+1/2}$. Based on the ENO method, Jiang and Shu [19] introduced the weighted ENO (WENO) method which considers different interpolation polynomials, based on different stencils, and combines them in a nonoscillatory manner to maximize the attainable accuracy. Further results on ENO and WENO methods can be found in [10, 11, 16].

C W E N O

The CWENO method is based on the WENO method and was introduced by Levy et al. [23] as a third order method. Further analysis and generalization to higher orders on general grids can be found in [6, 7].

Let us consider the standard semi-discrete formulation (2) with a monotone flux function $F(u, v)$. The goal is to construct a reconstruction $P_{rec,i}$ for each cell C_i based on the stencil $\{C_{i-k}, \ldots, C_{i+k}\}$ for $k \in \mathbb{N}$. In the smooth regions the algorithm should choose a polynomial of degree $2k$ which interpolates the central stencil $\bar{u}_{i-k}, \ldots, \bar{u}_{i+k}$ in the mean value sense. In case of a non-smooth solution it chooses a polynomial of degree k on one stencil $\{C_{i-k+l}, \ldots, C_{i+l}\}$ that avoids the discontinuity. Given the reconstruction, the high-order numerical flux is $F_{i+1/2} = F(P_{rec,i+1}(x_{i+1/2}), P_{rec,i}(x_{i+1/2}))$.

Specifically, let us consider P_{opt} as the polynomial of degree $2k$ that interpolates all data in the $2k + 1$ stencil and the polynomials P_l of degree k that interpolate the data on the stencil $\{C_{i-k+l-1}, \ldots, C_{i+l-1}\}$ for $l = 1, \ldots, k + 1$. Furthermore, the reconstruction depends on the choice of the positive real coefficients $d_0, \ldots, d_{k+1} \in [0, 1]$ such that $\sum_{l=0}^{k+1} d_l = 1, d_0 \neq 0$. Then, the reconstruction polynomial of degree $2k$ is

$$P_{rec}(x) = \sum_{l=0}^{k+1} \omega_l P_l(x), \tag{3}$$

with

$$P_0(x) = \frac{1}{d_0}\Big(P_{opt}(x) - \sum_{l=1}^{k+1} d_l P_l(x)\Big), \tag{4}$$

and the nonlinear coefficients ω_l that are defined as

$$\omega_l = \frac{\alpha_l}{\sum_{i=0}^{k+1} \alpha_i}, \qquad \alpha_l = \frac{d_l}{(I[P_l] + \bar{\epsilon})^t}, \tag{5}$$

where $I[P_l]$ indicates the smoothness of P_l, $1 \gg \bar{\epsilon} > 0$ and $t \geq 2$. A classical indicator of smoothness in the cell C for a polynomial is the Jiang–Shu indicator [19]

$$I[P] = \sum_{l>0} \operatorname{diam}(C)^{2l-1} \int_C \left(\frac{\mathrm{d}^l}{\mathrm{d}x^l} P(x) \right)^2 \mathrm{d}x. \tag{6}$$

The choice of $\bar{\epsilon}$ is of importance: if it is too small, it might affect the order of convergence. On the other hand if it is too big, spurious oscillations may occur. Cravero et al. [7] show that the choice $\bar{\epsilon} = \hat{\epsilon} h^p$ for $p = 1, 2$ leads to the maximal order of convergence. As proposed in [7] we define the coefficients d_j over the temporary weights

$$\hat{d}_j = \hat{d}_{k+2-j} = j, \qquad 1 \leq j \leq \frac{k+2}{2}, \tag{7}$$

and we choose $d_0 \in (0, 1)$ for the high-order polynomial. This gives us a possible choice for the coefficients

$$d_j = \frac{\hat{d}_j}{\sum_{i>0} \hat{d}_i} (1 - d_0). \tag{8}$$

The main difference with respect to the classical WENO method is that for the smooth case we are not constructing P_{opt} out of the polynomials P_l, but we build it independently by resolving an additional system of equations. This method has the advantage that it is easier to generalize on general grids in high dimensions, while maintaining high-order accuracy.

Radial Basis Functions

An alternative to the classical polynomial interpolation is the interpolation with radial basis functions (RBF). RBFs were proposed in the seminal work by Hardy [13]. They have been successfully applied in scattered data interpolation [4, 9, 17, 24, 27] and as a basis for a generalized finite difference method (RBF-FD) [5, 12]. The advantage is its flexibility in high dimensions and the possibility to reduce the risk of ill-conditioned point constellations. Its disadvantage is the ill-conditioning of the interpolation matrix for small grid sizes [8, 21, 26].

The RBF interpolation is based on a basis $\mathscr{B}$, obtained from a univariate continuous function $\phi : \mathbb{R}^d \to \mathbb{R}$, composed with the Euclidean norm centered at the data points

$$\phi(x - x_j) := \phi(\varepsilon \|x - x_j\|), \tag{9}$$

Table Commonly used RBFs with $\mathbb{N} \not\ni \nu > 0$, $k \in \mathbb{N}$ and $\varepsilon > 0$

RBF	$\phi(r)$	Order
Infinitely smooth RBFs		
Multiquadratics	$(1 + (\varepsilon r)^2)^{\nu}$	$\lceil \nu \rceil$
Inverse multiquadratics	$(1 + (\varepsilon r)^2)^{-\nu}$	0
Gaussians	$\exp(-(\varepsilon r)^2)$	0
Piecewise smooth RBFs		
Polyharmonic splines	r^{2k-d}	k
	$r^{2k-d}\log(r)$	k

with the shape parameter ε. Some common RBFs can be found in Table Thus, for given scattered data points $X = (x_1, \ldots, x_n)^T$ with $x_j \in \mathbb{R}^d$ and corresponding values $f_1, \ldots, f_n \in \mathbb{R}$ we look for

$$s(x) = \sum_{j=1}^{n} a_j \phi(x - x_j) + p(x), \tag{10}$$

with a polynomial $p \in \Pi_{m-1}(\mathbb{R}^d)$, $m \in \mathbb{N}$, the interpolation condition $s(x_j) = f_j$ and the additional constraints

$$\sum_{j=1}^{n} a_j q(x_j) = 0, \qquad \text{for all } q \in \Pi_{m-1}(\mathbb{R}^d), \tag{11}$$

with the coefficients $a_j \in \mathbb{R}$ for all $j = 1, \ldots, n$.

The same concept can be applied in the case of cell-averages. We seek functions

$$s(x) = \sum_{j=1}^{n} a_j \lambda_{C_j}^{\xi} \phi(x - \xi) + p(x), \qquad p \in \Pi_{m-1}(\mathbb{R}^d), \tag{12}$$

such that

$$\lambda_{C_j} s = \bar{u}_j, \qquad \text{for all } j = 1, \ldots, n, \tag{13a}$$

$$\sum_{j=1}^{n} a_j \lambda_C(p) = 0, \qquad \text{for all } C \in \{C_1, \ldots, C_n\}, \tag{13b}$$

with the averaging operator $\lambda_C^x f(x) = \frac{1}{|C|} \int_C f(x)\mathrm{d}x$. A well-known problem with RBFs is the high condition number of the interpolation matrix for small grid sizes or small shape parameters [8, 21, 26]. This problem can be resolved by using the vector-valued rational approximation method [28].

RBF-CWENO

Methods combining RBFs and essentially nonoscillatory methods have been proposed, e.g. RBFs with ENO [18, 25], RBFs with WENO [1–3]. The advantage of the CWENO method over the WENO method is its flexibility on general grids and its independence of the construction of a high-order interpolation function out of lower order ones. This facilitates the use of the whole grid in smooth regions and is important for non-polynomial interpolation functions which cannot be combined to an higher order function.

We propose the RBF-CWENO method which works as the classical CWENO method with the reconstruction function (3) and the weights (5), but as interpolation function we use RBFs instead of polynomials. Since the problem of the ill-conditioning can be solved by using the vector-valued rational approximation method [28], the main challenge for RBF methods is the choice of the smoothness indicator. For polyharmonic splines, Aboyar et al. [1] use the semi-norm of the Beppo-Levi space and Bigoni et al. [3] use a modified version of the Jiang-Shu indicator (6).

Smoothness Indicator

The smoothness indicator is the heart of the essentially nonoscillatory methods. We consider one based on the one introduced by Bigoni and Hesthaven [3]

$$
\begin{aligned}
I_i[s] = &\sum_{l=1}^{g+1} \Delta x_i^{2l-1} \int_{C_i} \left(\frac{\partial^l p(x)}{\partial x^l}\right)^2 \mathrm{d}x \\
&+ \Delta x_i^{2g+1} \int_{C_i} \left(\frac{\partial^{g+1}}{\partial x^{g+1}}\Big[\sum_{j=1}^{g+1} a_j \lambda_{C_j}^{\xi} \phi(\|x-\xi\|)\Big]\mathrm{d}x\right)^2,
\end{aligned}
\tag{14}
$$

where the first part is the sum of the derivatives of the polynomial part and the second term expresses the highest derivative of the RBF-part. The original Jiang-Shu indicator applied to (12) would include the lower derivatives of the RBF-part plus all mixed terms, but we find this to be less efficient. For simplicity the integrals can be approximated with a simple mid-point rule.

We face again the problem of ill-conditioning when recovering the coefficients a_i. Numerical examples indicate that small shape parameter improve the accuracy, but they do not affect the choice of the stencil using this smoothness indicator. Thus, we use a bigger shape parameter ε_R, that is smaller than the smallest distance to a singularity

$$\varepsilon_R = 0.95(\max_{i,j\leq N} \|x_i - x_j\|)^{-1}, \tag{15}$$

which ensures the solvability of the system of equations [28].

Numerical Results

We now discuss the numerical results of the RBF-CWENO method and compare it with the RBF-WENO method [3] and the classical ENO method [14]. All methods are using the Lax-Friedrichs numerical flux and integration in time is done using the SSPRK-5 method [15] with time step $dt = CFL \cdot \Delta x/\lambda_{max}$ and the maximal eigenvalue λ_{max} of $\nabla_u F$. Furthermore, we use the vector-valued rational approximation approach [28] to circumvent ill-conditioning of the interpolation matrix and a shape parameter $\varepsilon = 0.1$. For the nonlinear weights (5) we choose $\bar{\epsilon} = \hat{\epsilon} h^2$ with $\hat{\epsilon} = 0.1$.

Linear Advection Equation

Let us consider the linear advection equation

$$u_t + a u_x = 0, \qquad x \in [0, 1], \tag{16}$$

with wave speed $a = 1$, initial condition $u_0(x) = sin(2\pi x)$ and periodic boundary conditions [22]. Note that for $k = 3$ we expect the order of convergence to be 7, therefore we use the reduced time step $dt = CFL \cdot \Delta x^{7/5}/\lambda_{max}$ to recover the right order of convergence. The correct order of convergence of the RBF-CWENO method is shown in Table and it seems to be more accurate than the RBF-WENO method.

Burger's Equation

Considering the Burger's equation

$$u_t + \frac{1}{2}(u^2)_x = 0, \qquad x \in [0, 1], \tag{17}$$

Table Convergence rates of RBF-CWENO using multiquadratics for the linear advection equation at time $t = 0.05$

		RBF-CWENO						RBF-WENO	
		L_h^1		L_h^2		L_h^∞		L_h^2	
k	N	Error	Rate	Error	Rate	Error	Rate	Error	Rate
1	16	$5.6409e-04$	–	$2.1702e-04$	–	$1.5903e-04$	–	$1.5754e-02$	–
	32	$7.6612e-05$	2.75	$2.4817e-05$	2.99	$1.6221e-05$	3.15	$4.8924e-03$	1.69
	64	$1.0082e-05$	2.79	$2.5297e-06$	3.15	$1.3561e-06$	3.42	$1.2608e-03$	1.96
	128	$1.3812e-06$	2.74	$2.4032e-07$	3.24	$9.6982e-08$	3.63	$9.2931e-05$	3.76
	256	$2.1322e-07$	2.57	$2.3289e-08$	3.21	$6.5703e-09$	3.71	$2.3008e-06$	5.34
2	16	$2.3796e-05$	–	$7.3671e-06$	–	$4.1241e-06$	–	$5.4401e-04$	–
	32	$3.5783e-06$	2.61	$8.3093e-07$	3.01	$3.9675e-07$	3.22	$4.4938e-05$	3.60
	64	$2.8691e-07$	3.48	$5.9366e-08$	3.63	$3.6940e-08$	3.27	$3.4787e-06$	3.69
	128	$1.4563e-08$	4.11	$2.5775e-09$	4.32	$1.3965e-09$	4.51	$2.5956e-07$	3.74
	256	$6.8835e-10$	4.20	$9.6168e-11$	4.53	$4.4249e-11$	4.75	$1.9221e-08$	3.76
3	16	$3.8815e-05$	–	$1.3319e-05$	–	$7.7293e-06$	–	$2.2578e-04$	–
	32	$4.3423e-07$	6.48	$1.3452e-07$	6.63	$8.1494e-08$	6.57	$7.3483e-06$	4.94
	64	$5.1821e-09$	6.39	$1.4750e-09$	6.51	$8.8273e-10$	6.54	$1.4075e-07$	5.71
	128	$7.6636e-11$	6.08	$1.6792e-11$	6.46	$7.8655e-12$	6.81	$1.4510e-09$	6.60
	256	$1.1554e-12$	6.05	$1.5855e-13$	6.73	$6.9487e-14$	6.82	$2.0120e-11$	6.17

We use shape parameter $\varepsilon = 0.1$, CFL $= 0.01$

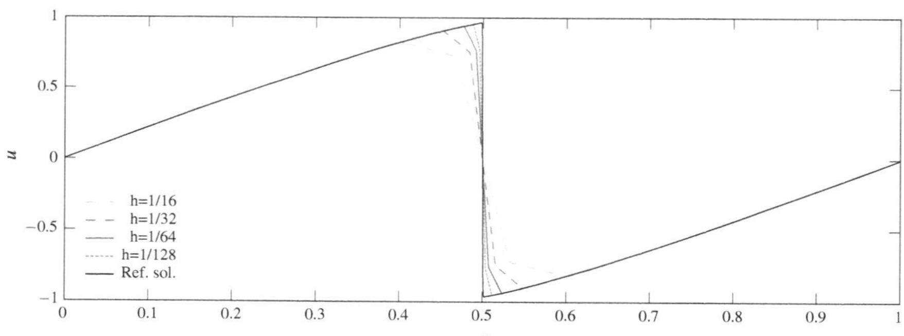

Fig. Burger's equation at $t = 0.3$ with $u_0 = \sin(2\pi x)$ solved by using RBF-CWENO method with MQ interpolants of order $k = 3$

we analyze its robustness with respect to discontinuities. In Fig. we report the results performed with $CFL = 0.5$ at $t = 0.3$. We observe no oscillations around the discontinuity at $x = 0.5$ and as expected an increasing accuracy for increasing number of elements.

Euler Equations

The one-dimensional Euler equations express conservation of mass, momentum and the total energy. They can be described by the density ρ, the mass flow m, the energy per unit volume E and the pressure p through

$$\begin{pmatrix} \rho \\ m \\ E \end{pmatrix}_t + \begin{pmatrix} m \\ \frac{m^2}{\rho} + p \\ \frac{m}{\rho}(E+p) \end{pmatrix}_x = 0, \tag{18}$$

with $p = \mathscr{R}\rho T = (\gamma - 1)(E - \frac{1}{2}\frac{m^2}{\rho})$ for an ideal gas with the ratio of specific heat $\gamma = 1.4$ [15]. For $k = 3$ we need to change the nonlinear weights (5) by using $\bar{\epsilon} = \hat{\epsilon}h^2$ with $\hat{\epsilon} = 10^{-6}$ to avoid oscillations.

Sod's Shock Tube Problem

The Sod's shock tube problem describes two colliding gases in [0, 1] with different densities given by the initial conditions

$$(\rho_0, m_0, p_0) = \begin{cases} (1, 0, 1) & \text{if } x < 0.5 \\ (0.125, 0, 0.1) & \text{if } x \geq 0.5 \end{cases}. \tag{19}$$

This results in a rarefaction wave followed by a contact and a shock discontinuity which separates the domain into four domains with constant variables. The RBF-CWENO method resolves it well, see Fig.. For $k = 3$, we observe minor

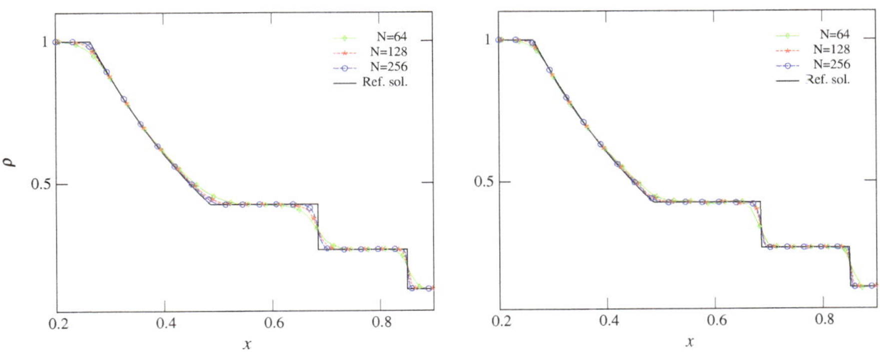

Fig. Results for the Sod shock tube problem at $t = 0.2$ solved by using RBF-CWENO with MQ interpolants of order $k = 2, 3$ on characteristic variables (left: $k = 2$, right: $k = 3$)

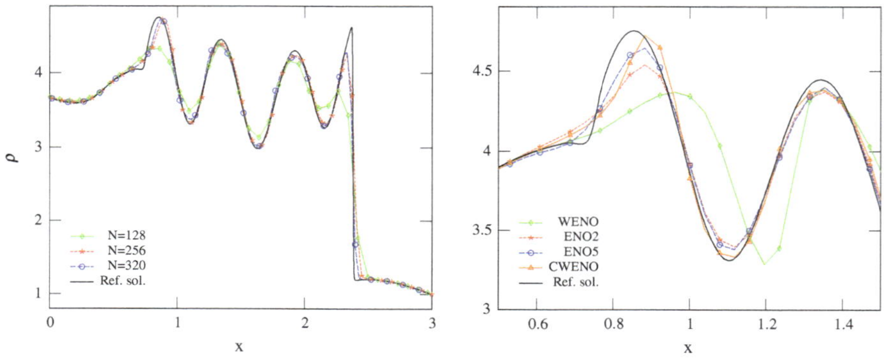

Fig. Results for the Euler shock entropy problem at $t = 1.8$ solved by using RBF-CWENO with MQ interpolants of order $k = 2$ on characteristic variables (Left) and a comparison with WENO, ENO2 and ENO5 for $N = 256$ cells (Right)

oscillations, but their amplitude decreases for increasing number of elements. Furthermore, we observe the increasing accuracy for $k = 3$ compared to $k = 2$.

Shu–Osher Shock-Entropy Wave Interaction Problem

The Shu–Osher problem describes the interaction of a discontinuity with a low frequency wave which introduces some high frequent waves. Its initial conditions are

$$(\rho_0, m_0, p_0) = \begin{cases} (3.857143, 2.629369, 10.33333) & \text{if } x < -4 \\ (1 + 0.2\sin(5x), 0, 1) & \text{if } x \geq -4 \end{cases}. \tag{20}$$

In Fig. we observe on the left side the increasing accuracy for increasing number of elements for $k = 2$. On the right side we see its good approximative behaviour compared to the existing methods ENO2, ENO5 and the corresponding WENO. In particular we observe that the performance of the RBF-CWENO ($k = 2$) is comparable to ENO5 and superior to WENO ($k = 2$).

Conclusion

In this work, we introduce the RBF-CWENO method that relies on the CWENO method [23] and the use of radial basis functions for the interpolation. We develop a smoothness indicator that is based on RBFs but works similarly to the one for polynomials. Furthermore, we tackle the problem about the choice of the weight

$1 \gg \bar{\epsilon} > 0$. For $\bar{\epsilon} = \hat{\epsilon} h^2$ with $\hat{\epsilon} = 0.1$ we get the right order of convergence, but for the 7th order method ($k = 3$) we choose $\hat{\epsilon} = 10^{-6}$ to reduce spurious oscillations for the Euler equations.

Moreover, we should point out that the choice of the linear weight d_0 can influence the result; indeed if it is too close to 1 then the reconstruction almost coincides with P_{opt}, which can lead to spurious oscillations in case of discontinuous solutions. We present multiple numerical examples to show the robustness of the method.

We can conclude that the RBF-CWENO method works comparable to the existing RBF-WENO and ENO methods in one dimension. The advantage of RBFs is clearer when considering unstructured grids in higher dimensions where polynomial reconstruction is complex.

References

1. Aboiyar, T., Georgoulis, E.H., Iske, A.: High order WENO finite volume schemes using polyharmonic spline reconstruction. In: Proceedings of the International Conference on Numerical Analysis and Approximation Theory NAAT2006, Cluj-Napoca (Romania). Department of Mathematics, University of Leicester (2006)
2. Aboiyar, T., Georgoulis, E.H., Iske, A.: Adaptive ADER methods using kernel-based polyharmonic spline WENO reconstruction. SIAM J. Sci. Comput. **32**(6), 3251–3277 (2010)
3. Bigoni, C., Hesthaven, J.S.: Adaptive WENO methods based on radial basis function reconstruction. J. Sci. Comput. **72**(3), 986–1020 (2017)
4. Buhmann, M.D.: Radial Basis Functions: Theory and Implementations. Cambridge University Press, Cambridge (2003)
5. Chandhini, G., Sanyasiraju, Y.: Local RBF-ŘFD solutions for steady convection–diffusion problems. Int. J. Numer. Methods Eng. **72**(3), 352–378 (2007)
6. Cravero, I., Semplice, M.: On the accuracy of WENO and CWENO reconstructions of third order on nonuniform meshes. J. Sci. Comput. **67**(3), 1219–1246 (2016)
7. Cravero, I., Puppo, G., Semplice, M., Visconti, G.: CWENO: uniformly accurate reconstructions for balance laws (2016). Preprint. arXiv:1607.07319
8. Driscoll, T.A., Fornberg, B.: Interpolation in the limit of increasingly flat radial basis functions. Comput. Math. Appl. **43**(3), 413–422 (2002)
9. Duchon, J.: Splines Minimizing Rotation-Invariant Semi-Norms in Sobolev Spaces, pp. 85–100. Springer, Berlin (1977)
10. Fjordholm, U.S., Ray, D.: A sign preserving WENO reconstruction method. J. Sci. Comput. 1–22 (2016)
11. Fjordholm, U.S., Mishra, S., Tadmor, E.: Arbitrarily high-order accurate entropy stable essentially nonoscillatory schemes for systems of conservation laws. SIAM J. Numer. Anal. **50**(2), 544–573 (2012)
12. Fornberg, B., Lehto, E., Powell, C.: Stable calculation of gaussian-based RBF-FD stencils. Comput. Math. Appl. **65**(4), 627–637 (2013)
13. Hardy, R.L.: Multiquadric equations of topography and other irregular surfaces. J. Geophys. Res. **76**(8), 1905–1915 (1971)
14. Harten, A., Engquist, B., Osher, S., Chakravarthy, S.R.: Uniformly high order accurate essentially non-oscillatory schemes, iii. J. Comput. Phys. **71**(2), 231–303 (1987)
15. Hesthaven, J.S.: Numerical Methods for Conservation Laws: From Analysis to Algorithms. Society for Industrial and Applied Mathematics, Philadelphia (2017)

16. Hu, C., Shu, C.-W.: Weighted essentially non-oscillatory schemes on triangular meshes. J. Comput. Phys. **150**(1), 97–127 (1999)
17. Iske, A.: Multiresolution Methods in Scattered Data Modelling, vol. 37. Springer, Berlin (2004)
18. Iske, A., Sonar, T.: On the structure of function spaces in optimal recovery of point functionals for ENO-schemes by radial basis functions. Numer. Math. **74**(2), 177–201 (1996)
19. Jiang, G.-S., Shu, C.-W.: Efficient implementation of weighted ENO schemes. J. Comput. Phys. **126**(1), 202–228 (1996)
20. Kröner, D.: Numerical Schemes for Conservation Laws. Wiley, Chichester (1997)
21. Larsson, E., Fornberg, B.: Theoretical and computational aspects of multivariate interpolation with increasingly flat radial basis functions. Comput. Math. Appl. **49**(1), 103–130 (2005)
22. LeVeque, R.J.: Numerical Methods for Conservation Laws. Springer Science & Business Media, New York (1992)
23. Levy, D., Puppo, G., Russo, G.: Central WENO schemes for hyperbolic systems of conservation laws. ESAIM: Math. Model. Numer. Anal. **33**(3), 547–571 (1999)
24. Micchelli, C.A.: Interpolation of scattered data: distance matrices and conditionally positive definite functions. Constr. Approx. **2**(1), 11–22 (1986)
25. Mönkeberg, F., Hesthaven, J.S.: Entropy stable essentially nonoscillatory methods based on RBF reconstruction. ESAIM: Math. Model. Numer. Anal. **53**, 925–958 (2019)
26. Schaback, R.: Multivariate interpolation by polynomials and radial basis functions. Constr. Approx. **21**(3), 293–317 (2005)
27. Wendland, H.: Scattered Data Approximation. Cambridge University Press, Cambridge (2004)
28. Wright, G.B., Fornberg, B.: Stable computations with flat radial basis functions using vector-valued rational approximations. J. Comput. Phys. **331**, 137–156 (2017)

9

On the use of Hermite Functions for the Vlasov–Poisson System

Lorella Fatone, Daniele Funaro, and Gianmarco Manzini

Introduction

A semi-Lagrangian spectral method has been proposed in [8] for the numerical approximation of the nonrelativistic Vlasov–Poisson equations, which describe the dynamics of a collisionless plasma of charged particles, coupled under the effect of their own electric field. We assume for simplicity that the development of the plasma is only due to electrons. Moreover, we just treat the case of a 1D-1V distribution function, defined in a phase space consisting of the two one-dimensional independent variables x (space) and v (velocity). The approximation introduced in [8] has been initially developed and tested on Fourier-Fourier periodic discretizations, for both variables in the phase space. In the successive paper [9], the approximation in the variable v has been approached with the help of Hermite functions, i.e., Hermite polynomials multiplied by the Gaussian weight $\exp(-v^2)$.

Semi-Lagrangian methods for plasma physics calculations were originally proposed in [5, 18] and more recently in [6, 15, 16]. By this approach, at different times, the solution is approximated at the nodes of a Cartesian grid covering the space-

L. Fatone
Dipartimento di Matematica, Università degli Studi di Camerino, Camerino, Italy
e-mail: lorella.fatone@unicam.it

D. Funaro
Dipartimento di Scienze Chimiche e Geologiche, Università degli Studi di Modena e Reggio Emilia, Modena, Italy
e-mail: daniele.funaro@unimore.it

G. Manzini (✉)
Group T-5, Applied Mathematics and Plasma Physics, Theoretical Division, Los Alamos National Laboratory, Los Alamos, NM, USA
e-mail: gmanzini@lanl.gov

velocity domain. The solution at each space-velocity node is traced back along the characteristic curve originating backward from that node. In [8] a high-order Taylor expansion of the characteristic curves is used to trace back the solution in time, which is then approximated by spectral interpolation. Such a method guarantees the conservation of the main physical quantities (charge, mass, and momentum).

The first attempt in using Hermite polynomials to solve the Vlasov equation dates back to the work [10], where the Hermite basis is used in the velocity variable to describe a plasma in a physical state near the thermodynamic equilibrium. Within this approach, exact discrete conservation laws can be constructed [7, 13, 14, 20, 21]. The weight function of the Hermite basis can be generalized by introducing a parameter α in such a way that it becomes $\exp(-\alpha^2 v^2)$. A proper choice of this parameter can significantly improve the convergence [2, 3, 19]. This fact was also confirmed in earlier works on plasmas physics based on Hermite spectral methods (see [11, 17] and more recently [4]).

The paper is organized as follows. In Section, we present the continuous model, i.e., the 1D-1V Vlasov equation. In Section, we introduce the spectral approximation in the phase space. We present the semi-Lagrangian schemes based on an approximation of the characteristic curves coupled with a second-order backward differentiation formula (BDF). In other section, we numerically assess the performance of the method for a standard test case, and we show how the solution's behavior can be affected by the choice of a certain parameter β, acting on the location of Hermite weight function.

The Continuous Model

We deal with the 1D-1V Vlasov equation defined in the domain $\Omega = \Omega_x \times \mathbb{R}$, with $\Omega_x \subseteq \mathbb{R}$. The unknown $f = f(t, x, v)$ denotes the probability of finding negative charged particles at the location x with velocity v. This is solution of the problem

$$\frac{\partial f}{\partial t} + v\frac{\partial f}{\partial x} - E(t, x)\,\frac{\partial f}{\partial v} = 0, \qquad t \in (0, T],\ x \in \Omega_x,\ v \in \mathbb{R}. \tag{1}$$

At time $t = 0$ we have the initial distribution $f(0, x, v) = \bar{f}(x, v)$. The problem is nonlinear, since the electric field E is coupled with f. Indeed, we set

$$\frac{\partial E}{\partial x}(t, x) = 1 - \rho(t, x) = 1 - \int_{\mathbb{R}} f(t, x, v)dv, \tag{2}$$

where ρ denotes the electron charge density. System (1)–(2) in the unknowns f and E is a simplification of the Vlasov–Poisson equations in two or three dimensional space domains. Uniqueness of the solution is ensured by imposing that

$$\int_{\Omega_x} E(t, x)dx = 0, \qquad \text{which implies that} \qquad \int_{\Omega_x} \rho(t, x)dx = |\Omega_x|, \tag{3}$$

where $|\Omega_x|$ is the size of Ω_x. We assume periodic boundary conditions in the variable x and a suitable exponential decay at infinity for the variable v. After integration and by using the boundary constraints, we obtain the conservation of mass

$$\frac{d}{dt}\int_{\Omega} f(t,x,v)\,dx\,dv = 0. \tag{4}$$

When f and E are smooth enough, for a sufficiently small $\delta > 0$, the local system of characteristics associated with (1) is given by the curves $(X(\tau), V(\tau))$ solving

$$\frac{dX}{d\tau} = -V(\tau), \qquad \frac{dV}{d\tau} = E(\tau, X(\tau)), \qquad \tau \in]t-\delta, t+\delta[. \tag{5}$$

with the condition that $(X(t), V(t)) = (x, v)$ when $\tau = t$. With this setting we have in mind that for $\tau > 0$ we proceed backward. Under suitable regularity assumptions, there exists a unique solution of the Vlasov–Poisson problem (1)–(2) which is formally obtained by propagating the initial condition along the characteristic curves described by (5), i.e. we have

$$f(t,x,v) = \bar{f}(X(t), V(t)), \tag{6}$$

where we recall that $\bar{f}$ is the initial datum. By using the first-order approximation

$$X(\tau) = x - v(\tau - t), \qquad V(\tau) = v + E(t,x)(\tau - t), \tag{7}$$

the Vlasov equation is satisfied up to an error decaying as $|\tau - t|$, for τ tending to t.

Phase-Space Discretization

We briefly recall the construction of the approximation method proposed in [8]. At each point of a given grid, the new value of the discrete solution is set up to be equal to the value obtained by going backward, by a suitably small amount, along the local characteristic lines. The algorithm follows from a Taylor expansion of arbitrary order, where the derivatives in the variable x and v are carried out with spectral accuracy. In particular, for the variable x we consider the domain $\Omega_x = [0, 2\pi[$. Given the positive integer N, we have the equispaced nodes $x_i = 2\pi i/N$, $i = 0, 1, \dots, N-1$. Regarding the direction v, when M is a given positive integer, the nodes v_j, $j = 0, 1, \dots, M-1$, are the zeros of H_M, which is the Hermite polynomial of degree M.

We introduce the polynomial Lagrangian basis functions for the x and v variables, that are $B_i^{(N)}(x_n) = \delta_{in}$ and $B_j^{(M)}(v_m) = \delta_{jm}$, where δ_{ij} is the usual Kronecker symbol. We recall that Hermite functions are obtained from Hermite polynomials after multiplication by the weight $\omega(v) = e^{-v^2}$. We also define the discrete spaces

$$\mathbf{X}_N = \text{span}\left\{B_i^{(N)}\right\}_{i=0,1,\dots,N-1}, \qquad \mathbf{Y}_{N,M} = \text{span}\left\{B_i^{(N)} B_j^{(M)} \omega\right\}_{\substack{i=0,1,\dots,N-1 \\ j=0,1,\dots,M-1}}. \tag{8}$$

Any function $f_{N,M}$ that belongs to $\mathbf{Y}_{N,M}$ can be represented as

$$f_{N,M}(x,v) = \sum_{i=0}^{N-1} \sum_{j=0}^{M-1} c_{ij}\, B_i^{(N)}(x)\, B_j^{(M)}(v)\, \omega(v), \tag{9}$$

where the coefficients of such an expansion are given by $c_{ij} = f_{N,M}(x_i, v_j)$.

In the following, the matrices $d_{ni}^{(N,s)}$ and $d_{mj}^{(M,s)}$ denote the s-th derivative of $B_i^{(N)}$ evaluated at point x_n and $(B_j^{(M)}\omega)$ evaluated at point v_m

$$d_{ni}^{(N,s)} = \frac{d^s B_i^{(N)}}{dx^s}(x_n) \quad \text{and} \quad d_{mj}^{(M,s)} = \frac{d^s\left(B_j^{(M)}\omega\right)}{dv^s}(v_m). \tag{10}$$

As a special case, we set $d_{ni}^{(N,0)} = \delta_{ni}$, $d_{mj}^{(M,0)} = \delta_{mj}$.

Now, let us assume that the one-dimensional function $E_N \in \mathbf{X}_N$ is known. Given $\Delta t > 0$, by taking $\tau = t - \Delta t$ in formula (7), we define the new set of points $\tilde{x}_{nm} = x_n - v_m\,\Delta t$ and $\tilde{v}_{nm} = v_m + E_N(x_n)\Delta t$. To evaluate a function $f_{N,M} \in \mathbf{Y}_{N,M}$ at the new points $(\tilde{x}_{nm}, \tilde{v}_{nm})$ through the coefficients c_{ij}, we use a Taylor expansion in time. By omitting the terms in Δt of order higher than one, we get

$$\begin{aligned} &B_i^{(N)}(\tilde{x}_{nm})\left(B_j^{(M)}\omega\right)(\tilde{v}_{nm}) \approx \\ &\qquad \delta_{in}\,\delta_{jm}\omega(v_m) - v_m\,\Delta t\,\delta_{jm}\,d_{ni}^{(N,1)}\omega(v_m) + E_N(x_n)\,\Delta t\,\delta_{in}\,d_{mj}^{(M,1)}. \end{aligned} \tag{11}$$

By substituting (11) in (9), we obtain the approximation

$$\begin{aligned} f_{N,M}(\tilde{x}_{nm}, \tilde{v}_{nm}) &= \sum_{i=0}^{N-1} \sum_{j=0}^{M-1} c_{ij}\, B_i^{(N)}(\tilde{x}_{nm})\, B_j^{(M)}(\tilde{v}_{nm})\, \omega(\tilde{v}_{nm}) \\ &\approx c_{nm}\omega(v_m) - v_m\omega(v_m)\,\Delta t \sum_{i=0}^{N-1} d_{ni}^{(N,1)} c_{im} + E_N(x_n)\,\Delta t \sum_{j=0}^{M-1} d_{mj}^{(M,1)} c_{nj}, \end{aligned} \tag{12}$$

which is the main building block for more advanced schemes.

Discretization of the Vlasov Equation

Given the time instants $t^k = k\Delta t = kT/K$ for any integer $k = 0, 1, \dots, K$, we consider the approximation of the unknowns f and E of problem (1)–(2), given by

$$\left(f_{N,M}^{(k)}(x, v),\ E_N^{(k)}(x)\right) \simeq \left(f(t^k, x, v),\ E(t^k, x)\right), \qquad x \in \Omega_x,\ v \in \mathbb{R}, \tag{13}$$

where the function $f_{N,M}^{(k)}$ belongs to $\mathbf{Y}_{N,M}$ and the function $E_N^{(k)}$ belongs to $\mathbf{X}_N$. Concerning the density function, we define

$$\rho_N^{(k)}(x) = \int_{\Omega_v} f_{N,M}^{(k)}(x, v)\, dv \ \simeq\ \rho(t^k, x). \tag{14}$$

Hence, at any time step k, we express $f_{N,M}^{(k)}$ in the following way

$$f_{N,M}^{(k)}(x, v) = \sum_{i=0}^{N-1} \sum_{j=0}^{M-1} c_{ij}^{(k)}\, B_i^{(N)}(x)\, B_j^{(M)}(v)\omega(v), \tag{15}$$

where $c_{ij}^{(k)} = f_{N,M}^{(k)}(x_i, v_j)$. At time $t = 0$, we use the initial condition $c_{ij}^{(0)} = f(0, x_i, v_j) = \bar{f}(x_i, v_j)$.

Suppose that $E_N^{(k)}$ is given at step k. According to [8], we write

$$E_N^{(k)}(x) = -\sum_{n=1}^{N/2} \frac{1}{n}\left[\hat{a}_n^{(k)} \sin(nx) - \hat{b}_n^{(k)} \cos(nx)\right], \tag{16}$$

where the discrete Fourier coefficients $\hat{a}_n^{(k)}$ and $\hat{b}_n^{(k)}$, $n = 1, 2, \dots, N/2$, are suitably related to those of $\rho_N^{(k)}$.

By taking $\tau = t - \Delta t$ in (7), we define $\tilde{x}_{nm} = x_n - v_m\,\Delta t$ and $\tilde{v}_{nm} = v_m + E_N^{(k)}(x_n)\Delta t$. The distribution function f is expected to remain constant along the characteristics. The most straightforward discretization method is obtained by advancing the coefficients according to the approximation

$$f_{N,M}^{(k+1)}(x_n, v_m) \approx f_{N,M}^{(k)}(\tilde{x}_{nm}, \tilde{v}_{nm}). \tag{17}$$

This states that the value of $f_{N,M}^{(k+1)}$, at the grid points and time step $(k+1)\Delta t$, is assumed to correspond to the previous value at time $k\Delta t$, recovered by going backwards along the characteristics. To compute $\tilde{v}_{nm}$, we should use $E_N^{(k+1)}(x_n)$ instead of $E_N^{(k)}(x_n)$. However, the distance between these two quantities is of the order of Δt, so that the replacement has no practical effects on the accuracy of first-order methods. Between each step k and the successive one, we need to update the electric field. This can be done by using the Gaussian quadrature formula in (14), so obtaining

$$\rho_N^{(k)}(x_i) = \sum_{j=0}^{M-1} \frac{1}{\omega(v_j)} f_{N,M}^{(k)}(x_i, v_j)\, w_j = \sum_{j=0}^{M-1} \frac{1}{\omega(v_j)} c_{ij}^{(k)}\, w_j, \tag{18}$$

where w_j, for $j = 1, \ldots, M-1$, are the quadrature weights. Afterwards, in order to compute the new point-values $E_N^{(k+1)}(x_n)$ of the electric field, it is necessary to integrate $\rho_N^{(k)}$. By using approximation (12) in (17), we end up with the first-order explicit scheme of Euler type:

$$c_{nm}^{(k+1)} = c_{nm}^{(k)} + \Delta t\, \Phi_{nm}^{(k)}, \tag{19}$$

where

$$\Phi_{nm}^{(k)} = -v_m \sum_{i=0}^{N-1} d_{ni}^{(N,1)} c_{im}^{(k)} + E_N^{(k)}(x_n) \sum_{j=0}^{M-1} d_{mj}^{(M,1)} c_{nj}^{(k)} \frac{1}{\omega(v_m)}. \tag{20}$$

The parameter Δt must satisfy a suitable CFL condition, which is obtained by requiring that the point $(\tilde{x}_{nm}, \tilde{v}_{nm})$ falls inside the box $]x_{n-1}, x_{n+1}[\times]v_{m-1}, v_{m+1}[$. A straightforward way to increase the time accuracy is to use a multistep discretization scheme as the second-order accurate two-step BDF scheme. We have

$$f_{N,M}^{(k+1)}(x_n, v_m) \approx \frac{4}{3} f_{N,M}^{(k)}(\tilde{x}_{nm}, \tilde{v}_{nm}) - \frac{1}{3} f_{N,M}^{(k-1)}(\tilde{\tilde{x}}_{nm}, \tilde{\tilde{v}}_{nm}), \tag{21}$$

where $(\tilde{x}_{nm}, \tilde{v}_{nm})$ is the point obtained from (x_n, v_m) going back of one step Δt along the characteristic lines. Similarly, the point $(\tilde{\tilde{x}}_{nm}, \tilde{\tilde{v}}_{nm})$ is obtained by going two steps back along the characteristic lines, i.e., by using $2\Delta t$ instead of Δt when computing $\tilde{x}_{nm}$ and $\tilde{v}_{nm}$. Despite the fact that a BDF scheme is commonly presented as an implicit technique, in our context (f constant along the characteristics) it assumes the form of an explicit method. In terms of the coefficients, we end up with the scheme

$$\begin{aligned} c_{nm}^{(k+1)} &= \frac{4}{3}\left(c_{nm}^{(k)} + \Delta t\, \Phi_{nm}^{(k)}\right) - \frac{1}{3}\left(c_{nm}^{(k-1)} + 2\Delta t\, \Phi_{nm}^{(k-1)}\right) \\ &= \frac{4}{3} c_{nm}^{(k)} - \frac{1}{3} c_{nm}^{(k-1)} + \frac{2}{3}\Delta t \left[-v_m \sum_{i=0}^{N-1} d_{ni}^{(N,1)} (2c_{im}^{(k)} - c_{im}^{(k-1)}) \right. \\ &\quad \left. + E_N^{(k)}(x_n) \sum_{j=0}^{M-1} d_{mj}^{(M,1)} (2c_{nj}^{(k)} - c_{nj}^{(k-1)}) \frac{1}{\omega(v_m)} \right]. \end{aligned} \tag{22}$$

From theoretical considerations and the experiments in [8], it turns out that the above method is actually second-order accurate in Δt. Higher order schemes can be obtained with similar principles. All the above schemes guarantee mass conservation (see (4) for the continuous case), which is a crucial physical property.

For practical purposes, it is advisable to make the change of variable $f(t,x,v) = p(t,x,v)\exp(-v^2)$ in the Vlasov equation, so obtaining

$$\frac{\partial p}{\partial t} + v\frac{\partial p}{\partial x} - E(t,x)\left[\frac{\partial p}{\partial v} - 2vp\right] = 0, \ t \in (0,T], \ x \in \Omega_x, \ v \in \mathbb{R}. \tag{23}$$

At time step k, the function $p(t^k,x,v)$ is approximated by a function $p_{N,M}^{(k)}(x,v)$ in such a way that $p_{N,M}^{(k)}e^{-v^2}$ belongs to the finite dimensional space $\mathbf{Y}_{N\,M}$.

A generalization consists in introducing a real parameter α and assuming that the weight function is $\omega(v) = \exp(-\alpha^2 v^2)$. The approximation scheme can be easily adjusted by modifying nodes and weights of the Gaussian formula, through a multiplication by suitable constants. The difficulty in the implementation is practically the same, but, as observed in [9], the results are quite sensitive to the variation of α.

Numerical Experiments

The numerical scheme here proposed is validated in the standard two-stream instability benchmark test. We consider the Vlasov–Poisson problem (1)–(2) where we set $\Omega_x = [0, 4\pi[$, $\Omega_v = [-5, 5]$. The initial solution is given by

$$\bar{f}(x,v) = \frac{1}{2a\sqrt{2\pi}}\Big[G_R(v) + G_L(v)\Big](1 + \epsilon\cos(\kappa x)), \tag{24}$$

where $G_R(v) = e^{-\alpha^2(v-\beta)^2}$ and $G_R(v) = e^{-\alpha^2(v+\beta)^2}$ are two Gaussians centered symmetrically at the points $v = \pm\beta$. The parameters for (24) are: $a = 1/\sqrt{8}$, $\epsilon = 10^{-3}$, $\kappa = 0.5$, $\alpha = \bar{\alpha} = 2$, $\beta = \bar{\beta} = 1$.

In all the experiments that follow, we integrate up to time $T = 30$ using the second-order BDF scheme with a suitably small time step, in order to guarantee stability and a good accuracy. In this way we can concentrate our attention to the spectral approximation in the variable x and v. A study of the convergence rate in time of the proposed numerical scheme can be found in [8]. First of all, in Fig. we show the results at time $T = 30$ of the solution recovered by the Fourier-Fourier method, by choosing $N = 2^5$, $M = 2^6$ and time step equal to $\Delta t = 0.00125$. This will be the referring figure for the successive comparisons. Besides we show the corresponding time evolution of $|\hat{a}_1^{(k)}|$, the first Fourier mode of the electric field $E_N^{(k)}$ in (16). The behavior of this last quantity is predicted by theoretical

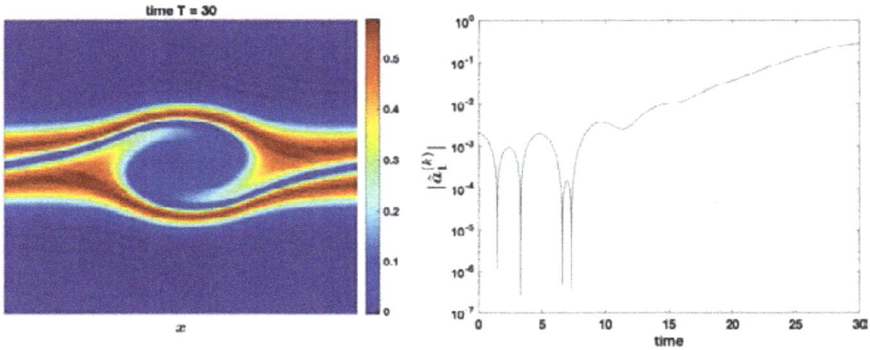

Fig. Two-stream instability test: approximated distribution function at time $T = 30$ obtained by using the Fourier-Fourier method with $N = 2^5$, $M = 2^6$, $\Delta t = 0.00125$, and the corresponding time evolution of the first Fourier mode of the electric field $E_N^{(k)}$, i.e. $|\hat{a}_1^{(k)}|$ in (16)

considerations, and the slope of the "segment" starting at $T = 15$ agrees with the expectancy.

As done in [9], we perform a series of experiments using less degrees of freedom than those actually necessary to resolve accurately the equation. In practice, we set $N = M = 2^4$. In this way, we could for instance detect what happens by varying the parameters α and β. Of course, if we increase the number of degrees of freedom, the numerical solution improves and cannot be distinguished from the referring one shown in Fig.. The purpose in [9] was to check what happens by varying the parameter α in the Hermite weight $\exp(-\alpha^2 v^2)$. The conclusions are that the approximate solution is very sensitive to the choice of α and that there are values of α that perform better than others. In general these values are those belonging to a neighbourhood of $\alpha = 1$. Moreover, in [9], we note that keeping α constantly equal to the value that better fits the initial datum (i.e. $\alpha = \bar{\alpha} = 2$ for (24)) may create instability as time increases. For such motivations, since at the moment a practical algorithm able to vary α in a dynamical way during the computations is not available, in the numerical experiments that follow we fix $\alpha = 1$, while play with β.

Due to the particular initial condition, we adopt a two-species decomposition of the Vlasov equation, where the distribution function is given by the sum of two electron distribution functions, i.e., $f = f_R + f_L$. These distribution functions refer to the two initial electron distributions, so that $f_R = p_R G_R$ and $f_L = p_L G_L$, where p_L and p_R are given polynomials. We consider the two systems of electrons described by the distribution functions f_L and f_R at the initial time as distinct plasma species that maintain their diversity throughout the whole numerical simulation. Therefore, we can split the Vlasov equation into two equations that are still of Vlasov type and are solvable independently, although they are coupled through the same electric field, which depends on the total charge density. This amounts to approximate two independent equations of the same type of that given

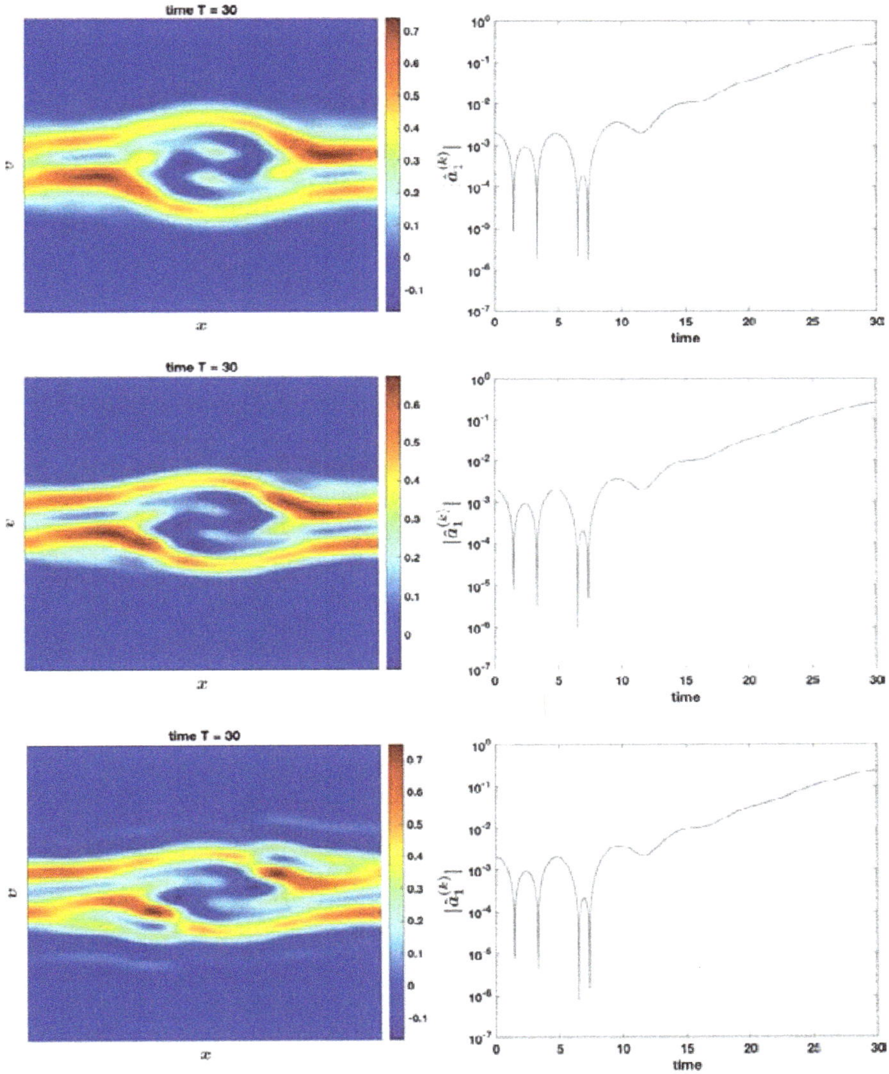

Fig. Two-stream instability test: approximated distribution function at time $T = 30$ obtained by using the Fourier–Hermite method with $N = M = 2^4$, $\Delta t = 0.01$, $\alpha = 1$ (left panel) and the corresponding time evolution of the first Fourier mode of the electric field $E_N^{(k)}$, i.e. $|\hat{a}_1^{(k)}|$ in (16) (right panel) when $\beta = 0.5$ (top), $\beta = 1$ (center) and $\beta = 1.5$ (bottom)

in (23), respectively shifted by $\pm\beta$, i.e.

$$\frac{\partial p_R}{\partial t} + (v-\beta)\frac{\partial p_R}{\partial x} - E(t,x)\left[\frac{\partial p_R}{\partial v} - 2\alpha^2(v-\beta)p_R\right] = 0, \tag{25}$$

$$\frac{\partial p_L}{\partial t} + (v+\beta)\frac{\partial p_L}{\partial x} - E(t,x)\left[\frac{\partial p_L}{\partial v} - 2\alpha^2(v+\beta)p_L\right] = 0. \tag{26}$$

The two unknowns are then coupled through the density function as in (2).

The plots of Fig. show the numerical distribution function at time $T = 30$ obtained by using the Fourier–Hermite method with $N = M = 2^4$, $\Delta t = 0.01$, $\alpha = 1$ and different values of the parameter β (i.e. $\beta = 0.5$, $\beta = 1$ and $\beta = 1.5$), together with the corresponding time evolution of the (log of the) first Fourier mode of the electric field $E_N^{(k)}$, i.e. $|\hat{a}_1^{(k)}|$ in (16).

The distribution functions presented in the left column of Fig. 2 are visibly and significantly different depending on β, while the first Fourier mode of the electric field shown in the right column seems to be less affected. These differences practically confirm that the choice of the Hermite weight functions $\omega(v) = \exp(-\alpha^2(v \pm \beta)^2)$ is a crucial aspect of the method (see also [11, 12, 17, 22]). This conclusion is heuristic. Unfortunately, there is no space enough for a deeper quantitative analysis in these pages. The question deserves however further investigation. Moreover, it would be advisable to develop appropriate algorithms allowing for the automatic adjustment of both parameters α and β during the time advancing procedure, in order to optimize the performance.

References

1. Bittencourt, J.A.: Fundamentals of Plasma Physics. Springer, New York (2004)
2. Boyd, J.P.: The rate of convergence of Hermite function series. Math. Comput. **35**, 1309–1316 (1980)
3. Boyd, J.P.: Asymptotic coefficients of Hermite function series. J. Comput. Phys. **54**, 382–410 (1984)
4. Camporeale, E., et al. On the velocity space discretization for the Vlasov–Poisson system: comparison between implicit Hermite spectral and Particle-in-Cell methods. Comput. Phys. Commun. **198**, 47–58 (2016)
5. Cheng, C.Z., Knorr, G.: The integration of the Vlasov equation in configuration space. J. Comput. Phys. **22**(3), 330–351 (1976)
6. Crouseilles, N., Respaud, T., Sonnendrücker, E.: A forward semi-Lagrangian method for the numerical solution of the Vlasov equation. Comput. Phys. Commun. **180**(10), 1730–1745 (2009)
7. Delzanno, G.L.: Multi-dimensional, fully-implicit, spectral method for the Vlasov–Maxwell equations with exact conservation laws in discrete form. J. Comput. Phys. **301**, 338–356 (2015)
8. Fatone, D., Funaro, L., Manzini, G.: Arbitrary-order time-accurate semi-Lagrangian spectral approximations of the Vlasov–Poisson system. J. Comput. Phys. **384**, 349–375 (2019)

9. Fatone, D., Funaro, L., Manzini, G.: A semi-Lagrangian spectral method for the Vlasov–Poisson system based on Fourier, Legendre and Hermite polynomials. Commun. Appl. Math. Comput. **1**, 333–360 (2019)
10. Grad, H.: On the kinetic theory of rarefied gases. Commun. Pure Appl. Math. **2**(4), 331–407 (1949)
11. Holloway, J.P.: Spectral velocity discretizations for the Vlasov-Maxwell equations. Transp. Theory Stat. Phys. **25**(1), 1–32 (1996)
12. Ma, H., Sun, W., Tang, T.: Hermite spectral methods with a time-dependent scaling for parabolic equations in unbounded domains. SIAM J. Numer. Anal. **43**, 58–75 (2005)
13. Manzini, G., et al. A Legendre-Fourier spectral method with exact conservation laws for the Vlasov–Poisson system. J. Comput. Phys. **317**, 82–107 (2016)
14. Manzini, G., Funaro, D., Delzanno, G.L.: Convergence of spectral discretizations of the Vlasov–Poisson system. SIAM J. Numer. Anal. **55**(5), 2312–2335 (2017)
15. Qiu, J.-M., Christlieb, A.: A conservative high order semi-Lagrangian WENO method for the Vlasov equation. J. Comput. Phys. **229**(4), 1130–1149 (2010)
16. Qiu, J.-M., Russo, G.: A high order multidimensional characteristic tracing strategy for the Vlasov–Poisson system. J. Sci. Comput. **71**, 414–434 (2017)
17. Schumer, J.W., Holloway, J.P.: Vlasov simulations using velocity-scaled Hermite representations. J. Comput. Phys. **144**(2), 626–661 (1998)
18. Sonnendrücker, E., et al.: The semi-Lagrangian method for the numerical resolution of the Vlasov equation. J. Comput. Phys. **149**(2), 201–220 (1999)
19. Tang, T.: The Hermite spectral method for Gaussian-type functions. SIAM J. Sci. Comput. **14**(3), 594–606 (1993)
20. Vencels, J., et al.: Spectral solver for multi-scale plasma physics simulations with dynamically adaptive number of moments. Proc. Comput. Sci. **51**, 1148–1157 (2015)
21. Vencels, J., et al.: SpectralPlasmaSolver: a spectral code for multiscale simulations of collisionless, magnetized plasmas. J. Phys. Conf. Series **719**(1), 012022 (2016)
22. Xiang, X.-M., Wang, Z.-Q.: Generalized Hermite approximations and spectral method for partial differential equations in multiple dimensions. J. Sci. Comput. **57**, 229–253 (2013)

10

Discrete Equivalence of Adjoint Neumann–Dirichlet div-grad and grad-div Equations in Curvilinear 3D Domains

Yi Zhang, Varun Jain, Artur Palha, and Marc Gerritsma

Introduction

In $\mathbb{R}^d$, given a bounded domain Ω with Lipschitz boundary $\partial\Omega$ and $\hat{\boldsymbol{\sigma}}_{\boldsymbol{n}} \in H^{-1/2}(\partial\Omega) = \mathrm{tr}\, H(\mathrm{div}, \Omega)$, $\omega \in H^1(\Omega)$ solves the Neumann problem,

$$\begin{cases} \dfrac{\partial\omega}{\partial\boldsymbol{n}} = \hat{\boldsymbol{\sigma}}_{\boldsymbol{n}} & \text{on } \partial\Omega \\ -\mathrm{div}\left(\mathrm{grad}\,\omega\right) + \omega = 0 & \text{in } \Omega \end{cases}, \tag{1}$$

if and only if $\boldsymbol{\sigma} \in H(\mathrm{div}, \Omega)$ which solves the Dirichlet problem,

$$\begin{cases} \boldsymbol{\sigma} \cdot \boldsymbol{n} = \hat{\boldsymbol{\sigma}}_{\boldsymbol{n}} & \text{on } \partial\Omega \\ -\mathrm{grad}\,(\mathrm{div}\,\boldsymbol{\sigma}) + \boldsymbol{\sigma} = 0 & \text{in } \Omega \end{cases}, \tag{2}$$

satisfies $\boldsymbol{\sigma} = \mathrm{grad}\,\omega$ [3]. This is obvious at the continuous level. The question is whether we can find a set of finite dimensional function spaces such that $\boldsymbol{\sigma}^h = \mathrm{grad}\,\omega^h$ holds if ω^h and $\boldsymbol{\sigma}^h$ solve the discrete Neumann and Dirichlet problems respectively. The answer is yes.

Y. Zhang (✉) · V. Jain · M. Gerritsma
Delft University of Technology, Delft, Netherlands
e-mail: y.zhang-14@tudelft.nl; v.jain@tudelft.nl; m.i.gerritsma@tudelft.nl

A. Palha
Eindhoven University of Technology, Eindhoven, Netherlands
e-mail: a.palha@tue.nl

Throughout this paper, we restrict ourselves to $\mathbb{R}^3$. We will first construct the primal polynomial spaces and their algebraic dual representations, and then use them to discretize problems (1) and (2) such that the identity $\boldsymbol{\sigma}^h = \text{grad}\,\omega^h$ holds at the discrete level in any curvilinear domain for any polynomial approximation degree. This work extends [7, 9], where similar dual Neumann–Dirichlet problems are considered, to 3-dimensional space. These primal spaces and their algebraic dual representations can be ideal for the so-called mimetic or structure-preserving discretizations [1, 4, 8, 11, 12]. Together with their trace spaces, they can be used for the hybrid finite element methods which first decompose the domains into discontinuous elements then connect them with Lagrange multipliers living in the trace spaces [2, 13, 14].

The outline of this paper is as follows: In other sect, we introduce the construction of polynomial spaces and their algebraic dual representations. The discrete formu-lations of the Neumann–Dirichlet problems and the proof of their equivalence at the discrete level follow in Sect. A 3-dimensional numerical test case is then presented in Sect.

Function Spaces

Primal Polynomial Spaces

Let $-1 = \xi_0^i < \xi_1^i < \cdots < \xi_{I^i}^i = 1, i = 1, 2, 3$, being three partitionings of $[-1, 1]$. The associated Lagrange polynomials are

$$h_j(\xi^i) = \prod_{m=0,m\neq j}^{I^i} \frac{\xi^i - \xi_m^i}{\xi_j^i - \xi_m^i}, \ j = 0, 1, \cdots, I^i.$$

They are polynomials of degree I^i which satisfy the Kronecker delta property, $h_j(\xi_k^i) = \delta_{jk}$. The associated edge functions can be derived as [6],

$$e_j(\xi^i) = -\sum_{k=0}^{j-1} \frac{\mathrm{d}h_k(\xi^i)}{\mathrm{d}\xi^i}, \ j = 1, 2, \cdots, I^i,$$

which are polynomials of degree $I^i - 1$. Edge functions also satisfy the Kronecker delta property, but in the integral sense,

$$\int_{\xi_{k-1}^i}^{\xi_k^i} e_j(\xi^i)\,\mathrm{d}\xi^i = \delta_{jk}.$$

Consider a reference domain $\Omega_{\text{ref}}|_{\xi^1,\xi^2,\xi^3} := [-1, 1]^3$. With the tensor product, we can construct finite dimensional scalar function space $\mathcal{P}^{I^1,I^2,I^3}$ spanned by polynomial basis functions

$$\left\{h_i(\xi^1)h_j(\xi^2)h_k(\xi^3)\right\},$$

and vector-valued function space $\mathcal{L}^{I^1,I^2,I^3}$ spanned by polynomial basis functions

$$\left\{e_i(\xi^1)h_j(\xi^2)h_k(\xi^3),\ h_i(\xi^1)e_j(\xi^2)h_k(\xi^3),\ h_i(\xi^1)h_j(\xi^2)e_k(\xi^3)\right\}.$$

Let $\omega^h \in \mathcal{P}^{I^1,I^2,I^3}$ be

$$\omega^h = \sum_{i=0}^{I^1}\sum_{j=0}^{I^2}\sum_{k=0}^{I^3} w_{i,j,k}h_i(\xi^1)h_j(\xi^2)h_k(\xi^3). \tag{3}$$

Due to the way of constructing the edge functions, we can easy derive $\boldsymbol{\rho}^h = \operatorname{grad}\omega^h \in \mathcal{L}^{I^1,I^2,I^3}$,

$$\boldsymbol{\rho}^h = \operatorname{grad}\omega^h = (\rho_1,\ \rho_2,\ \rho_3)^{\mathsf{T}},$$

where [6],

$$\rho_1 = \sum_{i=1}^{I^1}\sum_{j=0}^{I^2}\sum_{k=0}^{I^3}\left(w_{i,j,k} - w_{i-1,j,k}\right) e_i(\xi^1)h_j(\xi^2)h_k(\xi^3),$$

$$\rho_2 = \sum_{i=0}^{I^1}\sum_{j=1}^{I^2}\sum_{k=0}^{I^3}\left(w_{i,j,k} - w_{i,j-1,k}\right) h_i(\xi^1)e_j(\xi^2)h_k(\xi^3),$$

$$\rho_3 = \sum_{i=0}^{I^1}\sum_{j=0}^{I^2}\sum_{k=1}^{I^3}\left(w_{i,j,k} - w_{i,j,k-1}\right) h_i(\xi^1)h_j(\xi^2)e_k(\xi^3).$$

Let $\underline{\omega}$, $\underline{\boldsymbol{\rho}}$ be the vectors of expansion coefficients of ω^h, $\boldsymbol{\rho}^h$. We can obtain

$$\underline{\boldsymbol{\rho}} = \mathbb{E}\,\underline{\omega}, \tag{4}$$

where $\mathbb{E}$ is called the incidence matrix. The incidence matrix is very sparse, only consists of ± 1 as non-zero entries. If we squeeze, stretch or distort the domain, of course, the polynomial basis functions change, but the incidence matrix will remain the same. It only depends on the topology of the mesh and the numbering of the degrees of freedom. And it is exact. In other words, it introduces no extra error. All these features make it an excellent discrete counterpart of the grad operator. Examples of incidence matrices can be found in [8, 10–12].

For a comprehensive explanation of these polynomial basis functions, we refer to [6]. In isogeometric analysis, tensor-product B-splines with similar properties have been developed, see, for example [5]. For tetrahedral elements, an analogue development can be found in [15].

From (3), we can derive the trace of ω^h, for example, on the back boundary of Ω_{ref}, $\Gamma_{\mathrm{b}} = \left\{\xi^1 = -1,\ \xi^2,\ \xi^3 \in [-1,1]\right\}$,

$$\mathrm{tr}_{\mathrm{b}}\, \omega^h = \sum_{j=0}^{I^2} \sum_{k=0}^{I^3} w_{0,j,k} h_0(-1) h_j(\xi^2) h_k(\xi^3).$$

Let $\underline{\omega}_{\mathrm{b}}$ be the vector of expansion coefficients of $\mathrm{tr}_{\mathrm{b}}\, \omega^h$. Clearly, there exists a linear operator $\mathbb{N}_{\mathrm{b}}$ such that

$$\underline{\omega}_{\mathrm{b}} = \mathbb{N}_{\mathrm{b}}\, \underline{\omega}.$$

The same processes can be done for other boundaries. If we collect the traces of ω^h on all boundaries and combine their vectors of expansion coefficients and corresponding linear operators, we can eventually obtain

$$\underline{\omega}_{\mathrm{tr}} = \mathbb{N}\, \underline{\omega},$$

where the matrix $\mathbb{N}$, like $\mathbb{E}$, is sparse and only depends on the topology of the mesh and the numbering of the degrees of freedom. Furthermore, it contains only 1 as non-zero entries. An example of $\mathbb{N}$ can be found in [7]. Now, we can conclude that the trace space, $P^{I^1,I^2,I^3} = \mathrm{tr}\, \mathcal{P}^{I^1,I^2,I^3}$, is given as

$$P^{I^1,I^2,I^3} := P_{-1}^{I^2,I^3} \cup P_{1}^{I^2,I^3} \cup P_{-1}^{I^1,I^3} \cup P_{1}^{I^1,I^3} \cup P_{-1}^{I^1,I^2} \cup P_{1}^{I^1,I^2},$$

where $P_{-1}^{I^2,I^3}$ is the space spanned by $\left\{h_0(-1)h_j(\xi^2)h_k(\xi^3)\right\}$, $P_{1}^{I^2,I^3}$ is the space spanned by $\left\{h_{I^1}(1)h_j(\xi^2)h_k(\xi^3)\right\}$ and so on. Notice that the polynomial basis functions in $\left\{h_0(-1)h_j(\xi^2)h_k(\xi^3)\right\}$ are exactly the same as those in $\left\{h_{I^1}(1)h_j(\xi^2)h_k(\xi^3)\right\}$ because $h_0(-1) = h_{I^1}(1) = 1$. But here we still distinguish them because they represent basis functions at different boundaries.

Algebraic Dual Polynomial Spaces

We first consider the space $\mathcal{P}^{I^1,I^2,I^3}$. Let $\mathbb{M}_{\mathcal{P}}$ be the symmetric mass matrix, for example,

$$\mathbb{M}_{\mathcal{P}\, i+j(I^1+1)+k(I^1+1)(I^2+1),\, l+m(I^1+1)+n(I^1+1)(I^2+1)} := \iiint_{\Omega_{\text{ref}}} h_i(\xi^1)h_j(\xi^2)h_k(\xi^3)h_l(\xi^1)h_m(\xi^2)h_n(\xi^3)\, \mathrm{d}\xi^1\mathrm{d}\xi^2\mathrm{d}\xi^3.$$

The associated algebraic dual polynomial representations, or simply dual polynomials, are linear combinations of the polynomial basis functions, or simply primal polynomials, defined in the previous section,

$$\left[\widetilde{h_{0,0,0}}(\xi^1,\xi^2,\xi^3),\ \cdots,\ \widetilde{h_{I^1,I^2,I^3}}(\xi^1,\xi^2,\xi^3)\right] := \left[h_0(\xi^1)h_0(\xi^2)h_0(\xi^3),\ \cdots,\ h_{I^1}(\xi^1)h_{I^2}(\xi^2)h_{I^3}(\xi^3)\right]\mathbb{M}_{\mathcal{P}}^{-1}.$$

These dual polynomials are always well-defined. This is because the primal polynomials are linearly independent. So the mass matrix $\mathbb{M}_{\mathcal{P}}$ is injective and surjective, therefore invertible. Let the finite dimensional space spanned by $\left\{\widetilde{h_{i,j,k}}(\xi^1,\xi^2,\xi^3)\right\}$ be denoted by $\widetilde{\mathcal{P}}^{I^1,I^2,I^3}$. We say $\widetilde{\mathcal{P}}^{I^1,I^2,I^3}$ is the algebraic dual space of the primal space $\mathcal{P}^{I^1,I^2,I^3}$. Note that $\mathcal{P}^{I^1,I^2,I^3}$ and $\widetilde{\mathcal{P}}^{I^1,I^2,I^3}$ actually represent the same space. The change of basis functions only leads to a different representation. Therefore, we also call the algebraic dual space a dual representation. Let $\widetilde{\mathbb{M}}_{\mathcal{P}}$ be the mass matrix of $\widetilde{\mathcal{P}}^{I^1,I^2,I^3}$, we can easily see that

$$\widetilde{\mathbb{M}}_{\mathcal{P}}\mathbb{M}_{\mathcal{P}} = \mathcal{I}, \tag{5}$$

where $\mathcal{I}$ is the identity matrix. Similarly, we can derive the algebraic dual space $\widetilde{\mathcal{L}}^{I^1,I^2,I^3}$ of the primal space $\mathcal{L}^{I^1,I^2,I^3}$. Let $\widetilde{\mathbb{M}}_{\mathcal{L}}$ and $\mathbb{M}_{\mathcal{L}}$ be their mass matrices, we have

$$\widetilde{\mathbb{M}}_{\mathcal{L}}\mathbb{M}_{\mathcal{L}} = \mathcal{I}. \tag{6}$$

If $\boldsymbol{\rho}^h \in \mathcal{L}^{I^1,I^2,I^3}$, $\boldsymbol{\sigma}^h$, whose vector of expansion coefficients $\underline{\boldsymbol{\sigma}}$ satisfies

$$\underline{\boldsymbol{\sigma}} = \mathbb{M}_{\mathcal{L}}\, \underline{\boldsymbol{\rho}}, \tag{7}$$

will be the representation of $\boldsymbol{\rho}^h$ in the algebraic dual space $\widetilde{\mathcal{L}}^{I^1,I^2,I^3}$.

To explain how the algebraic dual space of the trace space P^{I^1,I^2,I^3} is derived, we take $P_{-1}^{I^2,I^3}$ as example. We already know that $P_{-1}^{I^2,I^3}$ is a space spanned by primal polynomials $\left\{h_0(-1)h_j(\xi^2)h_k(\xi^3)\right\}$. With these primal polynomials, we can compute its mass matrix, denoted by $\mathbb{M}_\mathrm{b}$. The dual polynomials are then computed by

$$\left[\widetilde{h_{0,0,0}}(-1,\xi^2,\xi^3),\ \cdots,\ \widetilde{h_{0,I^2,I^3}}(-1,\xi^2,\xi^3)\right]$$
$$=\left[h_0(-1)h_1(\xi^2)h_1(\xi^3),\ \cdots,\ h_0(-1)h_{I^2}(\xi^2)h_{I^3}(\xi^3)\right]\mathbb{M}_\mathrm{b}^{-1}.$$

The algebraic dual space $\widetilde{P}_{-1}^{I^2,I^3}$ is spanned by dual polynomials $\left\{\widetilde{h_{0,j,k}}(-1,\xi^2,\xi^3)\right\}$. The algebraic dual space of the trace space P^{I^1,I^2,I^3} eventually can be written as

$$\widetilde{P}^{I^1,I^2,I^3}=\widetilde{P}_{-1}^{I^2,I^3}\cup\widetilde{P}_{1}^{I^2,I^3}\cup\widetilde{P}_{-1}^{I^1,I^3}\cup\widetilde{P}_{1}^{I^1,I^3}\cup\widetilde{P}_{-1}^{I^1,I^2}\cup\widetilde{P}_{1}^{I^1,I^2}.$$

The divergence of $\boldsymbol{\sigma}^h\in\widetilde{\mathcal{L}}^{I^1,I^2,I^3}$ can be done with the help of the boundary value $\hat{\boldsymbol{\sigma}}^h\in\widetilde{P}^{I^1,I^2,I^3}$. With vector proxies, it can be written as

$$\underline{\operatorname{div}\boldsymbol{\sigma}^h}=\mathbb{N}^\mathsf{T}\,\underline{\hat{\boldsymbol{\sigma}}}^h-\mathbb{E}^\mathsf{T}\underline{\boldsymbol{\sigma}}^h. \tag{8}$$

A detailed introduction of algebraic dual polynomial spaces is given in [9].

Function Spaces in Curvilinear Domains

So far, all polynomial spaces are defined only in the reference domain $\Omega_\mathrm{ref}|_{\xi^1,\xi^2,\xi^3}=[-1,1]^3$. Consider an arbitrary domain Ω and a C^1 diffeomorphism $\Phi:\Omega_\mathrm{ref}|_{\xi^1,\xi^2,\xi^3}\to\Omega|_{x^1,x^2,x^3}$. In Ω, the primal polynomials change. Therefore, the mass matrices will also change. But the process of constructing dual polynomials does not change. And as we mentioned before, the metric-independent incidence matrix $\mathbb{E}$ and the matrix $\mathbb{N}$ remain the same. The way of converting polynomials in Cartesian domain into those in curvilinear domains follows the general coordinate transformation process, for example, see [16].

From now on, notations mentioned in this section not only refer to the reference domain Ω_ref, but also refer to the physical domain Ω.

Weak Formulations

Discrete Neumann Problem

With integration by parts, we can derive the weak formulation of the Neumann problem, (1), written as: For given $\hat{\boldsymbol{\sigma}} \in H^{-1/2}(\partial\Omega)$, find $\omega \in H^1(\Omega)$ such that

$$\left(\text{grad}\,\omega,\ \text{grad}\,\bar{\omega}\right)_{L^2} + (\omega,\ \bar{\omega})_{L^2} = \left\langle \text{tr}\,\bar{\omega},\ \hat{\boldsymbol{\sigma}}\right\rangle, \quad \forall \bar{\omega} \in H^1(\Omega). \tag{9}$$

Note that on the right hand side, we use $\langle\cdot,\cdot\rangle$ to represent the duality pairing between $\text{tr}\,\bar{\omega} \in H^{1/2}(\partial\Omega)$ and $\hat{\boldsymbol{\sigma}} \in H^{-1/2}(\partial\Omega)$. We use finite dimensional space $\mathcal{P}^{I^1,I^2,I^3}$ to approximate the space $H^1(\Omega)$ and use the algebraic dual trace space $\widetilde{P}^{I^1,I^2,I^3}$ to approximate the space $H^{-1/2}(\partial\Omega)$. Then we obtain

$$\left(\text{grad}\,\omega^h,\ \text{grad}\,\bar{\omega}^h\right)_{L^2} = \underline{\bar{\omega}}^{h,\mathsf{T}}\ \mathbb{E}^{\mathsf{T}}\mathbb{M}_{\mathcal{L}}\mathbb{E}\ \underline{\omega}^h,$$

$$\left(\omega^h,\ \bar{\omega}^h\right)_{L^2} = \underline{\bar{\omega}}^{h,\mathsf{T}}\ \mathbb{M}_{\mathcal{P}}\ \underline{\omega}^h,$$

and

$$\int_{\partial\Omega} \text{tr}\,\bar{\omega}^h\ \hat{\boldsymbol{\sigma}}^h\ \text{d}\Gamma = \underline{\bar{\omega}}^{h,\mathsf{T}}\ \mathbb{N}^{\mathsf{T}}\ \underline{\hat{\boldsymbol{\sigma}}}^h,$$

which eventually leads to the discrete formulation of (9),

$$\mathbb{E}^{\mathsf{T}}\mathbb{M}_{\mathcal{L}}\mathbb{E}\ \underline{\omega}^h + \mathbb{M}_{\mathcal{P}}\ \underline{\omega}^h = \mathbb{N}^{\mathsf{T}}\ \underline{\hat{\boldsymbol{\sigma}}}^h. \tag{10}$$

Discrete Dirichlet Problem

For the Dirichlet problem, (2), the weak formulation is given as: For given $\hat{\boldsymbol{\sigma}} \in H^{-1/2}(\partial\Omega)$, find $\boldsymbol{\sigma} \in H(\text{div},\Omega)$, $\text{tr}\,\boldsymbol{\sigma} = \hat{\boldsymbol{\sigma}}$ such that

$$(\text{div}\,\boldsymbol{\sigma},\ \text{div}\,\bar{\boldsymbol{\sigma}})_{L^2} + (\boldsymbol{\sigma},\ \bar{\boldsymbol{\sigma}})_{L^2} = 0, \quad \forall \bar{\boldsymbol{\sigma}} \in H_0(\text{div},\Omega). \tag{11}$$

We use algebraic dual space $\widetilde{\mathcal{L}}^{I^1,I^2,I^3}$ to approximate $H(\text{div},\Omega)$. With $\hat{\boldsymbol{\sigma}}^h \in \widetilde{P}^{I^1,I^2,I^3}$ given and (8), we obtain

$$\left(\text{div}\,\boldsymbol{\sigma}^h,\ \text{div}\,\bar{\boldsymbol{\sigma}}^h\right)_{L^2} = -\underline{\bar{\boldsymbol{\sigma}}}^{h,\mathsf{T}}\ \mathbb{E}\widetilde{\mathbb{M}}_{\mathcal{P}}\left(\mathbb{N}^{\mathsf{T}}\ \underline{\hat{\boldsymbol{\sigma}}}^h - \mathbb{E}^{\mathsf{T}}\ \underline{\boldsymbol{\sigma}}^h\right),$$

and

$$\left(\boldsymbol{\sigma}^h,\ \bar{\boldsymbol{\sigma}}^h\right)_{L^2} = \underline{\bar{\boldsymbol{\sigma}}}^{h,\mathsf{T}}\ \widetilde{\mathbb{M}}_{\mathcal{L}}\ \underline{\boldsymbol{\sigma}}^h.$$

Therefore, the discrete formulation of (11) is written as

$$\mathbb{E}\widetilde{\mathbb{M}}_{\mathcal{P}}\mathbb{E}^{\mathsf{T}}\,\underline{\boldsymbol{\sigma}}^h + \widetilde{\mathbb{M}}_{\mathcal{L}}\,\underline{\boldsymbol{\sigma}}^h = \mathbb{E}\widetilde{\mathbb{M}}_{\mathcal{P}}\mathbb{N}^{\mathsf{T}}\,\underline{\hat{\boldsymbol{\sigma}}}^h. \tag{12}$$

Equivalence Between Discrete Formulations

Now it is time to check if the equivalence between (1) and (2) holds at the discrete level. In other words, it is time to check if the statement that $\underline{\omega}^h$ solves (10) if and only if $\underline{\boldsymbol{\sigma}}^h = \operatorname{grad}\underline{\omega}^h$ solves (12) is correct.

From (4) and (7), we know that $\underline{\boldsymbol{\sigma}}^h$,

$$\underline{\boldsymbol{\sigma}}^h = \mathbb{M}_{\mathcal{L}}\mathbb{E}\,\underline{\omega}^h, \tag{13}$$

is the vector representation of grad ω^h in the dual space. If we insert (13) into (12), we obtain

$$\mathbb{E}\widetilde{\mathbb{M}}_{\mathcal{P}}\mathbb{E}^{\mathsf{T}}\mathbb{M}_{\mathcal{L}}\mathbb{E}\,\underline{\omega}^h + \widetilde{\mathbb{M}}_{\mathcal{L}}\mathbb{M}_{\mathcal{L}}\mathbb{E}\,\underline{\omega}^h = \mathbb{E}\widetilde{\mathbb{M}}_{\mathcal{P}}\mathbb{N}^{\mathsf{T}}\,\underline{\hat{\boldsymbol{\sigma}}}^h. \tag{14}$$

From (10), we know that

$$\mathbb{E}^{\mathsf{T}}\mathbb{M}_{\mathcal{L}}\mathbb{E}\,\underline{\omega}^h = -\mathbb{M}_{\mathcal{P}}\,\underline{\omega}^h + \mathbb{N}^{\mathsf{T}}\,\underline{\hat{\boldsymbol{\sigma}}}^h. \tag{15}$$

By inserting (15) into (14), we get

$$\mathbb{E}\widetilde{\mathbb{M}}_{\mathcal{P}}\left(-\mathbb{M}_{\mathcal{P}}\,\underline{\omega}^h + \mathbb{N}^{\mathsf{T}}\,\underline{\hat{\boldsymbol{\sigma}}}^h\right) + \widetilde{\mathbb{M}}_{\mathcal{L}}\mathbb{M}_{\mathcal{L}}\mathbb{E}\,\underline{\omega}^h = \mathbb{E}\widetilde{\mathbb{M}}_{\mathcal{P}}\mathbb{N}^{\mathsf{T}}\,\underline{\hat{\boldsymbol{\sigma}}}^h. \tag{16}$$

From (5) and (6), we know that (16) holds, which proves the equivalence.

If the equivalence holds, relation $\left\|\omega^h\right\|_{H^1(\Omega)} = \left\|\boldsymbol{\sigma}^h\right\|_{H(\mathrm{div},\Omega)}$ should also be satisfied. To prove this, we have

$$\begin{aligned}
\left\|\boldsymbol{\sigma}^h\right\|^2_{H(\mathrm{div},\Omega)} &\overset{(8)}{=} \underline{\boldsymbol{\sigma}}^{h,\mathsf{T}}\,\widetilde{\mathbb{M}}_{\mathcal{L}}\,\underline{\boldsymbol{\sigma}}^h + \left(\mathbb{N}^{\mathsf{T}}\,\underline{\hat{\boldsymbol{\sigma}}}^h - \mathbb{E}^{\mathsf{T}}\underline{\boldsymbol{\sigma}}^h\right)^{\mathsf{T}}\widetilde{\mathbb{M}}_{\mathcal{P}}\left(\mathbb{N}^{\mathsf{T}}\,\underline{\hat{\boldsymbol{\sigma}}}^h - \mathbb{E}^{\mathsf{T}}\underline{\boldsymbol{\sigma}}^h\right)\\
&\overset{(13)}{=} \left(\mathbb{M}_{\mathcal{L}}\mathbb{E}\,\underline{\omega}^h\right)^{\mathsf{T}}\widetilde{\mathbb{M}}_{\mathcal{L}}\left(\mathbb{M}_{\mathcal{L}}\mathbb{E}\,\underline{\omega}^h\right)\\
&\quad + \left[\mathbb{N}^{\mathsf{T}}\,\underline{\hat{\boldsymbol{\sigma}}}^h - \mathbb{E}^{\mathsf{T}}\left(\mathbb{M}_{\mathcal{L}}\mathbb{E}\,\underline{\omega}^h\right)\right]^{\mathsf{T}}\widetilde{\mathbb{M}}_{\mathcal{P}}\left[\mathbb{N}^{\mathsf{T}}\,\underline{\hat{\boldsymbol{\sigma}}}^h - \mathbb{E}^{\mathsf{T}}\left(\mathbb{M}_{\mathcal{L}}\mathbb{E}\,\underline{\omega}^h\right)\right]\\
&\overset{(10)}{=} \underline{\omega}^{h,\mathsf{T}}\,\mathbb{E}^{\mathsf{T}}\mathbb{M}_{\mathcal{L}}\mathbb{E}\,\underline{\omega}^h + \underline{\omega}^{h,\mathsf{T}}\,\mathbb{M}_{\mathcal{P}}\widetilde{\mathbb{M}}_{\mathcal{P}}\mathbb{M}_{\mathcal{P}}\,\underline{\omega}^h\\
&= \left\|\omega^h\right\|^2_{H^1(\Omega)},
\end{aligned}$$

where we constantly use (5) and (6) and the fact that mass matrices are symmetric.

Numerical Test

Consider the mapping Φ which maps the Cartesian reference domain $\Omega_{\text{ref}}|_{\xi^1,\xi^2,\xi^3} := [-1, 1]^3$ into the physical domain $\Omega|_{x^1,x^2,x^3} = [0, 1]^3$ by

$$x^i = \frac{1}{2} + \frac{1}{2}\left(\xi^i + c\prod_j \sin(\pi\xi^j)\right), \quad i = 1, 2, 3.$$

When the deformation coefficient $c = 0$, the domain Ω is Cartesian. Otherwise the domain is curvilinear, meaning that a curvilinear coordinate system parametrizes Ω. Examples of such curvilinear domains in $\mathbb{R}^2$ are shown in Fig.

A manufactured solution of the Neumann problem, (1), is

$$\omega_{\text{exact}} = e^{x^1} + e^{x^2} + e^{x^3}.$$

Clearly, $\boldsymbol{\sigma}_{\text{exact}} = \text{grad}\ \omega_{\text{exact}} = \left(e^{x^1},\ e^{x^2},\ e^{x^3}\right)^{\mathsf{T}}$ solves the Dirichlet problem, (2).

In the domains of different deformation coefficient c, with the boundary condition $\hat{\boldsymbol{\sigma}} = \text{tr}\ \boldsymbol{\sigma}_{\text{exact}}$ imposed, we solve the discrete formulations (10) and (12) using Gauss–Lobatto–Legendre (GLL) polynomial spaces of degree $I^1 = I^2 = I^3 = N$.

The results of the L^2-error of $\left(\boldsymbol{\sigma}^h - \text{grad}\ \omega^h\right)$ are shown in Fig.

we can see that the relation $\boldsymbol{\sigma}^h = \text{grad}\ \omega^h$ is preserved up to the machine precision. With the growth of the polynomial degree, the error increases slowly because of the accumulation of the machine error as the amount of degrees of freedom grows significantly.

In Table, the results of the H -norm of ω ^{h}and $H(\text{div})$-norm of $\boldsymbol{\sigma}$ ^{h} are presented. It is shown that the relation $\left\|\omega^h\right\|_{H^1(\Omega)} = \left\|\boldsymbol{\sigma}^h\right\|_{H(\text{div},\Omega)}$ holds for all polynomial degrees irrespective of whether we use the Cartesian domain, $c = 0$, or

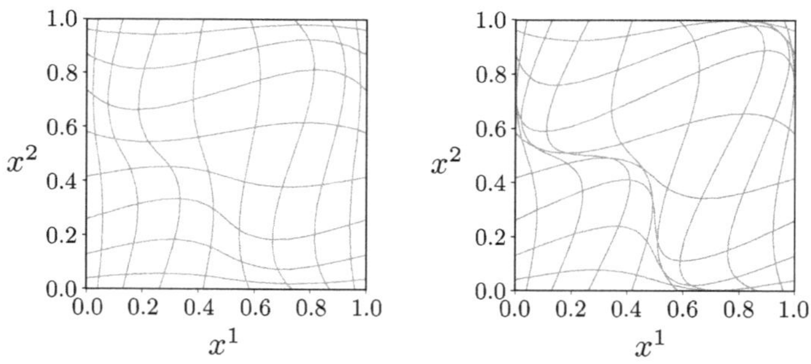

Fig. Curvilinear domains for $c = 0.15$ (Left) and $c = 0.3$ (Right) in $\mathbb{R}^2$. The gray lines illustrate the coordinate lines

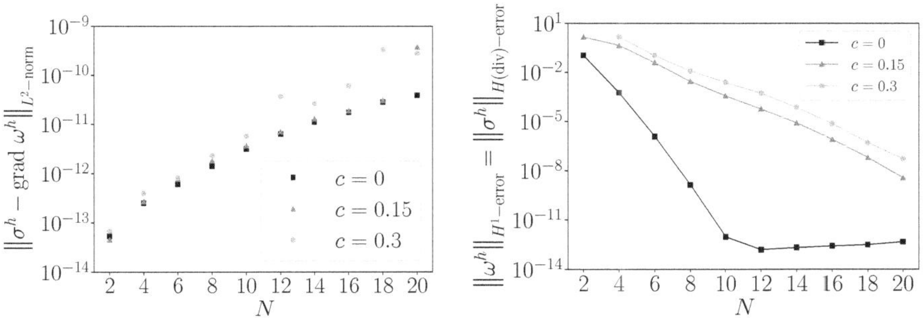

Fig. The L^2-error of $\left(\boldsymbol{\sigma}^h - \operatorname{grad}\omega^h\right)$ (Left) for $N = 2, 4, \cdots, 20$ and $c = 0, 0.15, 0.3$ and the p-convergence of the H^1-error of ω^h (Right)

Table The H^1-norm of ω^h and $H(\text{div})$-norm of $\boldsymbol{\sigma}^h$ for polynomial degree $N = 2, 4, \cdots, 20$ and deformation coefficient $c = 0, 0.15, 0.3$

	$c = 0$		$c = 0.15$		$c = 0.3$	
N	$\left\|\omega^h\right\|_{H^1}$	$\left\|\boldsymbol{\sigma}^h\right\|_{H(\text{div})}$	$\left\|\omega^h\right\|_{H^1}$	$\left\|\boldsymbol{\sigma}^h\right\|_{H(\text{div})}$	$\left\|\omega^h\right\|_{H^1}$	$\left\|\boldsymbol{\sigma}^h\right\|_{H(\text{div})}$
2	6.0720702909	6.0720702909	5.8899445673	5.8899445673	6.7381947027	6.7381947027
4	6.0730653395	6.0730653395	6.0567452129	6.0567452129	5.8849807780	5.8849807780
6	6.0730653668	6.0730653668	6.0729332275	6.0729332275	6.0721137212	6.0721137212
8	6.0730653668	6.0730653668	6.0730647051	6.0730647051	6.0730525346	6.0730525346
10	6.0730653668	6.0730653668	6.0730653557	6.0730653557	6.0730648440	6.0730648440
12	6.0730653668	6.0730653668	6.0730653665	6.0730653665	6.0730653428	6.0730653428
14	6.0730653668	6.0730653668	6.0730653667	6.0730653667	6.0730653663	6.0730653663
16	6.0730653668	6.0730653668	6.0730653668	6.0730653668	6.0730653667	6.0730653667
18	6.0730653668	6.0730653668	6.0730653668	6.0730653668	6.0730653668	6.0730653668
20	6.0730653668	6.0730653668	6.0730653668	6.0730653668	6.0730653668	6.0730653668

curvilinear domains, $c = 0.15, 0.3$. It is also seen that the results always converge to the analytical value $\|\omega_{\text{exact}}\|_{H^1} = \left\|\boldsymbol{\sigma}^h\right\|_{H(\text{div})} = 6.0730653668$. The p-convergence for the H^1-error of ω^h, therefore also for the $H(\text{div})$-error of $\boldsymbol{\sigma}^h$, is shown in Fig. (Right), which shows the exponential convergence of the method.

Conclusions

By constructing and using primal polynomial spaces and their algebraic dual representations both in the domain and on the boundary, we successfully preserve the equivalence of the div-grad Neumann problem and the grad-div Dirichlet problem at the discrete level in 3-dimensional curvilinear domains. This suggests the further usage of these spaces to structure-preserving methods and hybrid methods.

References

1. Bochev, P.B., Hyman, J.M.: Principles of mimetic discretizations of differential operators. In: Compatible Spatial Discretizations, pp. 89–119. Springer, New York (2006)
2. Brezzi, F., Fortin, M.: Mixed and Hybrid Finite Element Methods, vol. 15. Springer Science & Business Media, New York (2012)
3. Carstensen, C., Demkowicz, L., Gopalakrishnan, J.: Breaking spaces and forms for the DPG method and applications including Maxwell equations. Comput. Math. Appl. **72**(3), 494–522 (2016)
4. Castillo, J.E., Miranda, G.F.: Mimetic Discretization Methods. Chapman and Hall/CRC, London (2013)
5. Evans, J.A., Scott, M.A., Shepherd, K.M., Thomas, D.C., Vázquez Hernández, R.: Hierarchical B-spline complexes of discrete differential forms. IMA J. Numer. Anal. **40**(1), 422–473 (2020)
6. Gerritsma, M.: Edge functions for spectral element methods. In: Spectral and High Order Methods for Partial Differential Equations, pp. 199–207. Springer, Berlin (2011)
7. Gerritsma, M., Jain, V., Zhang, Y., Palha, A.: Algebraic dual polynomials for the equivalence of curl-curl problems (2018). arXiv:1805.00114
8. Gerritsma, M., Palha A., Jain, V., Zhang, Y.: Mimetic spectral element method for anisotropic diffusion. In: Numerical Methods for PDEs. Springer SEMA SIMAI Series, vol. 15, pp. 31–74. Springer, Berlin (2018)
9. Jain, V., Zhang, Y., Palha, A., Gerritsma, M.: Construction and application of algebraic dual polynomial representations for finite element methods (2017). arXiv:1712.09472
10. Jain, V., Zhang, Y., Fisser J., Palha, A., Gerritsma, M.: A conservative hybrid method for Darcy flow (ICOSAHOM 2018, accepted)
11. Kreeft, J., Gerritsma, M.: Mixed mimetic spectral element method for Stokes flow: a pointwise divergence-free solution. J. Comput. Phys. **240**, 284–309 (2013)
12. Palha, A., Rebelo, P.P., Hiemstra, R., Kreeft, J., Gerritsma, M.: Physics-compatible discretization techniques on single and dual grids, with application to the Poisson equation of volume forms. J. Comput. Phys. **257**, 1394–1422 (2014)
13. Pian, T.H.: Derivation of element stiffness matrices by assumed stress distributions. AIAA J. **2**(7), 1333–1336 (1964)
14. Pian, T.H., Tong, P.: Basis of finite element methods for solid continua. Int. J. Numer. Methods Eng. **1**(1), 3–28 (1969)
15. Rapetti, F.: High order edge elements on simplicial meshes. ESAIM: Math. Model. Numer. Anal. **41**(6), 1001–1020 (2007)
16. Steinberg, S.: Fundamentals of Grid Generation. CRC Press, Boca Raton (1993)

11

High-Order Finite Element Methods for Interface Problems: Theory and Implementations

Yuanming Xiao, Fangman Zhai, Linbo Zhang, and Weiying Zheng

Introduction

The interface problems which involve partial differential equations having discontinuous coefficients across certain interfaces are often encountered in fluid dynamics, electromagnetics and materials science. Because of the low global regularity and the irregular geometry of the interface, the standard numerical methods which are efficient for smooth solutions usually lead to loss in accuracy across the interface.

For arbitrarily shaped interface Γ, it is known that optimal or nearly optimal convergence rate can be recovered if body-fitted finite element meshes are used, see e.g. [6, 8, 20, 29]. Here, by "body-fitted meshes" we mean an element of the underlying mesh is required to intersect with the interface only through its boundaries (Fig.). Unfortunately, when the geometry is complex, this usually leads to a nontrivial interface meshing problem. Therefore, numerous modified finite difference methods based only on simple Cartesian grids have been proposed in the literature. We refer to the immersed boundary method [24], the immersed interface method [17, 18], the ghost fluid method [21], and the references therein. In the

Y. Xiao (✉)
Department of Mathematics, Nanjing University, Nanjing, China
e-mail: xym@nju.edu.cn

F. Zhai
Department of Applied Mathematics, Nanjing Forestry University, Nanjing, China

L. Zhang · W. Zheng
State Key Laboratory of Scientific and Engineering Computing, Academy of Mathematics and Systems Science, Chinese Academy of Sciences, Beijing, China

School of Mathematical Sciences, University of Chinese Academy of Sciences, Beijing, China
e-mail: zlb@lsec.cc.ac.cn; zwy@lsec.cc.ac.cn

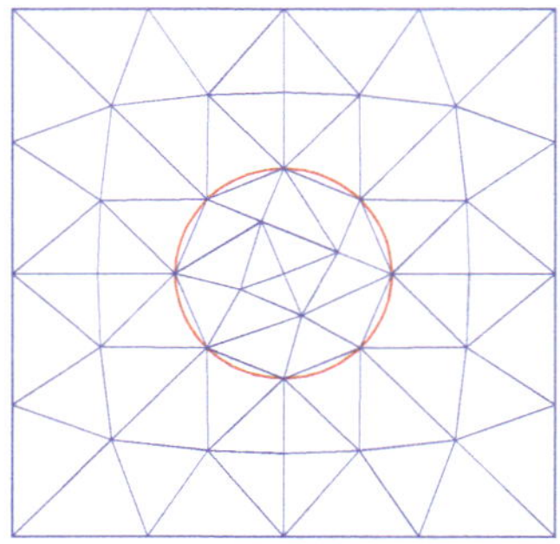

Fig. A body-fitted, shape regular mesh

finite element setting, we refer to the work of the immersed finite element method [7, 11, 19], the multiscale finite element method [9], the penalty finite element method [1].

In the past decade, a combination of the extended finite element method (XFEM) with the Nitsche scheme has become a popular discretization method. As the first attempt, an *unfitted finite element method* was proposed in [13] which can be viewed as a linear and consistent modification of [1]. This approach has motivated a number of works, e.g., the unfitted finite element method [4, 5, 12], the Ghost penalty method [2, 3], the unfitted discontinuous Galerkin methods [22]. Although significant progresses in the error analyses of some methods have been made, the development of high-order accurate unfitted FEMs with rigorous error analysis is still challenging. We refer to the work of [14–16, 22, 27, 28] which claim high order approximations. In [22], an *hp-unfitted discontinuous Galerkin method* for Problem (1) was considered, and optimal h-convergence for arbitrary p was shown for the two-dimensional case in the energy norm and in the L^2-norm. With an extra flux penalty term applied on the interface, [27] gave better hp a priori error estimates in both two and three dimensions. In [15, 16], an isoparametric finite element method with a high order geometrical approximation of level set domains was presented. The analysis reveals optimal order error bounds with respect to h for the geometry approximation and for the finite element approximation. In [14, 28], various issues related to unfitted methods was addressed, including the dependence of error estimates on the diffusion coefficients, the condition number of the discrete system, and the choice of stabilization parameters.

The Nitsche-XFEM can be interpreted as applying interior penalty (IP) methods on the interface, and our method falls into this category. The major step in our variant is an appropriate choice of the mesh and geometry dependent weights in the average (see (6)), which lead to trace and inverse inequalities for possibly degenerated sub-elements (see (9)). We note that in our approach, the penalization is applied only to the jump of the solution values across the interface (compared with the bilinear form in [27]). The optimal h-convergence rate for arbitrary high-order discretization in the energy and L_2-norm are proved regardless of the dimension. We refer to [14–16] for the similar estimates with respect to h and [27] for a refined version with respect to both h and p.

Efficient implementations of this method are then discussed in two aspects. We first consider an optimal multigrid solver for the generated linear system. We use the continuous FE space as a "background" subspace, with some smoothing operations added near the interface, to formulate a nested geometrical multigrid method. We prove the optimality of this special multiplicative multigrid method, which means the method converges uniformly with respect to the mesh size, and is independent of the location of the interface relative to the meshes. Since the assembling of the stiffness matrix will require integration over curved surfaces and volumes, we then implement a robust and arbitrarily high order numerical quadrature algorithm by transforming surface and volume integrals into multiple 1-D integrals. The code for the algorithm is freely available in the open source finite element toolbox Parallel Hierarchical Grid (PHG) [26]. We also refer to [23, 25] for different approaches to compute integrals on curved sub-elements and their curved boundaries.

The layout of this paper is as follows. In other Sect. we introduce the XFE spaces and reformulate the interface problem (1) in DG schemes. The H^1- and L^2- error estimates of both schemes—which attain the optimal order of the convergence rate in respect to mesh size h—are given. We give an optimal multigrid method for the aforementioned DG-XFE schemes. Numerical examples for both two and three dimensions are reported to illustrate the high accuracy of the algorithm.

XFE and DG Schemes for Interface Problems

We consider the following elliptic interface problem for u: Let $\Omega = \Omega_1 \cup \Gamma \cup \Omega_2$ be a bounded and convex polygonal or polyhedral domain in $\mathbb{R}^d$, $d = 2$ or 3, where Ω_1 and Ω_2 are two subdomains of Ω and are separated by a C^2-smooth interface Γ (see Fig. for an illustration of a unit square that contains a circle as an interface),

$$\begin{cases} -\nabla \cdot (\alpha(\mathbf{x})\nabla u) = f, & \text{in } \Omega_1 \cup \Omega_2, \\ [\alpha(\mathbf{x})\nabla u] = g_N, & \text{on } \Gamma, \\ [u] = \boldsymbol{g_D}, & \text{on } \Gamma, \\ u = 0, & \text{on } \partial\Omega. \end{cases} \tag{1}$$

Here $\alpha(\mathbf{x}) = \alpha_i$, $i = 1, 2$, is a piecewise constant function on the partition $\Omega_1 \cup \Omega_2$.

Denote by $\{\mathcal{T}_h\}$, a family of conforming, quasi-uniform, and regular partitions of Ω into triangles and parallelograms/tetrahedrons and parallelepipeds. As K is of regular shape, there is a constant γ_0 such that

$$h_K^d \le \gamma_0 |K|, \quad \forall K \in \mathcal{T}_h. \tag{2}$$

We define the set of all elements intersected by Γ as $\mathcal{T}_h^\Gamma = \{K \in \mathcal{T}_h : |K \cap \Gamma| \neq 0\}$. Each $\mathcal{T}_h^\Gamma$ induces a partition of interface Γ, which we denote by $\mathcal{E}_h^\Gamma = \{e_K : e_K =$

Fig. Domain $\Omega = \Omega_1 \cup \Gamma \cup \Omega_2$ with an unfitted mesh

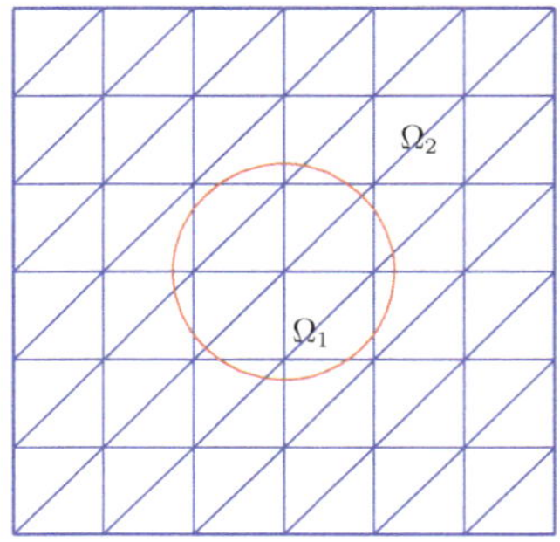

$K \cap \Gamma, K \in \mathcal{T}_h^\Gamma\}$. For any $K \in \mathcal{T}_h^\Gamma$, let $K_i = K \cap \Omega_i$ denote the part of K in Ω_i and $\boldsymbol{n}_i$ be the unit outward normal vector on ∂K_i with $i = 1, 2$. As Γ is of class C^2, it is easy to prove that (cf.[6, 31]) each interface segment/patch e_K is contained in a strip of width δ and satisfies

$$\delta \leq \gamma_1 h_K^2 \quad \text{and} \quad |\boldsymbol{n}_i(\mathbf{x}) - \boldsymbol{n}_i(\mathbf{y})| \leq \gamma_2 h_K, \ \forall \mathbf{x}, \mathbf{y} \in e_K. \tag{3}$$

We define the weighted average $\{\cdot\}$ and the jump $[\cdot]$ on $e \in \mathcal{E}_h^\Gamma$ by

$$\{v\} = \kappa_1 v_1 + \kappa_2 v_2, \qquad [v] = v_1 \boldsymbol{n}_1 + v_2 \boldsymbol{n}_2, \tag{4}$$

$$\{\boldsymbol{q}\} = \kappa_1 \boldsymbol{q}_1 + \kappa_2 \boldsymbol{q}_2, \qquad [\boldsymbol{q}] = \boldsymbol{q}_1 \cdot \boldsymbol{n}_1 + \boldsymbol{q}_2 \cdot \boldsymbol{n}_2. \tag{5}$$

For the stability analysis of our schemes, we define (κ_1, κ_2) on each element as follows:

$$\kappa_i = \begin{cases} 0, & \text{if } \frac{|K_i|}{|K|} < c_0 h_K, \\ 1, & \text{if } \frac{|K_i|}{|K|} > 1 - c_0 h_K, \\ \frac{|K_i|}{|K|}, & \text{otherwise}. \end{cases} \tag{6}$$

Clearly, $0 \leq \kappa_i \leq 1$ and $\kappa_1 + \kappa_2 = 1$ so that $\{\cdot\}$ is a convex combination along Γ. Roughly speaking, we adopt the weight $\kappa_i = \frac{|K_i|}{|K|}$ suggested in [13] for general sub-elements and we set $\kappa_i = 0$ for $|K_i| < ch_K^{d+1}$. Here, the user-defined constant $c_0 \geq 2\gamma_0\gamma_1$ and γ_0, γ_1 are constants defined in (2) and (3), respectively. The dependence of c_0 on these generic constants is elaborated in Lemma 1.

Let χ_i be the characteristic function on Ω_i with $i = 1, 2$. Given a mesh $\mathcal{T}_h$, let V_h be the continuous piecewise polynomial function space of degree $p \geq 1$ on the mesh. Let $V_h^0 := V_h \cap H_0^1(\Omega)$, $V_h^1 := V_h^0 \cdot \chi_1$ and $V_h^2 := V_h^0 \cdot \chi_2$. We define the XFE space as $V_h^\Gamma = V_h^1 + V_h^2$.

Then, the DG-XFE method for the interface problem is: Find $u_h \in V_h^\Gamma$ such that

$$B_h(u_h, v_h) = F_h(v_h), \quad \forall v_h \in V_h^\Gamma, \tag{7}$$

where

$$
\begin{aligned}
B_h(w,v) :=& \int_{\Omega_1\cup\Omega_2} \alpha(x)\nabla w\cdot\nabla v - \int_\Gamma \{\alpha(x)\nabla w\}\cdot[v] \\
&- \beta\int_\Gamma [w]\cdot\{\alpha(x)\nabla v\} + \sum_{K\in\mathcal{T}_h^\Gamma} \frac{\eta_\beta}{h_K}\int_{K\cap\Gamma} [w]\cdot[v], \\
F_h(v) :=& \int_\Omega f\, v + \int_\Gamma g_N(\kappa_1 v_2 + \kappa_2 v_1) \\
&- \beta\int_\Gamma \boldsymbol{g_D}\cdot\{\alpha(x)\nabla v\} + \sum_{K\in\mathcal{T}_h^\Gamma} \frac{\eta_\beta}{h_K}\int_{K\cap\Gamma} \boldsymbol{g_D}\cdot[v],
\end{aligned}
$$

For η_β sufficiently large, the norm corresponding to the bilinear form $B_h(\cdot,\cdot)$ is uniformly equivalent to $\|\cdot\|_{B_h}$, which is defined by

$$
\|v\|_{B_h}^2 = |v|_{1,\Omega_1\cup\Omega_2}^2 + \sum_{K\in\mathcal{T}_h^\Gamma} \eta_\beta h_K^{-1}\|[v]\|_{L^2(e_K)}^2 + \sum_{K\in\mathcal{T}_h^\Gamma} \eta_\beta^{-1} h_K\|\{\alpha(x)\nabla v\}\|_{L^2(e_K)}^2. \tag{8}
$$

The crucial component in regard to establishing this equivalence result and also the stability of bilinear forms is the control on the weighted normal derivatives, which is stated as a trace and inverse inequality in Lemma 1.

Lemma 1 ([27, 28]) *Let γ_0 and γ_1 be constants defined in* (2) *and* (3), *respectively. If we choose $c_0 \geq 2\gamma_0\gamma_1$ in the definition* (6) *of κ, there exists a positive constant h_0 such that for all $h \in (0, h_0]$ and any interface segment/patch $e_K = K\cap\Gamma \in \mathcal{E}_h^\Gamma$, the following estimates hold on both sub-elements of K:*

$$
\|\kappa_i^{1/2} v_i\|_{L^2(e_K)} \leq \frac{C}{h_K^{1/2}}\|v_i\|_{L^2(K_i)}, \quad v_i \in \mathcal{P}_p(K_i),\ \ i = 1, 2. \tag{9}
$$

The coercivity and boundedness of $B_h(\cdot,\cdot)$ in its norm $\|\cdot\|_{B_h}^2$ is then a direct consequence of the Cauchy–Schwarz inequality.

Lemma 2 *Let $V = H^2(\Omega_1\cup\Omega_2)$ and $V(h) = V_h^\Gamma + V$, we have*

$$
B_h(w,v) \leq C_b\|w\|_{B_h}\|v\|_{B_h}, \quad \forall\, w, v \in V(h), \tag{10}
$$

and

$$
B_h(v,v) \geq C_s\|v\|_{B_h}^2, \quad \forall\, v \in V_h^\Gamma, \tag{11}
$$

provided the penalty parameter η_β is chosen sufficiently large.

The XFE space has optimal approximation quality for piecewise smooth functions in $H^p(\Omega_1 \cup \Omega_2)$. The following theorem is proved in [28] as an analogue of Cea's lemma.

Theorem 1 *Assume that the interface* Γ *is* C^2 *smooth and that the solution of the elliptic interface problem* (1) *satisfies* $u \in H^s(\Omega_1 \cup \Omega_2)$, *where* $s \geq 2$ *is an integer. Let* $\mu = \min\{p+1, s\}$. *The following error estimates hold for any* $h \in (0, h_0]$: *If* η_β *is chosen sufficiently large (see* (11)*) and* u_h *is the solution to the first scheme of* (7)*, then*

$$\|u - u_h\|_{B_h} \lesssim h^{\mu-1}\|u\|_{H^s(\Omega_1 \cup \Omega_2)}, \qquad \forall 0 < h \leq h_0. \tag{12}$$

The hidden constants in the above estimates are dependent on the angle condition of the mesh $\mathcal{T}_h$, *the degree of the polynomials, the parameter in the scheme, and* $\alpha(\mathbf{x})$, *but are independent of the location of the interface relative to the mesh. Here, the constant* h_0 *is from Lemma 1.*

An Optimal Multigrid Method for (7)

In this section, we propose a two-level geometric multigrid solver of the finite element problem (7). It is well known that the element K with a "small" cut (i.e. $|K \cap \Omega_i|/|K| \ll 1$) would have adverse effect on the conditioning of the resulting stiffness matrices (see e.g. [3]). Our approach is based on the general theory of the successive subspace correction (SSC) method of solving on a linear vector space $\tilde{V} = \sum_{i=0}^{J} V_i$ with inner product $(\cdot, \cdot)$ the equation $(Au, v) = (f, v)$, where $A : \tilde{V} \to \tilde{V}$ is a symmetric positive definite operator.

We apply SSC for a relatively simple case of two subspaces (i.e. $J = 2$), that is, $\tilde{V} = V_h^\Gamma = V_1 \oplus V_2$, with $V_1 = V_h^0$ and $V_2 = \widetilde{V}_h^\Gamma$, where $\widetilde{V}_h^\Gamma \subset V_h^\Gamma$ is the space of nodal basis functions that vanish on $\widetilde{\mathcal{N}}_h := \{x_j : |\mathrm{supp}(\psi_j) \cap \Gamma| = 0\}$. With a slight abuse of notation, the DG-XFE scheme induces a symmetric positive definite operator B_h for $\beta = 1$. Let $\tilde{B}_h$ and $\tilde{B}_h^\Gamma$ be the restrictions of B_h on V_h^0 and $\widetilde{V}_h^\Gamma$, respectively. Let $\tilde{R}_h : V_h^0 \to V_h^0$ be approximately an inverse of $\tilde{B}_h$. We have this two-level successive subspace correction method (Algorithm 1). The similar idea has been employed in a special linear case in [32] and analyzed using the framework given in [30, 33].

Algorithm 1 The multigrid method for (7)

Implement this iterative procedure until converge:

1. do subspace correction on V_h^0 with an inexact solver $\tilde{R}_h$;
2. do subspace correction on $\widetilde{V}_h^\Gamma$ with an exact solver $(\tilde{B}_h^\Gamma)^{-1}$.

Obviously, Algorithm 1 defines an iterative method for solving $B_h u_h = f_h$. Denote by A_h the iterator of the method, then the error contract property is summarized as the following theorem.

Theorem 2 ([28]) *Assume that* $\|I - \tilde{R}_h \tilde{B}_h\|_{\tilde{B}_h} \leqslant \rho < 1$. *Then Algorithm 1 is uniformly convergent with respect to the mesh size with*

$$\|I - A_h B_h\|_{B_h}^2 \leqslant \frac{\Lambda}{1 - \rho^2 + \Lambda},$$

where Λ *is a constant independent of* h.

When $\mathcal{T}_h$ is a shape-regular grid with a geometrical multilevel structure, then a geometric multigrid process can be implemented on V_h^0, and the approximate inverse $\tilde{R}_h$ of $\tilde{B}_h$ can be chosen to be the iterator of V-cycle multigrid method.

Numerical Tests

In this section, we present some initial results to demonstrate the high-order accuracy and robustness of our method. A 2-D example was implemented in MATLAB. The numerical experiment for a 3-D case was carried out in the open source finite element toolbox PHG [26].

High-Order Numerical Quadratures on "cut" Elements

Assembling the local stiffness matrix and the corresponding RHS for $K \in \mathcal{T}_h^\Gamma$ requires integration over irregularly shaped manifolds:

$$I = \int_{K \cap \Omega_i} u(\mathbf{x}) \mathrm{d}\mathbf{x} \quad \text{and} \quad I = \int_{K \cap \Gamma} u(\mathbf{x}) \mathrm{d}\Gamma, \tag{13}$$

where Γ is defined by the zero level set of a piecewise smooth function.

Our implementation of (13) relies on a general-purpose and arbitrarily high order numerical quadrature algorithm proposed in [10]. The basic idea is to choose a local coordinate system with three orthogonal directions, decompose integrals in (13) into multiple 1-D integrals along these directions, and use 1-D Gaussian quadratures to compute these integrals. For 1-D Gaussian quadratures to work, the local coordinate system should be suitably chosen according to properties of K and Γ to prevent essential singularities from appearing in the 1-D integrands, and the integration intervals are divided into subintervals at the non essential singularities of the integrands. We note that the proposed algorithm only requires finding roots of univariate nonlinear functions in given intervals and evaluating the integrand, the level set function, and the gradient of the level set function at given points. It can achieve arbitrarily high order by increasing the orders of Gaussian quadratures, and does not need extra a priori knowledge about the integrand and the level set function.

This algorithm has been implemented in the file `src/quad-interface.c` and `include/phg/quad-interface.h` in PHG [26]. Extensive $h-$ and $p-$convergence tests have been performed in [10] and included in a sample code `test/quad_test2.c`.

2-D Numerical Examples

Let domain Ω be the unit square $(0, 1)^2$ and interface Γ be the zero level set of the function $\varphi(x) = (x_1 - 0.5)^2 + (x_2 - 0.5)^2 - 1/7$. The subdomain Ω_1 is characterized by $\varphi(x) < 0$ and Ω_2 by $\varphi(x) > 0$. The domain Ω is partitioned into grids of squares with the same size h. The exact solution is chosen as

$$u(x_1, x_2) = \begin{cases} 1/\alpha_1 \exp(x_1 x_2), & (x_1, x_2) \in \Omega_1, \\ 1/\alpha_2 \sin(\pi x_1) \sin(\pi x_2), & (x_1, x_2) \in \Omega_2. \end{cases}$$

The right-hand side can be computed accordingly.

We implement Algorithm 1, with V-cycle geometric multigrid based on the unfitted grid $\mathcal{T}_h$ playing as the coarse grid corrector. In each pre- and post-smoothing stage of V-cycle iterator, we perform Gauss-Seidel for two times. We record the numerical results in Table . In these examples, the initial guess is $\mathbf{0}$, and the stopping criterion is

$$\| f_h - B_h u_h^{(k)} \|_\infty / \| f_h - B_h u_h^{(0)} \|_\infty < 10^{-10}.$$

From Table, we can see that the multigrid method converges uniformly with respect to the mesh size, which confirms our theoretical results.

Table Numerical performance of Algorithm 1 (2-D example)

		h	2^{-2}	2^{-3}	2^{-4}	2^{-5}	2^{-6}
$\alpha_1 : \alpha_2 = 1 : 10$	$p = 1$	#iter	7	10	10	11	12
	$p = 2$	#iter	13	10	12	13	14
$\alpha_1 : \alpha_2 = 10 : 1$	$p = 1$	#iter	24	30	29	27	25
	$p = 2$	#iter	24	23	22	21	20

3-D Numerical Examples

The settings of this numerical experiment are as follows. The domain $\Omega = (0, 1)^3$. The interfaces are two touched spheres of radius 0.1 centered at (0.4, 0.5, 0.5) and (0.6, 0.5, 0.5). The exact solution is given by

$$u(x_1, x_2, x_3) = \begin{cases} \exp(x_1 + x_2 + x_3), & (x_1, x_2, x_3) \in \Omega_1, \\ \sin(x_1)\sin(x_2)\sin(x_3), & (x_1, x_2, x_3) \in \Omega_2. \end{cases}$$

The discontinuous coefficient function is defined such that $\alpha_1 = 1$ and $\alpha_2 = 100$.

A convergence study is performed on a series of meshes generated by uniform refinements of an initial mesh consisting of 6 congruent tetrahedra. Relative errors and convergence rates of numerical solutions for P_p elements for $p = 1, 2, 3$ and 4 are listed in Table, with the quadrature order $q = 2p + 3$. The convergence rates are optimal for both $H^1(\Omega)$-errors (order p) and $L^2(\Omega)$-errors (order $p+1$). For the

Table Errors and convergence orders of the numerical solutions (3-D example)

Number of elements	Degrees of freedom	Relative H^1 error		Relative L^2 error	
		Error	Order	Error	Order
P_1 element($p = 1, q = 2p + 3 = 5$)					
768	189	1.690e−01	–	1.686e−02	–
6144	1241	7.510e−02	1.17	3.403e−03	2.31
49,152	9009	3.514e−02	1.10	9.618e−04	1.82
393,216	68,705	1.658e−02	1.08	2.272e−04	2.08
3,145,728	536,769	8.145e−03	1.03	4.869e−05	2.22
P_2 element($p = 2, q = 2p + 3 = 7$)					
768	1241	9.041e−03	–	4.150e−04	–
6144	9009	2.026e−03	2.16	4.323e−05	3.26
49,152	68,705	4.973e−04	2.03	5.171e−06	3.06
393,216	536,769	1.234e−04	2.01	6.413e−07	3.01
3,145,728	4,243,841	3.070e−05	2.01	7.965e−08	3.01
P_3 element ($p = 3, q = 2p + 3 = 9$)					
768	3925	2.175e−03	–	8.394e−05	–
6144	29,449	3.793e−05	5.84	5.864e−07	7.16
49,152	228,241	4.743e−06	3.00	3.683e−08	3.99
393,216	1,797,409	5.932e−07	3.00	2.321e−09	3.99
3,145,728	14,266,945	7.414e−08	3.00	1.456e−10	3.99
P_4 element ($p = 4, q = 2p + 3 = 11$)					
768	9009	2.971e−03	–	9.606e−05	–
6144	68,705	6.042e−07	12.26	7.560e−09	13.63
49,152	536,769	3.778e−08	4.00	2.380e−10	4.99
393,216	4,243,841	2.362e−09	4.00	7.481e−12	4.99

time being, however, the design of multigrid solver for 3-D case is still on-going. The computations for P_1, P_2 and P_3 elements were done using the 64-bit double precision and the linear systems were solved using MUMPS, but for P_4 element, to eliminate influences of roundoff errors, the computations were done using the 80-bit extended double precision and the linear systems were solved using the GMRES method with MUMPS in double precision as its preconditioner. The performance of Algorithm 1 will be reported in a future work.

Acknowledgements The authors would like to acknowledge the funding support of this research by the National Key Research and Development Program of China under grant number 2017YFC0209804, National Natural Science Foundation of China under grant number 11101208, and National Center for Mathematics and Interdisciplinary Sciences of Chinese Academy of Sciences. The authors are grateful to Jinchao Xu and Haijun Wu for the fruitful discussions and suggestions, to Dr. Huaqing Liu for his valuable help on preparing the numerical examples.

References

1. Babuška, I.: The finite element method for elliptic equations with discontinuous coefficients. Computing **5**, 207–213 (1970)
2. Becker, R., Burman, E., Hansbo, P.: A Nitsche extended finite element method for incompressible elasticity with discontinuous modulus of elasticity. Comput. Methods Appl. Mech. Eng. **198**, 3352–3360 (2009)
3. Burman, E.: Ghost penalty. C.R. Math. **348**, 1217–1220 (2010)
4. Burman, E., Hansbo, P.: Fictitious domain finite element methods using cut elements: II. A stabilized Nitsche method. Appl. Num. Math. **62**, 328–341 (2012)
5. Burman, E., Guzman, J., Sanchez, M.A., Sarkis, M.: Robust flux error estimation of an unfitted Nitsche method for high-contrast interface problems. IMA J. Numer. Anal. **38**, 646–668 (2018)
6. Chen, Z., Zou, J.: Finite element methods and their convergence for elliptic and parabolic interface problems. Numer. Math. **79**, 175–202 (1998)
7. Chen, Z., Xiao, Y., Zhang, L.: The adaptive immersed interface finite element method for elliptic and Maxwell interface problems. J. Comput. Phys. **228**, 5000–5019 (2009)
8. Chen, Z., Wu, Z., Xiao, Y.: An adaptive immersed finite element method with arbitrary Lagrangian-Eulerian scheme for parabolic equations in time variable domains. Int. J. Numer. Anal. Model. **12**, 567–591 (2015)
9. Chu, C.-C., Graham, I.G., Hou, T.Y.: A new multiscale finite element method for high-contrast elliptic interface problems. Math. Comput. **79**, 1915–1955 (2010)
10. Cui, T., Leng, W., Liu, H., Zhang, L., Zheng, W.: High-order numerical quadratures in a tetrahedron with an implicitly defined curved interface, ACM Transactions on Mathematical Software, in press.
11. Gong, Y., Li, B., Li, Z.: Immersed-interface finite-element methods for elliptic interface problems with non-homogeneous jump conditions. SIAM J. Numer. Anal. **46**, 472–495 (2008)
12. Hansbo, P.: Nitsche's method for interface problems in computational mechanics. GAMM-Mitt. **47**, 183–206 (2005)
13. Hansbo, A., Hansbo, P.: An unfitted finite element method, based on Nitsche's method for elliptic interface problems. Comput. Methods Appl. Mech. Eng. **191**, 5537–5552 (2002)
14. Huang, P., Wu, H., Xiao, Y.: An unfitted interface penalty finite element method for elliptic interface problems. Comput. Methods Appl. Mech. Eng. **323**, 439–460 (2017)
15. Lehrenfeld, C., Reusken, A.: L^2-error analysis of an isoparametric unfitted finite element method for elliptic interface problems (2016). Preprint. arXiv:1604.04529

16. Lehrenfeld, C., Reusken, A.: Analysis of a high-order unfitted finite element method for elliptic interface problems. IMA J. Numer. Anal. **38**, 1351–1387 (2018)
17. LeVeque, R., Li, Z.: The immersed interface method for elliptic equations with discontinuous coefficients and singular sources. SIAM J. Numer. Anal. **31**, 1019–1044 (1994)
18. Li, Z., Ito, K.: The Immersed Interface Method: Numerical Solutions of PDEs Involving Interfaces and Irregular Domains. SIAM, Philadephia (2006)
19. Li, Z., Lin, T., Wu, X.: New Cartesian grid methods for interface problems using the finite element formulation. Numer. Math. **96**, 61–98 (2003)
20. Li, J., Melenk, J.M., Wohlmuth, B., Zou, J.: Optimal a priori estimates for higher order finite elements for elliptic interface problems. Appl. Numer. Math. **60**, 19–37 (2010)
21. Liu, X., Fedkiw, R.P., Kang, M.: A boundary condition capturing method for Poisson's equation on irregular domains. J. Comput. Phys. **160**, 151–178 (2000)
22. Massjung, R.: An unfitted discontinuous Galerkin method applied to elliptic interface problems. SIAM J. Numer. Anal. **50**(6), 3134–3162 (2012)
23. Müller, B., Kummer, F., Oberlack, M.: Highly accurate surface and volume integration on implicit domains by means of moment-fitting. Int. J. Numer. Methods Eng. **96**, 512–528 (2013)
24. Peskin, C.S.: Numerical analysis of blood flow in the heart. J. Comput. Phys. **25**, 220–252 (1977)
25. Saye, R.I.: High-order quadrature methods for implicitly defined surfaces and volumes in hyperrectangles. SIAM J. Sci. Comput. **37**, A993–A1019 (2015)
26. The toolbox Parallel Hierarchical Grid (PHG). http://lsec.cc.ac.cn/phg
27. Wu, H., Xiao, Y.: An unfitted hp-interface penalty finite element method for elliptic interface problems. J. Comput. Math. **37**, 316–339 (2019). http://arXiv.org/abs/1007.2893
28. Xiao, Y., Xu, J., Wang, F.: High-order extended finite element methods for solving interface problems. Comput. Methods Appl. Mech. Eng. **364**, 112964 (2020). https://doi.org/10.1016/j.cma.2020.112964
29. Xu, J.: Estimate of the convergence rate of finite element solutions to elliptic equations of second order with discontinuous coefficients (in Chinese). Nat. Sci. J. Xiangtan Univ. **1**, 1–5 (1982)
30. Xu, J.: Iterative methods by space decomposition and subspace correction. SIAM Rev. **34**, 581–613 (1992)
31. Xu, J.: Estimate of the convergence rate of finite element solutions to elliptic equations of second order with discontinuous coefficients (2013, preprint). arXiv:1311.4178
32. Xu, J., Zhang, S.: Optimal finite element methods for interface problems. In: Dickopf, T., et al. (eds.), Domain Decomposition Methods in Science and Engineering XXII. Lecture Notes in Computational Science and Engineering, vol. 104, pp. 77–91. Springer, Cham (2016)
33. Xu, J., Zikatanov, L.: The method of alternating projections and the method of subspace corrections in Hilbert space. J. Am. Math. Soc. **15**, 573–597 (2002)

12

Hermite Methods in Time

Rujie Gu and Thomas Hagstrom

Introduction

Over the past decade a number of works have appeared which exploit the unique properties of Hermite-Birkhoff interpolation in space to construct arbitrary-order discretization methods for hyperbolic [1, 2, 4–7, 10, 14, 16–19] as well as Schrödinger [3] equations. The precise form of the interpolant in a single cell, which here we write in one dimension labelled t, is

$$u(t) \approx \mathscr{I}u(t) \in \Pi^{2m+1}, \qquad t \in (t_{j-1}, t_j), \tag{1}$$

$$\frac{d^k}{dt^k}\mathscr{I}u(t_\ell) = \frac{d^k u}{dt^k}(t_\ell); \;\; k = 0, \ldots, m, \qquad \ell = j-1, j, \tag{2}$$

where Π^{2m+1} denotes the polynomials of degree $2m+1$. (In higher dimensions one uses a tensor-product cell interpolant based on vertex data consisting of mixed derivatives of order through m in each Cartesian coordinate.)

In contrast, there has been little work on analogous methods for time discretization. A recent exception is the manuscript by Liu et al. [15]. They develop methods for second-order semilinear hyperbolic equations using interpolants of the form (1)–(2) combined with a reformulation of the evolution problem using exact solutions of the linear part. They demonstrate excellent long-time performance.

The outline of the paper is as follows. In Sect. we list a few properties of piecewise Hermite-Birkhoff interpolation. In Sect. we construct the time-stepping

R. Gu · T. Hagstrom (✉)
Southern Methodist University, Dallas, TX, USA
e-mail: rujieg@smu.edu; thagstrom@smu.edu

schemes and establish some basic results, with a few numerical experiments.

Basic Properties of Hermite-Birkhoff Interpolation

Hermite interpolants have a number of interesting properties which make them very attractive for the solution of differential equations; see, e.g., [2]. Here we will mainly use the simplest. Precisely, for $t \in (t_{j-1}, t_j)$, the Peano representation of the local error can be easily derived by noting that $e = u - \mathscr{I}u$ solves the two point boundary value problem

$$\frac{d^{2m+2}e}{dt^{2m+2}} = \frac{d^{2m+2}u}{dt^{2m+2}}, \quad \frac{d^k e}{dt^k} = 0, \quad t = t_{j-1}, t_j, \quad k = 0, \dots m. \tag{3}$$

Thus

$$e(t) = \int_{t_{j-1}}^{t_j} K_j(t, s) \frac{d^{2m+2}u}{dt^{2m+2}}(s) ds, \tag{4}$$

where the kernel K_j is the Green's function for (3). Simple scaling arguments combined with the transformation $t = t_{j-1} + zh_j$ then show that $e = O(h_j^{2m+2})$ where $h_j = t_j - t_{j-1}$ is the time step. A fundamental feature of piecewise Hermite interpolation is the following orthogonality property. For **any** functions $v(t)$, $w(t)$

$$\int_{t_{j-1}}^{t_j} \frac{d^{m+1}\mathscr{I}v}{dt^{m+1}}(t) \cdot \frac{d^{m+1}\left(w - \mathscr{I}w\right)}{dt^{m+1}}(t) dt = 0, \tag{5}$$

which in particular implies that interpolation reduces the H^{m+1} seminorm.

Time-Stepping Methods

We begin by considering the initial value problem for a first-order system ordinary differential equations:

$$\frac{du}{dt} = f(u, t), \quad u(t_0) = u_0, \quad u(t) \in \mathbb{R}^d. \tag{6}$$

Given a discrete time sequence $t_j > t_{j-1}$, $j = 1, \dots, N$, with time steps $h_j = t_j - t_{j-1}$ we write down the Picard integral formulation of the time evolution over a single step

$$u(t_j) = u(t_{j-1}) + \int_{t_{j-1}}^{t_j} f(u(s), s)ds. \tag{7}$$

The construction of our time integration formula proceeds in three steps. We denote by v_j the approximation to $u(t_j)$.

1. Given v_{j-1} and assuming for the moment that v_j is known, use the differential equation to compute m scaled derivatives of its solution, $V_\ell(t)$, satisfying $V_\ell(t_\ell) = v_\ell$, $\ell = j, j-1$. Setting

$$F_\ell(t) = f(V_\ell(t), t), \tag{8}$$

these are recursively defined by the formula

$$\frac{d^k V_\ell}{dt^k}(t_\ell) = \frac{d^{k-1} F_\ell}{dt^{k-1}}(t_\ell), \quad k = 0, \dots m. \tag{9}$$

2. Construct the Hermite-Birkhoff interpolant of this data; that is the polynomial $P_{j-1/2}(t; v_{j-1}, v_j)$ of degree $2m+1$, satisfying

$$\frac{d^k P_{j-1/2}}{dt^k}(t_\ell; v_{j-1}, v_j) = \frac{d^k V_\ell}{dt^k}(t_\ell); \quad \ell = j-1, j, \quad k = 0, \dots, m.. \tag{10}$$

3. Approximate (7) by replacing $u(t_\ell)$ by v_ℓ and replacing the integral by a $q+1$-point quadrature rule with f evaluated at the Hermite interpolant:

$$v_j = v_{j-1} + h_j \sum_{k=0}^{q} w_k f(P_{j-1/2}(t_{j,k}; v_{j-1}, v_j)). \tag{11}$$

4. Solve (11) for v_j. Note that this is a system of d nonlinear equations for any m; that is, unlike standard implicit Runge–Kutta methods, the size of the nonlinear system is independent of the order.

We remark that we have not studied in detail the unique solvability of (11) in the stiff case. In our numerical experiments we used the solution at the current time step as an initial approximation for Newton iterations and simply accepted the solution to which the iterates converged.

To emphasize the ideas we write down some specific examples of methods with $m = 1$ and $m = 2$ making the simplifying assumption of autonomy; that is $f = f(u)$. The derivation of methods of arbitrary order is straightforward and the formulas can be trivially obtained using software capable of symbolic computations. To apply them at higher order one must evaluate higher derivatives of f, which is also possible using automatic differentiation tools [11].

Example (m = 1) Set

$$\tau = \frac{t - t_{j-1}}{h_j}.$$

Now the interpolant $P_{j-1/2}(t; v_{j-1}, v_j)$ is given by:

$$P_{j-1/2}(t; v_{j-1}, v_j) = \sum_{k=0}^{3} a_k \tau^k, \tag{12}$$

where

$$\begin{aligned} a_0 &= v_{j-1}, \quad a_1 = h_j f(v_{j-1}), \\ a_2 &= 3\left(v_j - v_{j-1}\right) - h_j\left(2f(v_{j-1}) + f(v_j)\right), \\ a_3 &= -2\left(v_j - v_{j-1}\right) + h_j\left(f(v_{j-1}) + f(v_j)\right). \end{aligned} \tag{13}$$

We next introduce a quadrature rule which is exact for polynomials of degree 3. Possible choices include the 2-point Gauss-Legendre (14) rules, or the 3-point Gauss-Radau (15) or Gauss-Lobatto rules. Note that by using two different rules we obtain a possible error indicator. Here are the two different methods used below. Note that the methods are identical if f is linear.

$$v_j = v_{j-1} + \frac{h_j}{2}\left(f\left(P_{j-1/2}(\alpha_-; v_{j-1}, v_j)\right) + f\left(P_{j-1/2}(\alpha_+; v_{j-1}, v_j)\right)\right), \tag{14}$$

$$\begin{aligned} v_j = v_{j-1} + \frac{h_j}{36}&\left(\beta_+ f\left(P_{j-1/2}(\gamma_-; v_{j-1}, v_j)\right) + \beta_- f\left(P_{j-1/2}(\gamma_+; v_{j-1}, v_j)\right)\right. \\ &\left. +4f\left(v_j\right)\right), \end{aligned} \tag{15}$$

$$\alpha_\pm = \left(1 \pm \frac{1}{\sqrt{3}}\right), \quad \gamma_\pm = \frac{4 \pm \sqrt{6}}{10}, \quad \beta_\pm = 16 \pm \sqrt{6}. \tag{16}$$

A time step is executed by solving the nonlinear system, (14) or (15), for v_j.

Example (m = 2) Now we also need the second time derivative of u,

$$\frac{d^2u}{dt^2} = \frac{d}{dt} f(u) = J(u) f(u), \tag{17}$$

where $J(u)$ is the Jacobian derivative. The Hermite interpolant can now be written:

$$P_{j-1/2}(t; v_{j-1}, v_j) = \sum_{k=0}^{5} a_k \tau^k, \tag{18}$$

where

$$a_0 = v_{j-1}, \quad a_1 = h_j f(v_{j-1}), \quad a_2 = \frac{h_j^2}{2} J(v_{j-1}) f(v_{j-1}), \tag{19}$$

$$a_3 = 10\left(v_j - v_{j-1}\right) - h_j\left(6f(v_{j-1}) + 4f(v_j)\right) + \frac{h_j^2}{2}\left(-3J(v_{j-1})f(v_{j-1}) + J(v_j)f(v_j)\right),$$

$$a_4 = -15\left(v_j - v_{j-1}\right) + h_j\left(8f(v_{j-1}) + 7f(v_j)\right) + \frac{h_j^2}{2}\left(3J(v_{j-1})f(v_{j-1}) - J(v_j)f(v_j)\right),$$

$$a_5 = 6\left(v_j - v_{j-1}\right) - 3h_j\left(f(v_{j-1}) + f(v_j)\right) + \frac{h_j^2}{2}\left(-J(v_{j-1})f(v_{j-1}) + J(v_j)f(v_j)\right).$$

Again we can now use, for example, the 3-point Gauss-Legendre or 4-point Gauss-Radau quadrature rules to produce the equation we must solve for v_j.

Stability and Consistency

The consistency of the method is a straightforward consequence of its construction, and its linear stability properties can also be established.

Theorem 1 *Assume that the quadrature rule has positive weights and is exact for polynomials of degree* $2m + 1$. *Then the implicit Hermite method is A-stable and accurate of order* $2m + 2$.

Proof Assume that f is smooth and that $u(t) \in C^{2m+2}(0, T)$. Using (4), standard estimates for quadrature errors, and the Picard formula (7) we find for the truncation error

$$\tau_j = \frac{u(t_j) - u(t_{j-1})}{h_j} - \sum_k w_k f(P_{j-1/2}(t_{j,k}; u(t_{j-1}), u(t_j))), \tag{20}$$

$$|\tau_j| = \frac{1}{h_j}\left|\int_{t_{j-1}}^{t_j} f(u(s))ds - h_j \sum_k w_k f(P_{j-1/2}(t_{j,k}; u(t_{j-1}), u(t_j)))\right|$$

$$\leq \frac{1}{h_j} \left| \int_{t_{j-1}}^{t_j} f(u(s))ds - h_j \sum_k w_k f(u(t_{j,k})) \right|$$
$$+ \left| \sum_k w_k \left(f(u(t_{j,k})) - f(P_{j-1/2}(t_{j,k}; u(t_{j-1}), u(t_j))) \right) \right|$$
$$\leq C h_j^{2m+2}. \tag{21}$$

Now consider the Dahlquist test problem, $f(u) = \lambda u$. In this case all quadrature rules which are exact for the Hermite interpolant produce the same method. As interpolation is linear, we have that the coefficients of the interpolant are linear combinations $h_j^k \lambda^k v_{j-1}$ and $h_j^k \lambda^k v_j$, $k = 0, \ldots m$. The Picard integral then increases the powers of $h_j \lambda$ by one so that the implicit system (11) can be rearranged to:

$$Q_+(h_j\lambda)v_j = Q_-(h_j\lambda)v_{j-1} \Rightarrow v_j = \frac{Q_-(h_j\lambda)}{Q_+(h_j\lambda)} v_{j-1}, \tag{22}$$

where $Q_\pm(h_j\lambda)$ are polynomials of degree $m+1$. Consistency implies

$$e^{h_j\lambda} = \frac{Q_-(h_j\lambda)}{Q_+(h_j\lambda)} + O\left((h_j\lambda)^{2m+3}\right). \tag{23}$$

The only rational function of the given degree with this accuracy is the diagonal Padé approximant. We thus conclude that our methods are A-stable [12].◇

Numerical Experiments

Our first experiments treat standard problems from the ode literature and are restricted to the fourth and sixth order methods described above with either Gauss-Legendre or Gauss-Radau quadrature. Our practical implementations employ the classical Aitken algorithm adapted to Hermite interpolation to directly evaluate $P_{j-1/2}(t_{j,k}, v_{j-1}, v_j)$ and solve (11) using Newton's method with the Jacobian of the implicit system approximated by finite differences. For adaptive computations we

1. Compute v_j using the Gauss-Radau-based formulas,
2. Compute a residual, ρ_j, by substituting v_j into the Gauss-Legendre-based formulas.

We then adjust the time step by the simple rule

$$h_{j+1} = \left(\frac{\text{tol}}{\rho_j}\right)^{1/(2m+3)} h_j, \tag{24}$$

while also imposing a minimum time step.

Our final experiment examines the use of the method for evolving spectral discretizations of initial-boundary value problems for the Schrödinger equation.

Arentsorf Orbit

We first consider the problem of computing a periodic solutions of the restricted three-body problem which we reformulate as a first-order system of four variables:

$$\frac{d^2 y_1}{dt^2} = y_1 + 2\frac{dy_2}{dt} - (1-\mu)\frac{y_1+\mu}{\left((y_1+\mu)^2+y_2^2\right)^{3/2}} - \mu\frac{y_1-(1-\mu)}{\left((y_1-(1-\mu))^2+y_2^2\right)^{3/2}},$$

$$\frac{d^2 y_2}{dt^2} = y_2 - 2\frac{dy_1}{dt} - (1-\mu)\frac{y_2}{\left((y_1+\mu)^2+y_2^2\right)^{3/2}} - \mu\frac{y_2}{\left((y_1-(1-\mu))^2+y_2^2\right)^{3/2}},$$

$$\mu = 0.012277471, \quad y_1(0) = .994, \quad \frac{dy_1}{dt}(0) = y_2(0) = 0,$$

$$\frac{dy_2}{dt}(0) = -2.01585106379082\ldots, \quad T = 17.06521656015796\ldots.$$

(For graphs of the solution see [13, Ch. II].)

We note that this problem is not considered to be stiff. The main difficulty is a need for very small time steps when the orbits approach the singularities of f. However, we use it to verify convergence at the design order when (woefully inefficient) uniform time steps are employed and to test the utility of our naive time step adaptivity algorithm.

Results for fixed (small) time steps are displayed in Table. We observe that convergence is at design order and that the results for the two quadrature formulas are comparable, though the fourth order Radau method is somewhat more accurate than Gauss-Legendre with roles reversed at sixth order. The sixth order methods are more accurate with larger time steps. The error is simply $\sqrt{(y_1(T)-y_1(0))^2+(y_2(T)-y_2(0))^2}$.

Results for adaptive computations with $m = 2$ are shown in Table. Obviously, the adaptive methods lead to a very significant reduction in the number of time steps; an accuracy of 10^{-7} is achieved with 264 steps of the adaptive method

Table Convergence with fixed time steps for the Arentsorf orbit problem

Gauss-Legendre $m = 1$			Gauss-Radau $m = 1$			Gauss-Legendre $m = 2$			Gauss-Radau $m = 2$		
h	Error	Rate	h	Error	Rate	h	Error	Rate	h	Error	Rate
5.69(−4)	2.30(−4)		5.69(−4)	9.65(−5)		8.53(−4)	2.98(−6)		8.53(−4)	4.46(−6)	
4.88(−4)	1.24(−4)	4.01	4.88(−4)	5.21(−5)	4.00	6.83(−4)	7.82(−7)	6.00	6.83(−4)	1.16(−6)	6.03
4.27(−4)	7.26(−5)	4.01	4.27(−4)	3.05(−5)	4.00	5.69(−4)	2.62(−7)	6.00	5.69(−4)	3.88(−7)	6.02
3.79(−4)	4.53(−5)	4.01	3.79(−4)	1.91(−5)	4.00	4.88(−4)	1.04(−7)	6.00	4.88(−4)	1.53(−7)	6.01
3.41(−4)	2.97(−5)	4.01	3.41(−4)	1.25(−5)	4.00	4.27(−4)	4.66(−8)	6.00	4.27(−4)	6.87(−8)	6.01
3.10(−4)	2.03(−5)	4.00	3.10(−4)	8.54(−6)	4.00	3.79(−4)	2.30(−8)	6.00	3.79(−4)	3.39(−8)	6.01
2.84(−4)	1.43(−5)	4.00	2.84(−4)	6.03(−6)	4.00	3.41(−4)	1.22(−8)	6.00	3.41(−4)	1.80(−8)	6.00

Table Time steps and error as a function of tolerance for adaptive solutions of the Arentsorf problem with $m = 2$

Tol	Error	Steps
10^{-6}	8.21(−3)	65
10^{-8}	1.85(−5)	136
10^{-10}	1.15(−8)	264

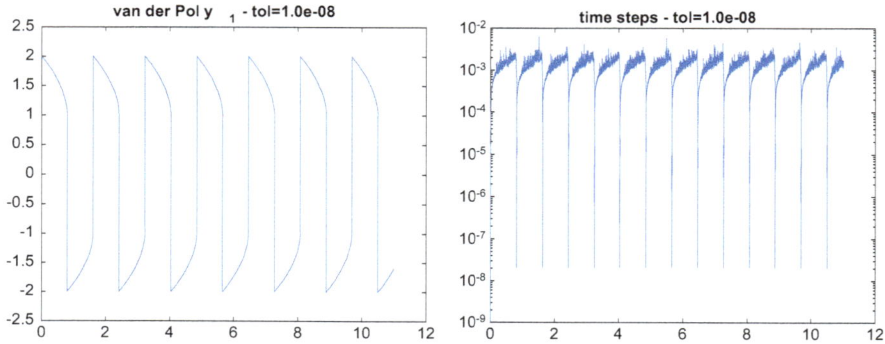

Fig Solution and time step history for the van der Pol oscillator with tolerance 10^{-8}

while 35,000 uniform steps are required. Due to the sensitivity of the problem, the global error is much larger than the error tolerance, but is reduced in proportion to it.

Van der Pol Oscillator

Our second example is the van der Pol oscillator problem, which again we rewrite as a first order system:

$$\frac{d^2y}{dt^2} = \epsilon^{-1}\left((1-y^2)\frac{dy}{dt} - y\right), \tag{25}$$

$$\epsilon = 10^{-6}, \quad y(0) = 2, \quad \frac{dy}{dt}(0) = 0.$$

We solve up to $T = 11$ using the adaptive method with $m = 2$. We plot the solution and the time step histories for a tolerance of 10^{-10} in Fig. . Note that very small steps are needed to resolve the fast transitions, while the problem is quite stiff in the regions where y is nearly constant. Plots for the other tolerances tested, 10^{-6} and 10^{-10}, are similar though the number of time steps required varies.

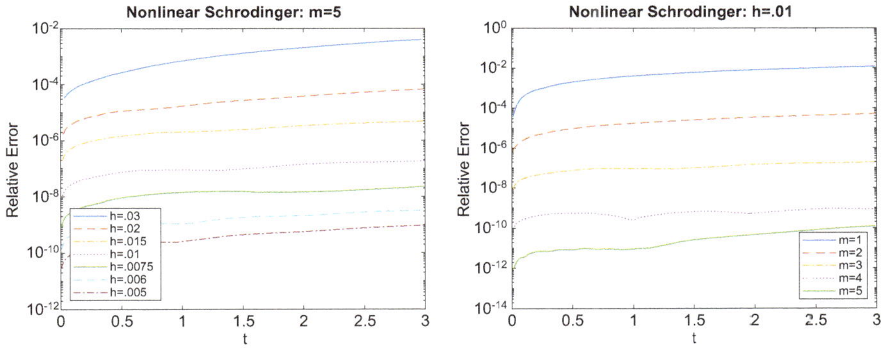

Fig. Left: Relative errors for NLS with various time steps and $m = 3$. Right: Relative errors for NLS with $h = .01$ and varying m

Schrödinger Equation

Lastly, we apply the method to evolve a Fourier pseudospectral discretization of the nonlinear Schrödinger equation. Precisely we consider the real problem

$$\frac{\partial v}{\partial t} = -\frac{\partial^2 w}{\partial x^2} - \left(v^2 + w^2\right) w, \quad \frac{\partial w}{\partial t} = \frac{\partial^2 v}{\partial x^2} + \left(v^2 + w^2\right) v, \tag{26}$$

for $x \in (-8, 8)$, $t \in (0, 3)$ with periodic boundary conditions $\frac{\partial v}{\partial x} = \frac{\partial w}{\partial x} = 0$ at $x = \pm 1$. We approximate the periodization of the exact solitary wave solution

$$\begin{aligned} v(x,t) &= \sqrt{50}\cos(rx - st) \cdot \text{sech}(5(x - ct)), \\ w(x,t) &= \sqrt{50}\sin(rx - st) \cdot \text{sech}(5(x - ct)), \end{aligned} \tag{27}$$

with $c = 2\pi$, $r = \pi$, $s = \pi^2 - 25$. We note that the amplitude of the solitary wave is reduced by about 17 digits at a distance of 8 from its peak so that the interaction with periodic copies is negligible over the simulation time. We use 512 Fourier modes in the computation of the derivatives and experiments show that this is sufficient to represent the solitary wave to machine precision. The implicit system was solved using Newton iterations each time step. In Fig. we present results for $m = 3$ (8th order) with varying time step and for m varying from 1 to 5 (order 4 through 12) with $h = 10^{-2}$. In both cases we observe rapid convergence. We also tabulate the errors at the final time and calculate the convergence rates when $m = 3$ in Table. The results are clearly consistent with the design order.

Table Relative errors for the Fourier pseudospectral discretization of the NLS (26) with solitary wave solution (27)

$m = 3$			$h = 0.01$	
h	Error	Rate	m	Error
3.0(−2)	4.0(−3)		1	1.2(−2)
2.0(−2)	6.7(−5)	10.1	2	5.0(−5)
1.5(−2)	4.9(−6)	9.1	3	1.8(−7)
1.0(−2)	1.8(−7)	8.1	4	8.6(−10)
7.5(−3)	2.2(−8)	7.3	5	1.2(−10)
6.0(−3)	3.2(−9)	8.7		
5.0(−3)	9.4(−10)	6.8		

Conclusions and Future Work

In conclusion, we have demonstrated that Hermite-Birkhoff interpolation can be used to develop singly-implicit A-stable timestepping methods of arbitrary order. A number of possible generalizations and improvements to the method are possible. These include

1. Stability analysis for variable coefficient or nonlinear problems using the projection properties (5);
2. Improved time step/order adaptivity;
3. Preconditioning of the implicit system for applications to partial differential equations such as spectral/pseudospectral discretizations of equations of Schrödinger type (e.g. integration preconditioners [8, 9]);
4. Development of IMEX schemes combining Hermite and Taylor polynomials.

Acknowledgements This work was supported in part by NSF Grant DMS-1418871. Any opinions, findings, and conclusions or recommendations expressed in this material are those of the author and do not necessarily reflect the views of the National Science Foundation.

References

1. Appelö, D., Hagstrom, T.: On advection by Hermite methods. Pac. J. Appl. Math. **4**,125–139 (2012)
2. Appelö, D., Hagstrom, T.: Solving PDEs with Hermite Interpolation. Lecture Notes in Computational Science, pp. 31–49. Springer, Berlin (2015)
3. Appelö, D., Kreiss, G., Wang, S.: An explicit Hermite-Taylor method for the Schrödinger equation. Commun. Comput. Phys **21**, 1207–1230 (2017)
4. Appelö, D., Hagstrom, T., Vargas, A.: Hermite methods for the scalar wave equation. SIAM J. Sci. Comput. **40**, A3902–A3927 (2018)
5. Chen, R., Hagstrom, T.: P-adaptive Hermite methods for initial value problems. ESAIM: Math. Model. Numer. Anal. **46**, 545–557 (2012)
6. Chen, X., Appelö, D., Hagstrom, T.: A hybrid Hermite-discontinuous Galerkin method for hyperbolic systems with application to Maxwell's equations. J. Comput. Phys. **257**, 501–520 (2014)

7. Chidwagyai, P., Nave, J.-C., Rosales, R., Seibold, B.: A comparative study of the efficiency of jet schemes. Int. J. Numer. Anal. Model.-B **3**, 297–306 (2012)
8. Coutsias, E., Hagstrom, T., Hesthaven, J., Torres, D.: Integration preconditioners for differential operators in spectral τ methods. Houst. J. Math. **1995**, 21–38 (1996). Special issue: ICOSAHOM
9. Coutsias, E., Hagstrom, T., Torres, D.: An efficient spectral method for ordinary differential equations with rational function coefficients. Math. Comp. **65**, 611–635 (1996)
10. Goodrich, J., Hagstrom, T., Lorenz, J.: Hermite methods for hyperbolic initial-boundary value problems. Math. Comp. **75**, 595–630 (2006)
11. Griewank, A.: Evaluating Derivatives: Principles and Techniques of Algorithmic Differentiation. SIAM, Philadelphia (2000)
12. Hairer, E., Wanner, G.: Solving Ordinary Differential Equations II, Stiff and Differential-Algebraic Problems. Springer, New York (1996)
13. Hairer, E., Norsett, S., Wanner, G.: Solving Ordinary Differential Equations I, Nonstiff Problems. Springer, New York (1992)
14. Kornelus, A., Appelö, D.: Flux-conservative Hermite methods for simulation of nonlinear conservation laws. J. Sci. Comput. **76**, 24–47 (2018)
15. Liu, C., Iserles, A., Wu, W.: Symmetric and arbitrarily high-order Birkhoff-Hermite time integrators and their long-time behaviour for solving nonlinear Klein-Gordon equations. J. Comput. Phys. **356**, 1–30 (2018)
16. Seibold, B., Rosales, R., Nave, J.-C.: Jet schemes for advection problems. Discrete Contin. Dyn. Syst. Ser. B **17**, 1229–1259 (2012)
17. Vargas, A., Chan, J., Hagstrom, T., Warburton, T.: GPU Acceleration of Hermite Methods for Simulation of Wave Propagation. Lecture Notes in Computational Science, pp. 357–368. Springer, Berlin (2017)
18. Vargas, A., Chan, J., Hagstrom, T., Warburton, T.: Variations on Hermite methods for wave propagation. Commun. Comput. Phys. **22**, 303–337 (2017)
19. Vargas, A., Hagstrom, T., Chan, J., Warburton, T.: Leapfrog time-stepping for Hermite methods. J. Sci. Comput. **80**, 289–314 (2019)

13

High-Order Mesh Generation Based on Optimal Affine Combinations of Nodal Positions

Mike Stees and Suzanne M. Shontz

Introduction

The advantage of high-order numerical methods for solving partial differential equations is their higher degree of accuracy compared to low-order numerical methods. A major hurdle in the usage of these methods in the presence of complex geometries is the absence of robust high-order mesh generation methods [23]. In other words, these methods need a high-order mesh that accurately captures the features of the geometry to achieve their full potential [1, 10].

The typical approach for generating high-order meshes is to transform a coarse linear mesh [2–6, 9, 11, 12, 14, 16, 17, 19–22, 24]. At a high-level, these transformations usually consist of the following three steps: (1) the low-order mesh is enriched with additional nodes; (2) the new nodes that lie along the boundary of the mesh are moved to the true boundary; (3) the interior nodes are moved based on the boundary deformation. The main challenge of these methods arises from step (2). In particular, the curving of the elements along the boundary can result in invalid mesh elements. With that in mind, these high-order mesh generation methods use different approaches in step (3) in an effort to obtain a valid high-order mesh. Methods for transforming the linear mesh usually fall into two groups. The first

M. Stees (✉)
Department of Electrical Engineering and Computer Science, Information and Telecommunication Technology Center, University of Kansas, Lawrence, KS, USA
e-mail: mstees@ku.edu

S. M. Shontz
Department of Electrical Engineering and Computer Science, Bioengineering Program, Information and Telecommunication Technology Center, University of Kansas, Lawrence, KS, USA
e-mail: shontz@ku.edu

group of methods transform the mesh based on the solution to a partial differential equation [3, 12, 14, 24]. The second group of methods are based on optimization of an objective function [2, 4, 5, 9, 16–21].

In this paper, we describe an optimization-based approach for generating high-order meshes based on affine combinations of nodal positions. The remainder of this paper is organized as follows. we present our new method for high-order mesh generation. In other sect. we demonstrate the performance of our proposed method on several aerospace engineering geometries. Finally, in some concluding remarks and possible directions for our future work.

High-Order Mesh Generation Based on Affine Combinations of Nodal Positions

In this section, we present our optimization-based method for high-order mesh generation. Our proposed method uses affine combinations of nodal positions to determine the movement of the interior nodes after deforming the boundary. Our method consists of three steps. First, for each interior node in the high-order straight-sided mesh, an optimization problem is solved to calculate a set of weights that relates the interior node to its neighbors. Second, the boundary nodes are moved to the true boundary. Third, the new positions of the interior nodes are calculated by solving a linear system of equations using the weights and the new boundary positions. In spirit, this method is similar to the weight-based method that we proposed in [19] with two major differences. The first difference is that we propose an affine combination of nodal positions in this work, as opposed to a convex combination. This change allows us to remove the inequality constraint and log-barrier term, leaving only the equality constraints. We also propose an alternative objective function that when combined with the equality constraints allows us to directly solve the optimization problem via a QR factorization. This change results in simplified computational complexity and faster execution time.

To frame our discussion of the method, we introduce the following notation for the 2D formulation of the problem; the 3D formulation is similar. Let the x- and y-coordinates of the ith interior node be represented as (x_i, y_i). In addition, define the x- and y-coordinates of the vertices adjacent to node i as $\{(x_j, y_j) : j \in N_i\}$, where N_i is the set of neighbors of node i. For each interior node i, this information can be represented as the following linear system, where w_{ij} are the weights:

$$\sum_{j \in N_i} w_{ij} x_j = x_i$$

$$\sum_{j \in N_i} w_{ij} y_j = y_i,$$

where

$$N_i = \{\text{high-order nodes of the patch to which } i \text{ belongs}\}.$$

There are several potential choices for the local neighboring set based on use of the low-order nodes, high-order nodes, or both. We include only the high-order nodes as neighbors, as only the high-order boundary nodes move during the boundary deformations. Using either the low-order nodes or both the low- and high-order nodes would dampen the effect the boundary deformation has on the interior nodes, which might lead to tangling near the boundary. Including additional nodes as neighbors would also result in a less sparse matrix when solving (7). Another important consideration is that while the weight calculation is based on only the local neighbors, the position of an interior node is indirectly affected by the deformation of all the interior nodes through the solution of (7).

Adding the additional constraint that the weights sum to one results in the following linear system $Aw = b$ for finding an affine combination of the x- and y-coordinates of the vertices adjacent to node i:

$$\begin{bmatrix} x_1 \; x_2 \; \ldots \; x_n \\ y_1 \; y_2 \; \ldots \; y_n \\ 1 \;\; 1 \;\; \ldots \;\; 1 \end{bmatrix} \begin{bmatrix} w_{i1} \\ w_{i2} \\ \vdots \\ w_{in} \end{bmatrix} = \begin{bmatrix} x_i \\ y_i \\ 1 \end{bmatrix},$$

where $n = |N_i|$. Based on the set of neighbors, this linear system will be underdetermined (i.e., $A = m \times n$ with $m < n$) in general. If we assume that A has full rank, we can find one particular solution to our problem by requiring that w has the smallest norm of any solution. This results in the following optimization problem:

$$\min_w ||w||_2^2 \tag{1}$$

$$\text{subject to } Aw = b. \tag{2}$$

From the Karush–Kuhn–Tucker (KKT) theory [13], we know that the following conditions must hold for a solution (w^*, λ^*) to our problem to be optimal:

$$\nabla_w \mathcal{L}(w^*, \lambda^*) = 0 \tag{3}$$

$$Aw^* - b = 0 \tag{4}$$

$$\lambda^*(Aw^* - b) = 0. \tag{5}$$

The Lagrangian of our problem is given by:

$$\mathcal{L}(w, \lambda) = w^T w - \lambda^T (Aw - b),$$

where λ are the Lagrange multipliers.

Using (3)–(5), we can find the following solution pair (w^*, λ^*) as follows:

$$\nabla_w \mathcal{L}(w, \lambda) = 2w - A^T \lambda.$$

$$\nabla_w \mathcal{L}(w^*, \lambda^*) = 0 \Rightarrow w^* = \frac{1}{2} A^T \lambda^*.$$

$$Aw^* - b = 0 \Rightarrow A(\frac{1}{2} A^T \lambda^*) - b = 0.$$

$$\lambda^* = 2(AA^T)^{-1} b$$

$$w^* = \frac{1}{2} A^T \lambda^* = \frac{1}{2} A^T 2(AA^T)^{-1} b = A^T (AA^T)^{-1} b.$$

Although we have verified that (w^*, λ^*) is a stationary point, we cannot yet claim that it is a minimum. To do so, we must investigate $\nabla_w^2 \mathcal{L}(w^*, \lambda^*)$:

$$\nabla_w^2 \mathcal{L}(w, \lambda) = 2I_{|N_i| \times |N_i|}$$
$$\nabla_w^2 \mathcal{L}(w^*, \lambda^*) = 2I_{|N_i| \times |N_i|}.$$

From the second-order sufficient conditions, if w^* satisfies (3)–(5) and the following condition is satisfied:

$$z^T \nabla_w^2 \mathcal{L}(w^*, \lambda^*) z > 0, \text{ for all } z \in C(w^*, \lambda^*), z \neq 0, \tag{6}$$

where $C(w^*, \lambda^*) = \{z \mid \nabla_w c(w^*)^T z = 0\}$ is the critical cone and $c(w) = Aw - b$, then our solution is a minimum. Since $\nabla_w^2 \mathcal{L}(w^*, \lambda^*)$ is symmetric positive definite, the inequality in (6) is satisfied for any choice of z. Thus we can conclude that our solution w^* is a minimum of (1)–(2).

Now that we have established that w^* is our solution, we will discuss calculating it via a reduced QR factorization. Suppose that $A^T = QR$, where $Q_{n \times m}$, $R_{n \times m}$ is upper triangular, and $Q^T Q = I_{m \times m}$. Substituting in the QR factorization of A^T into w^*, we get the following:

$$\begin{aligned} w^* = A^+ b = A^T (AA^T)^{-1} b &= QR(R^T Q^T QR)^{-1} b = QR(R^T R)^{-1} b \\ &= QRR^{-1} R^{-T} b = QR^{-T} b. \end{aligned}$$

Rearranging this into linear system form, we have:

$$R^T Q^T w^* = b.$$

If we let $t = Q^T w^*$, then $R^T t = b$ and $w^* = Qt$. Thus calculating w^* involves a QR decomposition of A^T, solving the lower triangular system $R^T t = b$ by forward substitution, and calculating the matrix-vector product Qt.

After calculating the weights, a boundary deformation is applied. The final step is to solve for the new locations of the interior nodes $[\hat{x}_I, \hat{y}_I]$ by solving the following global linear system:

$$A_I[\hat{x}_I, \hat{y}_I] = -A_B[\hat{x}_B, \hat{y}_B], \tag{7}$$

where $\hat{x}_B$ and $\hat{y}_B$ are the new x- and y-coordinates for the boundary nodes, and A_I and A_B contain the weights for the interior nodes and boundary nodes, respectively. In this global linear system, each row of the weight matrix corresponds to an interior node with nonzero entries for the node's neighbors and zero entries for the remainder of the row. The resulting global weight matrix is very sparse with irregular structure. In an effort to shift the nonzero entries closer to the diagonal, we apply the sparse reverse Cuthill–McKee ordering provided in Matlab. In Fig. , we show the matrix sparsity plots for the natural node ordering and the updated node ordering for the first two examples in Sect. After applying the matrix reordering, the linear system is solved using a sparse LU factorization.

Numerical Experiments

In this section, we demonstrate the results from applying our method to generate several high-order meshes. We use Gmsh [7, 8, 15, 21] to generate the initial straight-sided high-order meshes. Our method then uses this mesh to calculate the weights (step 1). Next, we curve the boundary nodes (step 2) using Gmsh. The positions of the interior nodes in the resulting curved high-order mesh are then updated (step 3) by our method. For each example, we show the mesh which results from our method (with high-order nodes visible), the mesh element distortion for the curved boundary mesh generated using Gmsh, and the distortion for the mesh resulting from our method. When reporting the mesh distortion, we list the minimum distortion, maximum distortion, average distortion computed over all elements (referred to as Avg1 in figures), and average distortion computed over curved elements (referred to as Avg2 in figures). The ideal distortion value is 1, indicating that the element is straight. We also list the execution times needed for steps 1 and 3 of our method (excluding I/O) in Table. The code was run using Matlab R2018a, and the wall-clock execution times were measured on a machine with 16GB of RAM and a Ryzen 7 1700 CPU. All mesh visualizations and distortion evaluations were done using Gmsh. Our first example is a third-order mesh of a square region around

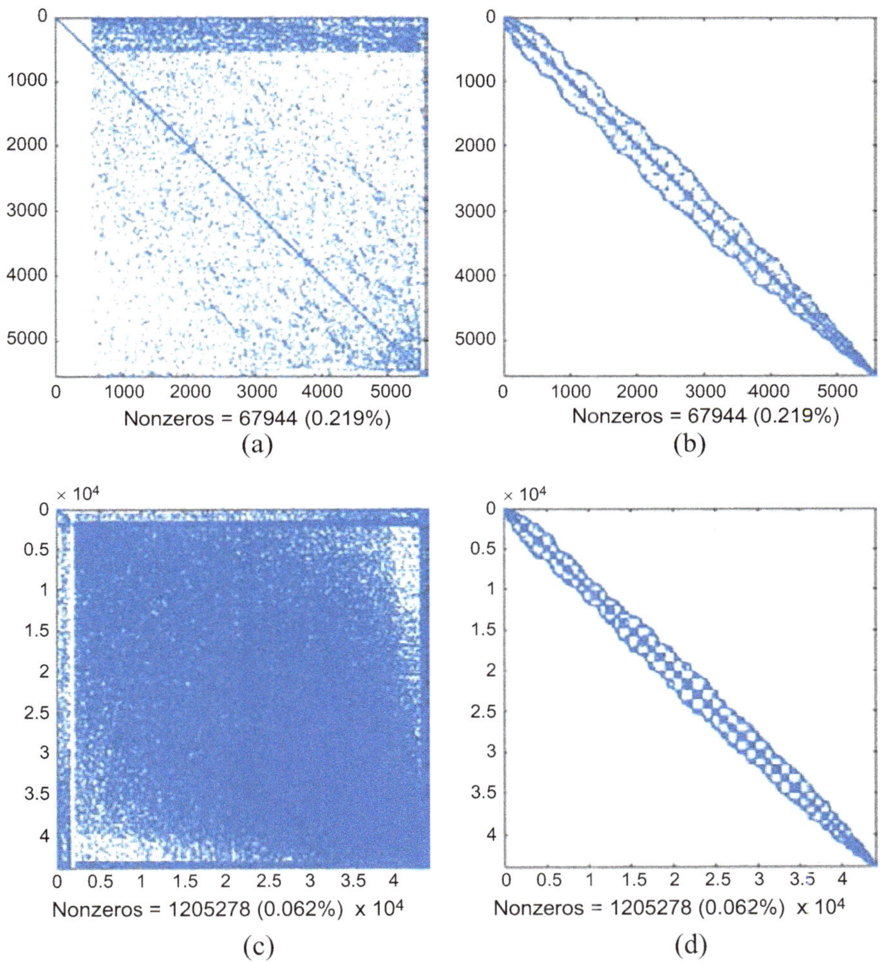

Fig. Sparsity plots for the first two examples in Sect.: (**a** and **c**) the nonzero entries using the original node ordering; (**b** and **d**) the nonzero entries after applying the sparse reverse Cuthill–McKee ordering

a NACA0012 airfoil. In Fig. a, b, we show the mesh resulting from our method and a table of the mesh quality values as measured by the distortion metric. In this example, our method increased the minimum distortion from 0.744 to 0.799, while causing only minor changes in the average distortion.

In our second example, we extrude the NACA0012 airfoil and create a third-order mesh of the resulting region. In Fig. a, b, we show the mesh resulting from our method, and a table of the mesh quality values. For this example, our method improved the minimum distortion by 0.125, increasing it from 0.317 to 0.442.

Table The number of elements and the wall clock times for steps 1 and 3 of our method (excluding I/O) for each example

Example	Number of elements	Execution time (s) Original ordering	Execution time (s) New reordering
NACA0012 airfoil	1312	11.51	3.61
Extruded NACA0012 airfoil	13,895	4826.09	958.14
Airbus A319	50,400	13,956.43	693.11

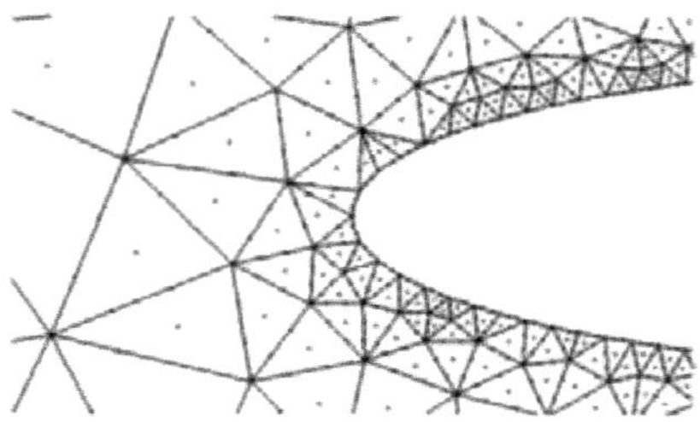

(a)

Example	Distortion Min	Max	Avg1	Avg2
original mesh	0.744	1.000	0.999	0.994
resulting mesh	0.799	1.000	0.997	0.997

(b)

Fig. NACA0012 airfoil example: (**a**) the mesh resulting from our method and (**b**) the mesh quality as measured by the element distortion metric

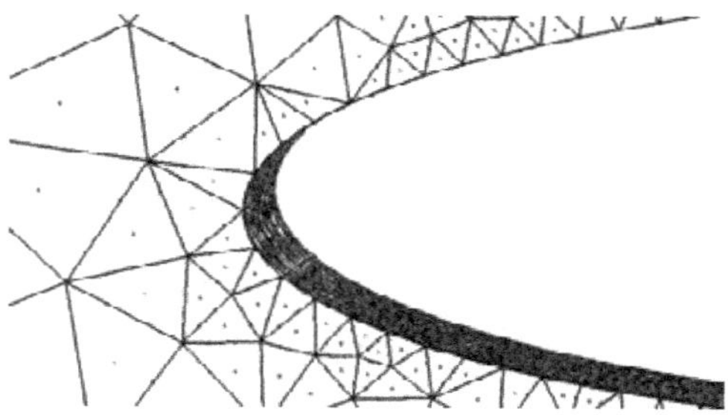

(a)

Example	Distortion Min	Max	Avg1	Avg2
original mesh	0.317	1.000	0.997	0.994
resulting mesh	0.442	1.000	0.995	0.995

(b)

Fig. Extruded NACA0012 airfoil example: (**a**) the mesh resulting from our method and (**b**) the mesh quality as measured by the element distortion metric

Our third and final example is a second-order mesh of an Airbus A319 aircraft. Unlike our previous examples, this geometry resulted in tangled elements after curving the boundary. Although our method still increased the minimum quality, it was not able to untangle the mesh. To address this, we applied the high-order

Fig. Airbus A319 example: (**a**) surfaces of the mesh resulting from our method; (**b**) a view of a cut through the interior volume mesh, and (**c**) the quality of the mesh with only boundary curving, the quality of the mesh resulting from our method, and the quality of both meshes after applying the regularization scheme available in Gmsh

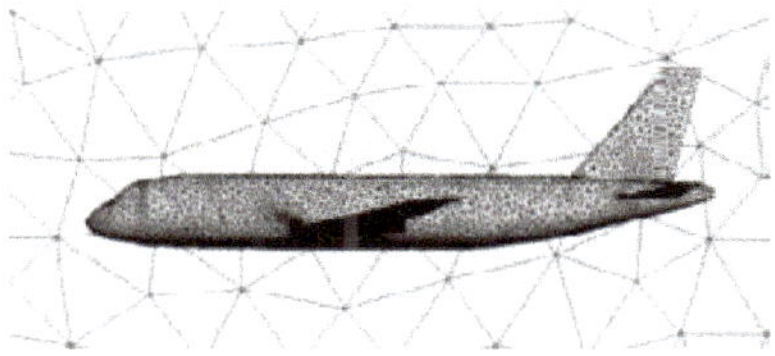

(a)

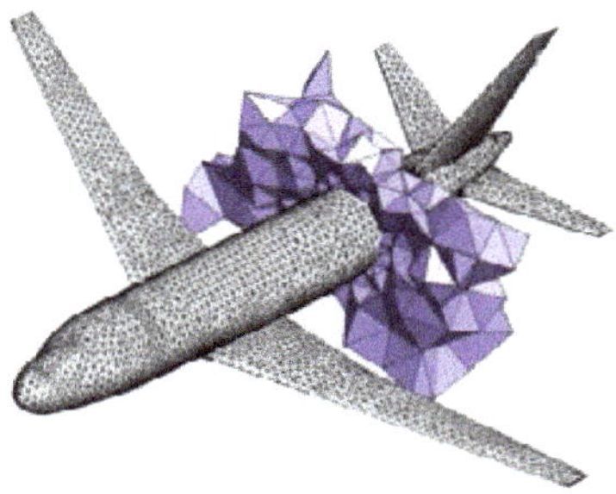

(b)

	Distortion			
Example	**Min**	**Max**	**Avg1**	**Avg2**
original mesh	–0.878	0.975	0.943	1.000
resulting mesh	–0.449	0.970	0.970	1.000
original mesh after untangling	0.206	0.975	0.945	1.000
resulting mesh after untangling	0.211	0.971	0.970	1.000

(c)

regularization scheme available in Gmsh as a post-processing step. Aside from changing the target Jacobian range to 0.3–2 and fixing the boundary nodes, all other parameters were left at their default values. The untangling for the original mesh took 14.14 s, while untangling the mesh resulting from our method required only 1.64 s. In Fig. a–c, we show the surfaces of the mesh resulting from our method, a view of a cut through the interior volume mesh, and a table of the mesh quality values. In Fig. c, we list the distortion for the mesh after curving the boundary, the distortion for the mesh resulting from our method, and the distortions of both meshes after applying the regularization scheme in Gmsh.

Aside from the third test case, all of these examples were relatively straightforward. In each case, our method increased the minimum distortion when compared to only curving the nodes along the boundary. While additional testing is necessary to confirm this, our results for the third example seem to indicate that our method could be used to reduce the severity of the mesh tangling and thus simplify the work for an untangling method during post-processing.

Concluding Remarks and Future Work

We have presented a new optimization-based method for generating high-order meshes. Our examples have shown that the proposed method based on affine combinations of nodal positions tends to improve the quality of the most distorted elements, while causing minor changes to the least distorted elements. While our approach is optimization-based, we have demonstrated that the optimization problem can be solved directly using a QR factorization as opposed to the typical iterative optimization approach. This change results in lessened computational complexity and reduced execution time.

As part of our future work, we will consider other definitions for the set of neighbors of an interior node. We will also investigate other aspects of the linear system including other node reordering schemes and solvers. Finally, we will apply the untangling method that we proposed in [20] after extending it to 3D.

Acknowledgements The work of the first author was funded in part by the Madison and Lila Self Graduate Fellowship and NSF CCF grant 1717894. The work of the second author was supported in part by NSF grants CCF 1717894 and OAC 1808553. The authors wish to thank the anonymous referee for his or her careful reading of the paper and for the helpful suggestions which strengthened it.

References

1. Bassi, F., Rebay, S.: High-order accurate discontinuous finite element solution of the 2D Euler equations. J. Comput. Phys. **138**(2), 251–285 (1997)
2. Dey, S., Shephard, M.S.: Curvilinear mesh generation in 3D. In: Proceedings of the 8th International Meshing Roundtable (1999)
3. Fortunato, M., Persson, P.-O.: High-order unstructured curved mesh generation using the Winslow equations. J. Comput. Phys. **307**(2016), 1–14 (2016)
4. Gargallo-Peiró, A., Roca, X., Peraire, J., Sarrate, J.: Distortion and quality measures for validating and generating high-order tetrahedral meshes. Eng. Comput. **31**(3), 423–437 (2015)
5. Gargallo-Peiró, A., Roca, X., Peraire, J., Sarrate, J.: Optimization of a regularized distortion measure to generate curved high-order unstructured tetrahedral meshes. Int. J. Numer. Methods Eng. **103**(5), 342–363 (2015)
6. George, P.L., Borouchaki., H.: Construction of tetrahedral meshes of degree two. Int. J. Numer. Methods Eng. **90**(9), 1156–1182 (2012)
7. Geuzaine, C., Remacle, J.-F.: Gmsh: a 3-D finite element mesh generator with built-in pre-and post-processing facilities. Int. J. Numer. Methods Eng. **79**(11), 1309–1331 (2009)
8. Johnen, A., Remacle, J.-F., Geuzaine, C.: Geometrical validity of high-order triangular finite elements. Eng. Comput. **30**(3), 375–382 (2014)
9. Karman, S.L., Erwin, J.T., Glasby, R.S., Stefanski, D.: High-order mesh curving using WCN mesh optimization. In: 46th AIAA Fluid Dynamics Conference, p. 3178 (2016)
10. Luo, X., Shephard, M.S., Remacle, J.-F.: The influence of geometric approximation on the accuracy of high order methods. Rensselaer SCOREC report, 1, 2001
11. Moxey, D., Green, M., Sherwin, S., Peiró, J.: An isoparametric approach to high-order curvilinear boundary-layer meshing. Comput. Methods Appl. Mech. Eng. **283**, 636–650 (2015)

12. Moxey, D., Ekelschot, D., Keskin, Ü., Sherwin, S.J., Peiró, J.: High-order curvilinear meshing using a thermo-elastic analogy. Comput. Aided Des. **72**, 130–139 (2016)
13. Nocedal, J., Wright, S.: Numerical Optimization, 2nd edn. Springer Series in Operations Research and Financial Engineering. Springer, Berlin (2006)
14. Persson, P.-O., Peraire, J.: Curved mesh generation and mesh refinement using Lagrangian solid mechanics. In: Proceedings of the 47th AIAA Aerospace Sciences Meeting Including the New Horizons Forum and Aerospace Exposition, p. 949 (2009)
15. Remacle, J.-F., Chevaugeon, N., Marchandise, E., Geuzaine, C.: Efficient visualization of high-order finite elements. Int. J. Numer. Methods Eng. **69**(4), 750–771 (2007)
16. Roca, X., Gargallo-Peiró, A., Sarrate, J.: Defining quality measures for high-order planar triangles and curved mesh generation. In: Proceedings of the 20th International Meshing Roundtable, pp. 365–383. Springer, Berlin (2012)
17. Ruiz-Gironés, E., Sarrate, J., Roca, X.: Generation of curved high-order meshes with optimal quality and geometric accuracy. In: Proceedings of the 25th International Meshing Roundtable, vol. 163, pp. 315–327. Procedia Engineering (2016)
18. Sherwin, S.J., Peiró, J.: Mesh generation in curvilinear domains using high-order elements. Int. J. Numer. Methods Eng. **53**(1), 207–223 (2001)
19. Stees, M., Shontz, S.M.: A high-order log barrier-based mesh generation and warping method. In: Proceedings of the 26th International Meshing Roundtable, vol. 203, pp. 180–192. Procedia Engineering (2017)
20. Stees, M., Shontz, S.M.: An angular approach to untangling high-order curvilinear triangular meshes. In: Proceedings of the 27th International Meshing Roundtable (2018)
21. Toulorge, T., Geuzaine, C., Remacle, J.-F., Lambrechts, J.: Robust untangling of curvilinear meshes. J. Comput. Phys. **254**, 8–26 (2013)
22. Turner, M., Moxey, D., Peiró, J., Gammon, M., Pollard, C.R., Bucklow, H.: A framework for the generation of high-order curvilinear hybrid meshes for CFD simulations. In: Proceedings of the 26th International Meshing Roundtable, vol. 203, pp. 206–218. Procedia Engineering (2017)
23. Wang, Z.J., Fidkowski, K., Abgrall, R., Bassi, F., Caraeni, D., Cary, A., Deconinck, H., Hartmann, R., Hillewaert, K., Huynh, H.T., et al. High-order CFD methods: current status and perspective. Int. J. Numer. Methods Fluids **72**(8), 811–845 (2013)
24. Xie, Z.Q., Sevilla, R., Hassan, O., Morgan, K.: The generation of arbitrary order curved meshes for 3D finite element analysis. Comput. Mech. **51**(3), 361–374 (2013)

14

A Conservative Hybrid Method for Darcy Flow

Varun Jain, Joël Fisser, Artur Palha, and Marc Gerritsma

Introduction

Hybrid formulations [1, 3, 10] are classical domain decomposition methods which reduce the problem of solving one global system to many small local systems. The local systems can then be efficiently solved independently of each other in parallel.

In this work we present a hybrid mimetic spectral element formulation to solve Darcy flow. We follow [8] which render the constraints on divergence of mass flux, the pressure gradient and the inter-element continuity metric free. The resulting system is extremely sparse and shows a reduced growth in condition number as compared to a non-hybrid system.

This document is structured as follows: In Sections we define the weak formulation for Darcy flow. The basis functions are introduced. The evaluation of weighted inner product and duality pairings are discussed in Sect. In Sect. we discuss the formulation of discrete algebraic system. In Sect.we present results for a test case taken from [7].

V. Jain (✉) · J. Fisser · A. Palha · M. Gerritsma
Faculty of Aerospace Engineering, TU Delft, Delft, The Netherlands
e-mail: V.Jain@tudelft.nl; A.PalhaDaSilvaClerigo@tudelft.nl; M.I.Gerritsma@tudelft.nl

Darcy Flow Formulation

For $\Omega \in \mathbb{R}^d$, where d is the dimension of the domain, the governing equations for Darcy flow, are given by,

$$\begin{cases} \boldsymbol{u} + \mathbb{A}\,\nabla p = 0 \\ \nabla \cdot \boldsymbol{u} \quad\;\; = f \end{cases} \text{in } \Omega \quad \text{and} \quad \begin{cases} \partial\Omega \;\; = \Gamma_D \cup \Gamma_N & \\ p \quad\;\; = \hat{p} & \text{on } \Gamma_D \\ \boldsymbol{u}\cdot\boldsymbol{n} = \hat{\boldsymbol{u}}_{\boldsymbol{n}} & \text{on } \Gamma_N \end{cases},$$

where, $\boldsymbol{u}$ is the velocity, p is the pressure, f the prescribed RHS term. $\mathbb{A}$ is a $d \times d$ symmetric positive definite matrix, $\hat{p}$ and $\hat{\boldsymbol{u}}_{\boldsymbol{n}}$ are the prescribed pressure and flux boundary conditions, respectively.

Notations

For $f, g \in L^2(\Omega)$, $(f, g)_\Omega$ denotes the usual L^2-inner product.

For vector-valued functions in L^2 we define the weighted inner product by,

$$(\boldsymbol{u}, \boldsymbol{v})_{\mathbb{A}^{-1},\Omega} = \int_\Omega \left(\boldsymbol{u}, \mathbb{A}^{-1}\boldsymbol{v}\right) d\Omega\,, \tag{1}$$

where $(\cdot\,,\cdot)$ denotes the pointwise inner product.

Duality pairing, denoted by $\langle\cdot,\cdot\rangle_\Omega$, is the outcome of a linear functional on $L^2(\Omega)$ acting on elements from $L^2(\Omega)$.

Let Ω_K be a disjoint partitioning of Ω with total number of elements K, and K_i is any element in Ω_K, such that, $K_i \in \Omega_K$. We define the following broken Sobolev spaces [2], $H\left(\text{div}; \Omega_K\right) = \prod_i H\left(\text{div}; K_i\right)$, and $H^{1/2}(\partial\Omega_K) = \prod_i H^{1/2}(\partial K_i)$.

Weak Formulation

The Lagrange functional for Darcy flow is defined as,

$$\begin{aligned}\mathcal{L}(\boldsymbol{u}, p, \lambda; f) = \tfrac{1}{2}\int_{\Omega_K} \boldsymbol{u}^T\mathbb{A}^{-1}\boldsymbol{u}\,\mathrm{d}\Omega_K - \int_{\Omega_K} p\left(\nabla\cdot\boldsymbol{u} - f\right)\mathrm{d}\Omega_K \\ + \int_{\partial\Omega_K\setminus\Gamma_D} \lambda\,(\boldsymbol{u}\cdot\boldsymbol{n})\,\mathrm{d}\Gamma + \int_{\Gamma_D} \hat{p}\,(\boldsymbol{u}\cdot\boldsymbol{n})\,\mathrm{d}\Gamma - \int_{\Gamma_N} \lambda\left(\hat{\boldsymbol{u}}_{\boldsymbol{n}}\right)\mathrm{d}\Gamma\end{aligned}.$$

The variational problem is then given by: For given $f \in L^2(\Omega_K)$, $\hat{p} \in H^{1/2}(\Gamma_D)$ and $\hat{\boldsymbol{u}}_{\boldsymbol{n}} \in H^{-1/2}(\Gamma_N)$ find $\boldsymbol{u} \in H(\text{div};\Omega_K)$, $p \in L^2(\Omega_K)$, $\lambda \in H^{\frac{1}{2}}(\partial\Omega_K)$, such that,

$$\begin{cases} (\boldsymbol{v},\boldsymbol{u})_{\mathbb{A}^{-1},\Omega_K} - \langle \nabla\cdot\boldsymbol{v}, p\rangle_{\Omega_K} + \langle (\boldsymbol{v}\cdot\boldsymbol{n}), \lambda\rangle_{\partial\Omega_K\setminus\Gamma_D} = -\langle \boldsymbol{v}\cdot\boldsymbol{n}, \hat{p}\rangle_{\Gamma_D} & \forall\, \boldsymbol{v}\in H(\text{div};\Omega_K) \\ -\langle q, \nabla\cdot\boldsymbol{u}\rangle_{\Omega_K} = -\langle q, f\rangle_{\Omega_K} & \forall\, q\in L^2(\Omega_K) \\ \langle \mu, (\boldsymbol{u}\cdot\boldsymbol{n})\rangle_{\partial\Omega_K\setminus\Gamma_D} = \langle \mu, \hat{\boldsymbol{u}}_{\boldsymbol{n}}\rangle_{\Gamma_N} & \forall\, \mu\in H^{\frac{1}{2}}(\partial\Omega_K) \end{cases} . \tag{2}$$

Basis Functions

Primal and Dual Nodal Degrees of Freedom

Let ξ_j, $j = 0, 1, \ldots, N$, be the $N+1$ Gauss–Lobatto–Legendre (GLL) points in $I \in [-1, 1]$. The Lagrange polynomials $h_i(\xi)$ through ξ_j, of degree N, given by,

$$h_i(\xi) = \frac{\left(\xi^2 - 1\right) L_N'(\xi)}{N(N+1) L_N(\xi_i)(\xi - \xi_i)} ,$$

form the 1D primal nodal polynomials which satisfy, $h_i(\xi_j) = \delta_{ij}$.

Let a^h and b^h be two polynomials expanded in terms of $h_i(\xi)$. The L^2—inner product is then given by,

$$\left(a^h, b^h\right)_I = \mathbf{a}^T \mathbb{M}^{(0)} \mathbf{b} , \quad \text{where} \quad \mathbb{M}^{(0)}_{i,j} = \int_{-1}^{1} h_i(\xi)\, h_j(\xi)\, \mathrm{d}\xi ,$$

and, $\mathbf{a} = [\mathsf{a}_0\ \mathsf{a}_1\ \ldots\ \mathsf{a}_N]$ and $\mathbf{b} = [\mathsf{b}_0\ \mathsf{b}_1\ \ldots\ \mathsf{b}_N]$ are the nodal degrees of freedom. We define the algebraic *dual* degrees of freedom, $\widetilde{\mathbf{a}}$, such that the duality pairing is simply the vector dot product between primal and dual degrees of freedom,

$$\left\langle a^h, b^h\right\rangle_I = \widetilde{\mathbf{a}}^T \mathbf{b} := \mathbf{a}^T \mathbb{M}^{(0)} \mathbf{b} \quad \Rightarrow \quad \widetilde{\mathbf{a}} = \mathbb{M}^{(0)} \mathbf{a} .$$

Thus, the dual degrees of freedom are linear functionals of primal degrees of freedom.

Primal and Dual Edge Degrees of Freedom

The edge polynomials, for the N edges between $N+1$ GLL points (ξ_{j-1}, ξ_j), of polynomial degree $N-1$, are defined as [4],

$$e_j(\xi) = -\sum_{k=0}^{j-1} \frac{\mathrm{d}h_k}{\mathrm{d}\xi}(\xi)\ , \quad \text{such that} \quad \int_{\xi_{j-1}}^{\xi_j} e_i(\xi) = \delta_{ij}$$

Let p^h and q^h be two polynomials expanded in edge basis functions. The inner product in L^2 space is given by,

$$\left(p^h, q^h\right)_I = \mathbf{p}^T \mathbb{M}^{(1)} \mathbf{q}\ , \quad \text{where} \quad \mathbb{M}^{(1)}_{i,j} = \int_{-1}^{1} e_i(\xi)\, e_j(\xi)\, \mathrm{d}\xi\ ,$$

and, $\mathbf{p} = \left[\mathsf{p}_1\ \mathsf{p}_2\ \ldots\ \mathsf{p}_N\right]$ and $\mathbf{q} = \left[\mathsf{q}_1\ \mathsf{q}_2\ \ldots\ \mathsf{q}_N\right]$ are the edge degrees of freedom. As before, we define the *dual* degrees of freedom such that,

$$\left\langle p^h, q^h\right\rangle_I = \widetilde{\mathbf{p}}^T \mathbf{q} := \mathbf{p}^T \mathbb{M}^{(1)} \mathbf{q} \quad \Rightarrow \quad \widetilde{\mathbf{p}} = \mathbb{M}^{(1)} \mathbf{p}\ .$$

A similar construction can be used for dual degrees of freedom in higher dimensions. For construction of the dual degrees of freedom in 2D see [8] and for 3D see [9].

Differentiation of Nodal Polynomial Representation

Let $a^h(\xi)$ be expanded in Lagrange polynomials, then

$$\frac{d}{d\xi} a^h(\xi) = \frac{d}{d\xi} \sum_{i=0}^{N} \mathsf{a}_i h_i(\xi) = \sum_{i=1}^{N} \left(\mathsf{a}_i - \mathsf{a}_{i-1}\right) e_i(\xi)\ . \tag{3}$$

Therefore, taking the derivative of a polynomial involves two steps: First, take the difference of degrees of freedom; second, change of basis from nodal to edge [4].

Discrete Inner Product and Duality Pairing

For 2D domains, the higher dimensional primal basis are constructed using the tensor product of the 1D basis.

For the weak formulation (2) we expand the velocity $\boldsymbol{u}^h$ in primal edge basis as,

$$\boldsymbol{u}^h\left(\xi,\eta\right)=\sum_{i=0}^{N}\sum_{j=1}^{N}\mathsf{u}_{\mathsf{x}i,j}\;h_i(\xi)\;e_j(\eta)\;\hat{\boldsymbol{\imath}}+\sum_{i=1}^{N}\sum_{j=0}^{N}\mathsf{u}_{\mathsf{y}i,j}\;e_i(\xi)\;h_j(\eta)\;\hat{\boldsymbol{\jmath}}\;, \tag{4}$$

where $\mathsf{u}_{x\,i,j}$ denotes the flux, $\int \boldsymbol{u}\cdot\boldsymbol{n}$, over the vertical edges and $\mathsf{u}_{y\,i,j}$ the flux over the horizontal edges, see Fig.

Weighted Inner Product

Using (1) and the expansions in (4), the weighted inner product is evaluated as,

$$\left(\boldsymbol{v}^h,\boldsymbol{u}^h\right)_{\mathbb{A}^{-1},\Omega_K}=\sum_{K_i}\mathbf{v}_{K_i}^T\;\mathbb{M}^{(1)}_{\mathbb{A}^{-1},K_i}\;\mathbf{u}_{K_i}\;,$$

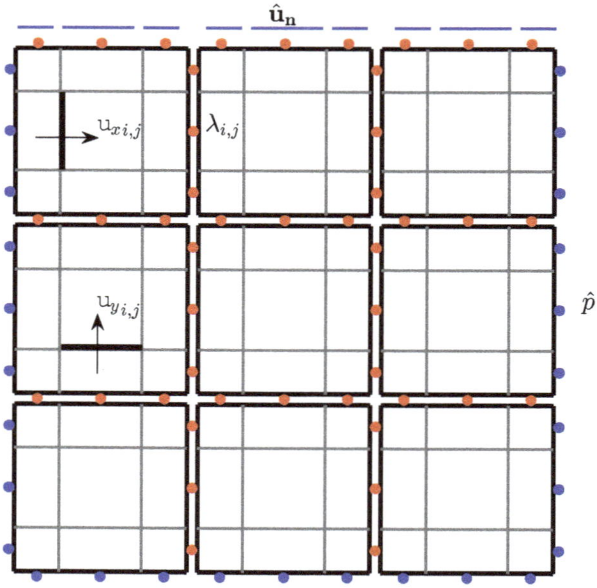

Fig. Discretized domain for $K = 3\times 3$, $N = 3$. The blue dots represent the pressure boundary condition $\hat{p}$, and the blue edges represent the velocity boundary condition $\hat{\boldsymbol{u}}_{\boldsymbol{n}}$

where, $\mathbf{u}_{K_i}$ are the degrees of freedom in element K_i, and

$$\mathbb{M}^{(1)}_{\mathbb{A}^{-1},K_i} = \int_{K_i} \begin{pmatrix} h_i(\xi)\, e_j(\eta) \\ e_i(\xi)\, h_j(\eta) \end{pmatrix}^T \mathbb{A}^{-1}(\xi,\eta) \begin{pmatrix} h_i(\xi)\, e_j(\eta) \\ e_i(\xi)\, h_j(\eta) \end{pmatrix} \mathrm{d}K_i \,.$$

For mapping of elements please refer to [6].

Divergence of Velocity

Divergence of velocity, $\nabla \cdot \boldsymbol{u}^h$, is evaluated using (3), but now for 2D,

$$\begin{aligned} \nabla \cdot \boldsymbol{u}^h &= \tfrac{\partial}{\partial x} \textstyle\sum_{i=0}^{N} \sum_{j=1}^{N} \mathsf{u_x}_{i,j}\, h_i(\xi) e_j(\eta) + \tfrac{\partial}{\partial y} \sum_{i=1}^{N} \sum_{j=0}^{N} \mathsf{u_y}_{i,j}\, e_i(\xi) h_j(\eta) \\ &= \textstyle\sum_{i,j=1}^{N} \left(\mathsf{u_x}_{i,j} - \mathsf{u_x}_{i-1,j} + \mathsf{u_y}_{i,j} - \mathsf{u_y}_{i,j-1} \right) e_i(\xi)\, e_j(\eta) \end{aligned} \, . \tag{5}$$

For pressure we will use dual degrees of freedom. Therefore the weak constraint on divergence of velocity is a duality pairing evaluated as,

$$\left\langle q^h, \nabla \cdot \boldsymbol{u}^h \right\rangle_{\Omega_K} = \sum_{K_i} \tilde{\mathbf{q}}^T_{K_i}\, \mathbb{E}^{2,1}\, \mathbf{u}_{K_i} \,,$$

where $\mathbb{E}^{2,1}$ represents the discrete divergence operator. It is an incidence matrix that is metric-free and topological, and remains the same for each element in Ω_K. For an extensive discussion on the incidence matrix, see for instance [6]. For an element of degree $N = 3$,

$$\mathbb{E}^{2,1} = \begin{bmatrix} -1 & 0 & 0 & 1 & 0 & 0 & 0 & 0 & 0 & 0 & 0 & 0 & -1 & 0 & 0 & 1 & 0 & 0 & 0 & 0 & 0 & 0 & 0 & 0 \\ 0 & 0 & 0 & -1 & 0 & 0 & 1 & 0 & 0 & 0 & 0 & 0 & 0 & -1 & 0 & 0 & 1 & 0 & 0 & 0 & 0 & 0 & 0 & 0 \\ 0 & 0 & 0 & 0 & 0 & 0 & -1 & 0 & 0 & 1 & 0 & 0 & 0 & 0 & -1 & 0 & 0 & 1 & 0 & 0 & 0 & 0 & 0 & 0 \\ 0 & -1 & 0 & 0 & 1 & 0 & 0 & 0 & 0 & 0 & 0 & 0 & 0 & 0 & 0 & -1 & 0 & 0 & 1 & 0 & 0 & 0 & 0 & 0 \\ 0 & 0 & 0 & 0 & -1 & 0 & 0 & 1 & 0 & 0 & 0 & 0 & 0 & 0 & 0 & 0 & -1 & 0 & 0 & 1 & 0 & 0 & 0 & 0 \\ 0 & 0 & 0 & 0 & 0 & 0 & 0 & -1 & 0 & 0 & 1 & 0 & 0 & 0 & 0 & 0 & 0 & -1 & 0 & 0 & 1 & 0 & 0 & 0 \\ 0 & 0 & -1 & 0 & 0 & 1 & 0 & 0 & 0 & 0 & 0 & 0 & 0 & 0 & 0 & 0 & 0 & 0 & -1 & 0 & 0 & 1 & 0 & 0 \\ 0 & 0 & 0 & 0 & 0 & -1 & 0 & 0 & 1 & 0 & 0 & 0 & 0 & 0 & 0 & 0 & 0 & 0 & 0 & -1 & 0 & 0 & 1 & 0 \\ 0 & 0 & 0 & 0 & 0 & 0 & 0 & 0 & -1 & 0 & 0 & 1 & 0 & 0 & 0 & 0 & 0 & 0 & 0 & 0 & -1 & 0 & 0 & 1 \end{bmatrix} .$$

Connectivity Matrix

The connectivity matrix ensures continuity of the velocity flux across the elements. $\boldsymbol{\lambda}$ is the interface variable defined between the elements, shown as red dots in Fig. . λ acts as Lagrange multiplier that imposes the continuity constraint given by,

$$\left\langle \mu^h, \boldsymbol{u}^h \cdot \boldsymbol{n} \right\rangle_{\partial\Omega_K \backslash \Gamma_D} = \sum_K \widetilde{\boldsymbol{\mu}}_{K_i}^T \, \mathbb{N} \, \mathbf{u}_{K_i} = \widetilde{\boldsymbol{\mu}}^T \, \mathbb{E}_{\mathbb{N}} \, \mathbf{u} \ ,$$

where $\mathbb{N}$ is the discrete trace operator. It is a sparse matrix that consists of 1, -1 and 0 only. For construction of $\mathbb{N}$ please refer to [5]. $\mathbb{E}_{\mathbb{N}}$ is the assembled $\mathbb{N}$ for all elements. For, $K = 2 \times 2$, $N = 2$, $\mathbb{E}_{\mathbb{N}}$ is shown in (6). The matrix size of $\mathbb{E}_{\mathbb{N}}$ is 8×64, but it has only 16 non-zero entities. It is an extremely sparse matrix that is metric free and the location of $\pm$ valued entries depend only on the connection between different elements.

$$\mathbb{E}_{\mathbb{N}} = \left[\begin{array}{*{64}{c}}
\cdot & \cdot & \cdot & \cdot & 1 & \cdot & \cdot & \cdot & \cdot & \cdot & \cdot & \cdot & \cdot & \cdot & \cdot & \cdot & -1 & \cdot & \cdot & \cdot & \cdot & \cdot & \cdot & \cdot & \cdot & \cdot & \cdot & \cdot & \cdot & \cdot & \cdot & \cdot & \cdot & \cdot & \cdot & \cdot & \cdot & \cdot & \cdot & \cdot & \cdot & \cdot & \cdot & \cdot & \cdot & \cdot & \cdot & \cdot & \cdot & \cdot & \cdot & \cdot & \cdot & \cdot & \cdot & \cdot & \cdot & \cdot & \cdot & \cdot & \cdot & \cdot & \cdot & \cdot \\
\cdot & \cdot & \cdot & \cdot & \cdot & 1 & \cdot & \cdot & \cdot & \cdot & \cdot & \cdot & \cdot & \cdot & \cdot & \cdot & \cdot & -1 & \cdot & \cdot & \cdot & \cdot & \cdot & \cdot & \cdot & \cdot & \cdot & \cdot & \cdot & \cdot & \cdot & \cdot & \cdot & \cdot & \cdot & \cdot & \cdot & \cdot & \cdot & \cdot & \cdot & \cdot & \cdot & \cdot & \cdot & \cdot & \cdot & \cdot & \cdot & \cdot & \cdot & \cdot & \cdot & \cdot & \cdot & \cdot & \cdot & \cdot & \cdot & \cdot & \cdot & \cdot & \cdot & \cdot \\
\cdot & \cdot & \cdot & \cdot & \cdot & \cdot & \cdot & \cdot & \cdot & \cdot & \cdot & \cdot & \cdot & \cdot & \cdot & \cdot & \cdot & \cdot & \cdot & \cdot & \cdot & \cdot & \cdot & \cdot & \cdot & \cdot & \cdot & \cdot & \cdot & \cdot & \cdot & \cdot & \cdot & \cdot & \cdot & \cdot & 1 & \cdot & \cdot & \cdot & \cdot & \cdot & \cdot & \cdot & \cdot & \cdot & \cdot & \cdot & -1 & \cdot & \cdot & \cdot & \cdot & \cdot & \cdot & \cdot & \cdot & \cdot & \cdot & \cdot & \cdot & \cdot & \cdot & \cdot \\
\cdot & \cdot & \cdot & \cdot & \cdot & \cdot & \cdot & \cdot & \cdot & \cdot & \cdot & \cdot & \cdot & \cdot & \cdot & \cdot & \cdot & \cdot & \cdot & \cdot & \cdot & \cdot & \cdot & \cdot & \cdot & \cdot & \cdot & \cdot & \cdot & \cdot & \cdot & \cdot & \cdot & \cdot & \cdot & \cdot & \cdot & 1 & \cdot & \cdot & \cdot & \cdot & \cdot & \cdot & \cdot & \cdot & \cdot & \cdot & \cdot & -1 & \cdot & \cdot & \cdot & \cdot & \cdot & \cdot & \cdot & \cdot & \cdot & \cdot & \cdot & \cdot & \cdot & \cdot \\
\cdot & \cdot & \cdot & \cdot & \cdot & \cdot & \cdot & \cdot & \cdot & \cdot & 1 & \cdot & \cdot & \cdot & \cdot & \cdot & \cdot & \cdot & \cdot & \cdot & \cdot & \cdot & \cdot & \cdot & \cdot & \cdot & \cdot & \cdot & \cdot & \cdot & \cdot & \cdot & \cdot & \cdot & \cdot & \cdot & \cdot & \cdot & -1 & \cdot & \cdot & \cdot & \cdot & \cdot & \cdot & \cdot & \cdot & \cdot & \cdot & \cdot & \cdot & \cdot & \cdot & \cdot & \cdot & \cdot & \cdot & \cdot & \cdot & \cdot & \cdot & \cdot & \cdot & \cdot \\
\cdot & \cdot & \cdot & \cdot & \cdot & \cdot & \cdot & \cdot & \cdot & \cdot & \cdot & 1 & \cdot & \cdot & \cdot & \cdot & \cdot & \cdot & \cdot & \cdot & \cdot & \cdot & \cdot & \cdot & \cdot & \cdot & \cdot & \cdot & \cdot & \cdot & \cdot & \cdot & \cdot & \cdot & \cdot & \cdot & \cdot & \cdot & \cdot & -1 & \cdot & \cdot & \cdot & \cdot & \cdot & \cdot & \cdot & \cdot & \cdot & \cdot & \cdot & \cdot & \cdot & \cdot & \cdot & \cdot & \cdot & \cdot & \cdot & \cdot & \cdot & \cdot & \cdot & \cdot \\
\cdot & \cdot & \cdot & \cdot & \cdot & \cdot & \cdot & \cdot & \cdot & \cdot & \cdot & \cdot & \cdot & \cdot & \cdot & \cdot & \cdot & \cdot & \cdot & \cdot & \cdot & \cdot & \cdot & \cdot & \cdot & \cdot & 1 & \cdot & \cdot & \cdot & \cdot & \cdot & \cdot & \cdot & \cdot & \cdot & \cdot & \cdot & \cdot & \cdot & \cdot & \cdot & \cdot & \cdot & \cdot & \cdot & \cdot & \cdot & \cdot & \cdot & \cdot & \cdot & \cdot & \cdot & -1 & \cdot & \cdot & \cdot & \cdot & \cdot & \cdot & \cdot & \cdot & \cdot \\
\cdot & \cdot & \cdot & \cdot & \cdot & \cdot & \cdot & \cdot & \cdot & \cdot & \cdot & \cdot & \cdot & \cdot & \cdot & \cdot & \cdot & \cdot & \cdot & \cdot & \cdot & \cdot & \cdot & \cdot & \cdot & \cdot & \cdot & 1 & \cdot & \cdot & \cdot & \cdot & \cdot & \cdot & \cdot & \cdot & \cdot & \cdot & \cdot & \cdot & \cdot & \cdot & \cdot & \cdot & \cdot & \cdot & \cdot & \cdot & \cdot & \cdot & \cdot & \cdot & \cdot & \cdot & \cdot & -1 & \cdot & \cdot & \cdot & \cdot & \cdot & \cdot & \cdot & \cdot
\end{array} \right] . \tag{6}$$

Discrete Formulation

Using the weighted inner product and duality pairings discussed in Sect. 4, we can write the discrete form of weak formulation in (2) as,

$$\begin{bmatrix} \mathbb{B} & {\mathbb{E}_{\mathbb{N}}}^T \\ \mathbb{E}_{\mathbb{N}} & 0 \end{bmatrix} \begin{bmatrix} \mathbf{X} \\ \boldsymbol{\lambda} \end{bmatrix} = \begin{bmatrix} \mathbf{F} \\ 0 \end{bmatrix} , \tag{7}$$

where, $\mathbb{B}$ is an invertible block diagonal matrix given by,

$$\mathbb{B} = \begin{bmatrix} \mathbb{M}^{(1)}_{\mathbb{A}^{-1},K_1} & \mathbb{E}^{2,1\,T} & & & & & & \\ \mathbb{E}^{2,1} & 0 & & & & & & \\ & & \mathbb{M}^{(1)}_{\mathbb{A}^{-1},K_2} & \mathbb{E}^{2,1\,T} & & & & \\ & & \mathbb{E}^{2,1} & 0 & & & & \\ & & & & \ddots & \ddots & & \\ & & & & \ddots & \ddots & & \\ & & & & & & \mathbb{M}^{(1)}_{\mathbb{A}^{-1},K_K} & \mathbb{E}^{2,1\,T} \\ & & & & & & \mathbb{E}^{2,1} & 0 \end{bmatrix}, \tag{8}$$

$\mathbb{E}_{\mathbb{N}}$ is as given in (6), $\mathbf{X} = \sum_i \begin{bmatrix} \mathbf{u} \\ \mathbf{p} \end{bmatrix}_{K_i}$, and $\mathbf{F} = \sum_i \begin{bmatrix} \hat{\mathbf{p}} \\ \mathbf{f} \end{bmatrix}_{K_i}$, where $\mathbf{f}$ are the expansion coefficients of $f^h\left(x, y\right) = \sum_{i,j}^{N} \mathsf{f}_{ij}\, e_i\left(x\right) e_j\left(y\right)$.

In (8), the mass matrix $\mathbb{M}^{(1)}_{\mathbb{A}^{-1},K_i}$ is the only dense matrix and also the only matrix that changes with each local element, K_i. $\mathbb{E}_{\mathbb{N}}$ is a sparse incidence matrix for the global system and $\mathbb{E}^{2,1}$ is a sparse incidence matrix for the local systems that remains the same for each element.

Using the Schur complement method, the global system (7) can be reduced to solve for $\boldsymbol{\lambda}$, [1],

$$\boldsymbol{\lambda} = \left(\mathbb{E}_{\mathbb{N}}\mathbb{B}^{-1}\,\mathbb{E}_{\mathbb{N}}{}^{T}\right)^{-1} \cdot \left(\mathbb{E}_{\mathbb{N}}\mathbb{B}^{-1}\mathbf{F}\right). \tag{9}$$

To evaluate $\boldsymbol{\lambda}$ in (9) we need $\mathbb{B}^{-1}$ that can be calculated efficiently by taking inverse of each block of $\mathbb{B}$ separately. This part is trivially parallelized. Once the $\boldsymbol{\lambda}$ is determined the solution in each element, K_i, can be evaluated independent of each other.

The system (9) solves for interface degrees of freedom between the elements and will always be smaller than the full global system. For a comparison of the size of λ system with full system see Table (for 2D), and 3D).
In the Tables we see that, for constant K, increasing the order of polynomial basis the growth in size of λ system is less than the growth in size of full system. Thus, hybrid formulations are beneficial for high order methods where local degrees of freedom of an element are much higher than interface degrees of freedom.

Table For 2D

N	Full system	λ only	λ/full
5	825	60	0.07
10	3000	120	0.04
15	6525	180	0.03
20	11,400	240	0.02
25	17,625	300	0.02

K	Full system	λ only	λ/full
400	15,480	2280	0.15
1600	62,160	9360	0.15
3600	140,040	21,240	0.15
6400	249,120	37,920	0.15
10,000	389,400	59,400	0.15

Left: Number of total unknowns as a function of N, for $K = 3 \times 3$. Right: Number of total unknowns as a function of the number of elements K, for $N = 3$

Table For 3D

N	Full system	λ only	λ/full
5	16,875	1350	0.08
10	121,500	5400	0.04
15	394,875	12,150	0.03
20	918,000	21,600	0.02
25	1,771,875	33,750	0.02

K	Full system	λ only	λ/full
8000	1,285,200	205,200	0.16
64,000	10,324,800	1,684,800	0.16
216,000	34,894,800	5,734,800	0.16
512,000	82,771,200	13,651,200	0.16
1,000,000	161,730,000	26,730,000	0.17

Left: Number of total unknowns as a function of N, for $K = 3 \times 3 \times 3$. Right: Number of total unknowns as a function of the number of elements K, for $N = 3$

On the right of the Tables we see that, for constant N, the λ system is smaller than the full system, although the growth ratio of the size of λ and full systems do not change significantly.

Results

In this section we present the results for a test problem from [7] by solving system (7). The domain of the test problem is, $\Omega \in [0, 1]^2$. The RHS term is defined as,

$$f_{ex} = \nabla \cdot (-\mathbb{A}\nabla p_{ex}) \,, \quad \text{where} \,,$$

$$\mathbb{A} = \frac{1}{x^2+y^2+\alpha} \begin{pmatrix} 10^{-3}x^2 + y^2 + \alpha & \left(10^{-3} - 1\right) xy \\ \left(10^{-3} - 1\right) xy & x^2 + 10^{-3}y^2 + \alpha \end{pmatrix} ; \qquad \alpha = 0.1 \,,$$

$$p_{ex} = \sin(2\pi x) \sin\left(2\pi y\right)$$

and Dirichlet boundary conditions are imposed along the entire boundary, $\Gamma_{\mathrm{D}} = \partial\Omega$ and $\Gamma_{\mathrm{N}} = \emptyset$. We solve this problem on an orthogonal and a curved mesh, see Fig.

The same problem was earlier addressed in [6], but for a method with continuous elements and *primal* basis functions only. For the configuration $K = 3 \times 3$, $N = 6$, we compare the sparsity structure of the two approaches in Fig.. On the left we see

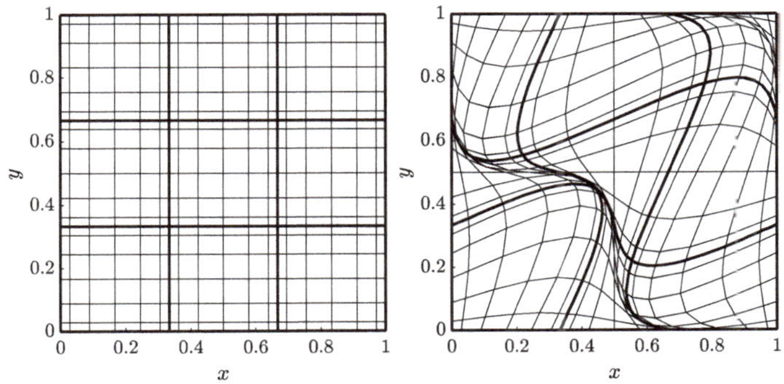

Fig. Mesh configuration: $K = 3 \times 3$, $N = 6$. Left: orthogonal. Right: curved

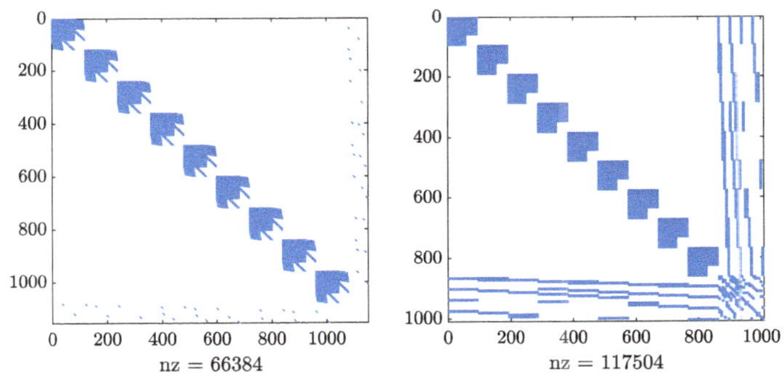

Fig. Sparsity plots $K = 3 \times 3$, $N = 6$. Left: hybrid elements method. Right: continuous element method

the hybrid formulation, and on the right we see the continuous elements formulation [6]. The number of non zero entries are almost half in the hybrid formulation, 66,384, as compared to the continuous element formulation, 117,504. Here, the sparsity is due to use of algebraic dual degrees of freedom and is not because of hybridization of the scheme.

In Fig., on the left we compare the growth in condition number, for the $\boldsymbol{\lambda}$ system (9) with full continuous element system, for $N = 7$ on the curved mesh, with increasing number of elements, K. We observe similar growth rates for hybrid and continuous formulation, however the condition number for continuous elements formulation is almost $O\left(10^2\right)$ higher. On the right we see the growth in condition number with increasing polynomial degree for $K = 9 \times 9$ on the curved mesh. A reduced growth rate in condition number for hybrid formulation is observed. Thus hybrid formulations are beneficial for high order methods.

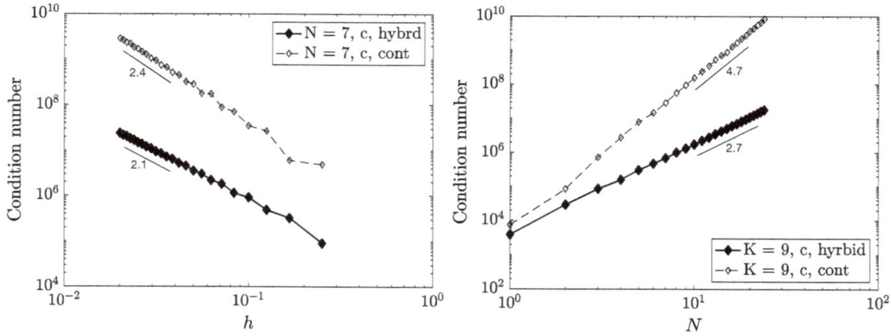

Fig. Growth in condition number for hybrid elements in dark line, and continuous elements in dotted line. Left: h-refinement; Right: N-refinement. 'c' refers to the curved mesh

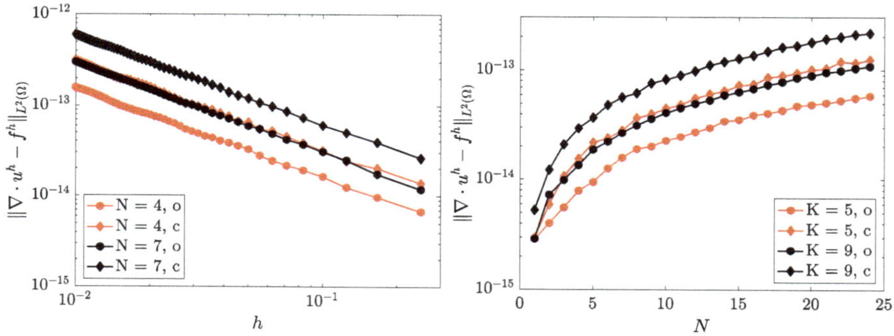

Fig. L^2-error in divergence of velocity: Left: h-refinement; Right: N-refinement. 'o' refers to the orthogonal mesh and 'c' to the curved mesh

In Fig. we show the L^2-error for $\|\nabla \cdot \boldsymbol{u}^h - f^h\|$. On the left side as a function of element size, $h = 1/\sqrt{K}$, and on the right side as a function of polynomial degree of the basis functions. In both cases the maximum error observed is of $O\left(10^{-12}\right)$.

In Figure, on the top two figures we show the error in the $H\left(\text{div};\, \Omega\right)$ norm for the velocity; and at the bottom two figures we show the error in $L^2\,(\Omega)$ norm for the pressure. On the left we have h-convergence plots, and on the right we have N-convergence plots. In all the figures, for the same number of elements, K, and polynomial degree, N, the error is higher for the curved mesh.

On the left we see that the error decreases with the element size. The slope of error rate of convergence is N, which is optimal for both curved and orthogonal meshes. On the right we see exponential convergence of the error with increasing polynomial degree of basis for both orthogonal and curved meshes.

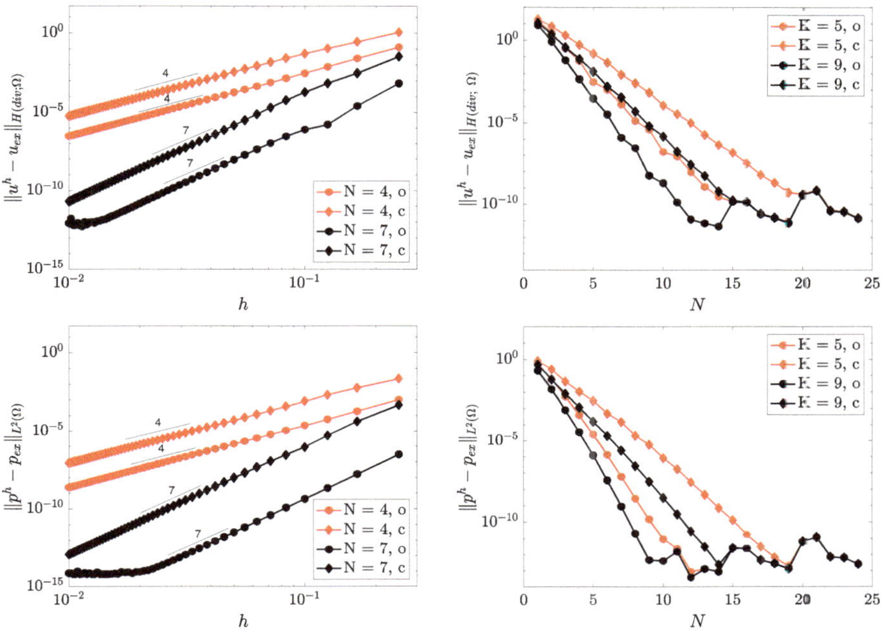

Fig. Top row: error in $H(\text{div};\Omega)$ norm for velocity; Bottom row: L^2-error in pressure. Left: h-refinement; Right: N-refinement. 'o' refers to the orthogonal mesh and 'c' to the curved mesh

References

1. Boffi, D., Brezzi, F., Fortin, M.: Mixed Finite Elements Methods and Applications. Springer Series in Computational Mechanics. Springer, Berlin (2010)
2. Carstensen, C., Demkowicz, L., Gopalakrishnan, J.: Breaking spaces and forms for the DPG method and applications including Maxwell equations. Comput. Math. Appl. **72**, 494–522 (2016)
3. Cockburn, B.: Static Condensation, Hybridization, and the Devising of the HDG Methods. Lecture Notes in Computational Science and Engineering, vol. 114. Springer, Berlin (2015)
4. Gerritsma, M.: Edge functions for spectral element methods. In: Spectral and High Order Methods for Partial Differential Equations, pp. 199–208. Springer, Berlin (2011)
5. Gerritsma, M., Jain, V., Zhang, Y., Palha, A.: Algebraic dual polynomials for the equivalence of curl-curl problems (2018). arXiv:1805.00114
6. Gerritsma, M., Palha, A., Jain, V., Zhang, Y.: Mimetic spectral element method for anisotropic diffusion. In: Numerical Methods for PDEs, pp. 31–74. Springer, Berlin (2018)
7. Herbin, R., Hubert, F.: Benchmark on Discretization Schemes for Anisotropic Diffusion Problems on General Grids. ISTE, Finite Volumes for Complex Applications V, pp. 659–692. Wiley, London (2008)
8. V. Jain, Y. Zhang, A. Palha, M. Gerritsma, Construction and application of algebraic dual polynomial representations for finite element methods (2017). arXiv:1712.09472

9. Zhang, Y., Jain, V., Palha, A., Gerritsma, M.: Discrete equivalence of adjoint Neumann-Dirichlet div-grad and grad-div equations in curvilinear 3D domains. In: Spectral and High Order Methods for Partial Differential Equations ICOSAHOM 2018. Springer, Cham (2020). https://doi.org/10.1007/978-3-030-39647-3_3
10. Zhang, Y., Jain, V., Palha, A., Gerritsma, M.: The discrete Steklov-Poincaré operator using algebraic dual polynomials. Comput. Methods Appl. Math. **19**(3), 645–661. https://doi.org/10.1515/cmam-2018-0208

15

Stability of Wall Boundary Condition Procedures for Discontinuous Galerkin Spectral Element Approximations of the Compressible Euler Equations

Florian J. Hindenlang, Gregor J. Gassner, and David A. Kopriva

Introduction

The ingredients for a reliable numerical method for the approximation of partial differential equations, e.g. one that will not blow up, include stable inter-element and physical boundary condition implementations. The recognition that the discontinuous Galerkin spectral element method (DGSEM) with Gauss-Lobatto quadratures satisfies a summation-by-parts (SBP) operators [4, 7] has allowed for the analysis of these schemes and to connect them with penalty collocation and SBP finite difference schemes. For instance, in [5], we showed that a split form approximation of the compressible Navier–Stokes equations was both linearly and entropy stable provided that the boundary conditions were properly imposed.

The importance of stable boundary condition procedures for hyperbolic equations has long been studied, especially in relation to finite difference methods, e.g. [3, 9, 10]. Only recently have they been studied for discontinuous Galerkin approximations. In [12], the authors showed that the reflection approach is stable when using an entropy conserving flux and an additional entropy stable dissipation

F. J. Hindenlang
Max Planck Institute for Plasma Physics, Garching, Germany
e-mail: florian.hindenlang@ipp.mpg.de

G. J. Gassner (✉)
Department for Mathematics and Computer Science, Center for Data and Simulation Science, University of Cologne, Cologne, Germany
e-mail: ggassner@math.uni-koeln.de

D. A. Kopriva
Florida State University, Tallahassee, FL, USA

San Diego State University, San Diego, CA, USA
e-mail: kopriva@math.fsu.edu

term (EC-ES). In [2], the authors show that the reflection condition is stable if the numerical flux is either the Godunov or HLL flux.

In this paper, we analyze both the linear and entropy stability of two types of commonly used wall boundary condition procedures used with the DGSEM applied to the compressible Euler equations. In both cases, wall boundary conditions are implemented through a numerical flux. The boundary condition might be implemented through a special wall numerical flux that includes the boundary condition, or a fictitious external state applied to a Riemann solver approximation. We show how to construct special wall numerical fluxes that are stable, and study the behavior of the approximations. In particular, we show that the use of Riemann solvers at the boundaries introduce numerical dissipation in an amount that depends on the size of the normal Mach number at the wall.

The Compressible Euler Equations and the Wall Boundary Condition

We write the Euler equations as

$$\mathbf{u}_t + \sum_{i=1}^{3} \frac{\partial \mathbf{f}_i}{\partial x_i} = 0. \tag{1}$$

The state vector contains the conservative variables

$$\mathbf{u} = \left[\varrho \ \ \varrho\vec{v} \ \ E\right]^T = \left[\varrho \ \ \varrho v_1 \ \ \varrho v_2 \ \ \varrho v_3 \ \ E\right]^T. \tag{2}$$

In standard form, the components of the advective fluxes are

$$\mathbf{f}_1 = \begin{bmatrix} \varrho v_1 \\ \varrho v_1^2 + p \\ \varrho v_1 v_2 \\ \varrho v_1 v_3 \\ (E+p)v_1 \end{bmatrix} \quad \mathbf{f}_2 = \begin{bmatrix} \varrho v_2 \\ \varrho v_2 v_1 \\ \varrho v_2^2 + p \\ \varrho v_2 v_3 \\ (E+p)v_2 \end{bmatrix} \quad \mathbf{f}_3 = \begin{bmatrix} \varrho v_3 \\ \varrho v_3 v_1 \\ \varrho v_3 v_2 \\ \varrho v_3^2 + p \\ (E+p)v_3 \end{bmatrix}, \tag{3}$$

Here, ϱ, $\vec{v} = (v_1, v_2, v_3)^T$, p, E are the mass density, fluid velocities, pressure and total energy. We close the system with the ideal gas assumption, which relates the total energy and pressure

$$p = (\gamma - 1)\left(E - \frac{1}{2}\varrho \left\|\vec{v}\right\|^2\right),$$

where γ denotes the adiabatic coefficient. For a compact notation that simplifies the analysis, we define *block vectors* (with the double arrow)

$$\overleftrightarrow{\mathbf{f}} = \left[\mathbf{f}_1 \ \ \mathbf{f}_2 \ \ \mathbf{f}_3\right]^T , \tag{5}$$

so that the system of equations can be written in the compact form

$$\mathbf{u}_t + \vec{\nabla}_x \cdot \overleftrightarrow{\mathbf{f}} = 0. \tag{6}$$

The linear Euler equations are derived by linearizing about a constant mean state $(\bar{\varrho}, \bar{v}_1, \bar{v}_2, \bar{v}_3, \bar{p})$. We follow [11] for the symmetrization of the linearized equations, with the constants

$$a = \sqrt{\frac{\gamma - 1}{\gamma}}\bar{c}, \quad b = \frac{\bar{c}}{\sqrt{\gamma}}, \bar{c} = \sqrt{\frac{\gamma \bar{p}}{\bar{\varrho}}}, \tag{7}$$

where $\bar{c}$ is the sound speed of the constant mean state. The state variables become

$$\mathbf{u} = \left[\varrho' \ \ v_1 \ \ v_2 \ \ v_3 \ \ p'\right]^T , \tag{8}$$

where $\vec{v}$ is the velocity perturbation from the mean state, and we introduce

$$\varrho' = b\frac{\tilde{\varrho}}{\bar{\varrho}}, \quad p' = \frac{1}{\bar{\varrho}a}\tilde{p} - \frac{1}{\sqrt{\gamma - 1}}\varrho', \tag{9}$$

which depend on the density and pressure perturbations $\tilde{\varrho}$, $\tilde{p}$. The flux vectors are

$$\mathbf{f}_i = \underline{A}_i\mathbf{u}, \quad \overleftrightarrow{\mathbf{f}} = \vec{\underline{A}}\mathbf{u} = \left(\underline{A}_1\hat{x} + \underline{A}_2\hat{y} + \underline{A}_3\hat{z}\right)\mathbf{u}, \tag{10}$$

where [11]

$$\underline{A}_1 = \begin{bmatrix} \bar{v}_1 & b & 0 & 0 & 0 \\ b & \bar{v}_1 & 0 & 0 & a \\ 0 & 0 & \bar{v}_1 & 0 & 0 \\ 0 & 0 & 0 & \bar{v}_1 & 0 \\ 0 & a & 0 & 0 & \bar{v}_1 \end{bmatrix}, \quad \underline{A}_2 = \begin{bmatrix} \bar{v}_2 & 0 & b & 0 & 0 \\ 0 & \bar{v}_2 & 0 & 0 & 0 \\ b & 0 & \bar{v}_2 & 0 & a \\ 0 & 0 & 0 & \bar{v}_2 & 0 \\ 0 & 0 & a & 0 & \bar{v}_2 \end{bmatrix}, \quad \underline{A}_3 = \begin{bmatrix} \bar{v}_3 & 0 & 0 & b & 0 \\ 0 & \bar{v}_3 & 0 & 0 & 0 \\ 0 & 0 & \bar{v}_3 & 0 & 0 \\ b & 0 & 0 & \bar{v}_3 & a \\ 0 & 0 & 0 & a & \bar{v}_3 \end{bmatrix} \tag{11}$$

are constant symmetric matrices.

The linear equations have the property that the L^2 norm of the solution over a domain Ω is bounded by terms of the boundary data on $\partial\Omega$, only. Let

$$\langle \mathbf{v}, \mathbf{w} \rangle = \int\limits_{\Omega} \mathbf{v}^T \mathbf{w}\, dxdydz, \quad \left\langle \overleftrightarrow{\mathbf{f}}, \overleftrightarrow{\mathbf{g}} \right\rangle = \int\limits_{\Omega} \sum_{i=1}^{3} \mathbf{f}_i^T \mathbf{g}_i\, dxdydz. \tag{12}$$

represent the L^2 inner product of two state vectors $\mathbf{v}$ and $\mathbf{w}$ and two block vectors $\overleftrightarrow{\mathbf{f}}$ and $\overleftrightarrow{\mathbf{g}}$, respectively. Since the coefficient matrices are constant the product rule and symmetry of $\vec{\underline{\mathrm{A}}}$ implies

$$\left\langle \vec{\nabla}_x \cdot \overleftrightarrow{\mathbf{f}}, \mathbf{u} \right\rangle = \left\langle \vec{\nabla}_x \cdot \left(\vec{\underline{\mathrm{A}}} \mathbf{u} \right), \mathbf{u} \right\rangle = \left\langle \nabla_x \mathbf{u}, \overleftrightarrow{\mathbf{f}} \right\rangle. \tag{13}$$

Then it follows from Gauss' law (integration by parts) that

$$\left\langle \vec{\nabla}_x \cdot \overleftrightarrow{\mathbf{f}}, \mathbf{u} \right\rangle = \frac{1}{2} \int_{\partial\Omega} \mathbf{u}^T \overleftrightarrow{\mathbf{f}} \cdot \vec{n}\, dS, \tag{14}$$

where $\vec{n}$ is the outward normal to the surface of Ω. The norm of the solution therefore satisfies

$$\frac{d}{dt} ||\mathbf{u}||^2 = - \int_{\partial\Omega} \mathbf{u}^T \overleftrightarrow{\mathbf{f}} \cdot \vec{n}\, dS. \tag{15}$$

Replacing the boundary terms by boundary conditions leads to a bound on the solution in terms of the boundary data. The argument of the boundary integral on the right of (15) is

$$\mathbf{u}^T \overleftrightarrow{\mathbf{f}} \cdot \vec{n} = \mathbf{u}^T \left(\vec{\underline{\mathrm{A}}} \cdot \vec{n} \right) \mathbf{u} = 2 \left(\varrho b + ap \right) v_n + (\vec{\bar{v}} \cdot \vec{n})(\varrho^2 + |\vec{v}|^2 + p^2), \tag{16}$$

where v_n is the wall normal velocity, $v_n = \vec{v} \cdot \vec{n}$. Note that here, the mean flow must be chosen such that the normal flow vanishes at the wall boundary $\vec{\bar{v}} \cdot \vec{n} = 0$, so that the boundary condition makes physical sense.

Therefore, with the no penetration wall condition $v_n = 0$ applied,

$$\frac{d}{dt} ||\mathbf{u}||^2 = 0, \tag{17}$$

and the (energy) norm of the solution is bounded for all time by its initial value.

The nonlinear equations, on the other hand, satisfy a bound on the entropy that depends only on the boundary data. For what follows, we assume that the solution is smooth so that we don't have to consider entropy generated at shock waves. We introduce the entropy density (scaled with $(\gamma - 1)$ for convenience) as

$$s(\mathbf{u}) = -\frac{\varrho\varsigma}{(\gamma - 1)}, \tag{18}$$

where $\varsigma = \ln(p) - \gamma \ln(\varrho)$ is the physical entropy. (The minus sign is conventional in the theory of hyperbolic conservation laws to ensure a decreasing entropy function.) The entropy flux for the Euler equations is

$$\vec{f}^{\,\varsigma}(\mathbf{u}) = \vec{v}\,s = -\frac{\varrho\varsigma\vec{v}}{(\gamma - 1)}. \tag{19}$$

Finally the entropy variables are

$$\mathbf{w} = \frac{\partial s(\mathbf{u})}{\partial \mathbf{u}} = \begin{bmatrix} \frac{\gamma-\varsigma}{\gamma-1} - \beta||\vec{v}||^2, \\ 2\beta\vec{v} \\ -2\beta \end{bmatrix}, \quad \beta = \frac{\varrho}{2p}. \tag{20}$$

The entropy pair contracts the solution and fluxes, meaning that it satisfies the relations

$$\mathbf{w}^T\,\mathbf{u}_t = \left(\frac{\partial s}{\partial \mathbf{u}}\right)^T \mathbf{u}_t = s_t(\mathbf{u}), \quad \mathbf{w}^T\,\vec{\nabla}_x \cdot \overleftrightarrow{\mathbf{f}} = \vec{\nabla}_x \cdot \vec{f}^{\,\varsigma}. \tag{21}$$

When we multiply (6) with the entropy variables and integrate over the domain,

$$\left\langle \mathbf{w}(\mathbf{u}), \mathbf{u}_t \right\rangle + \left\langle \mathbf{w}(\mathbf{u}), \vec{\nabla}_x \cdot \overleftrightarrow{\mathbf{f}} \right\rangle = 0. \tag{22}$$

Next we use the properties of the entropy pair to contract (22) and use integration by parts to get

$$\left\langle s_t(\mathbf{u}), 1 \right\rangle = -\left\langle \vec{\nabla}_x \cdot \vec{f}^{\,\varsigma}, 1 \right\rangle = -\int\limits_{\partial\Omega} \left(\vec{f}^{\,\varsigma} \cdot \vec{n}\right) \mathrm{d}S \tag{23}$$

showing that, in the continuous case, the total entropy in the domain can only change via the boundary conditions.

In the case of a zero-mass flux boundary condition, with $v_n = \vec{v} \cdot \vec{n} = 0$, the entropy is not changed by the slip-wall boundary condition, since

$$-\vec{f}^{\,\varsigma} \cdot \vec{n} = \frac{\varrho\varsigma}{(\gamma - 1)} v_n = 0. \tag{24}$$

Stability Bounds for the DGSEM

The DGSEM is described in detail in [5] and elsewhere [1, 6]. We will only quickly summarize the approximation here. The domain, Ω is subdivided into non-overlapping, conforming, hexahedral elements. Each element is mapped to the reference element $E = [-1, 1]^3$. Associated with the transformation from the reference element is a set of contravariant coordinate vectors, $\vec{a}^i$, and transformation Jacobian, $\mathcal{J}$. Equation (6) transform to another conservation law on the reference element as

$$\mathcal{J}\mathbf{u}_t + \vec{\nabla}_\xi \cdot \overset{\leftrightarrow}{\tilde{\mathbf{f}}} = 0, \tag{25}$$

where $\overset{\leftrightarrow}{\tilde{\mathbf{f}}}$ is the contravariant flux vector with components $\tilde{\mathbf{f}}^i = \mathcal{J}\vec{a}^i \cdot \overset{\leftrightarrow}{\mathbf{f}}$.

The approximation of (25) proceeds as follows: A weak form is created by taking the inner product of the equation with a test function. The Gauss law is applied to the divergence term to separate the boundary from the interior contributions. The resulting weak form is then approximated: The solution vector is approximated by a polynomial of degree N interpolated at the Legendre–Gauss–Lobatto points. In the following, we will represent the true continuous solutions by lower case letter. Upper case letters will denote their polynomial approximations, except for the density, where the approximation is denoted by ρ. The volume fluxes are replaced by two-point numerical fluxes. In the linear case, the two point fluxes are immediately relatable to a split form of the equations. Integrals are replaced by Legendre–Gauss–Lobatto quadratures. Finally, the boundary fluxes are replaced by a numerical flux. See [5] and [8] for details.

The result is an approximation that is energy stable for the linearized equations if at every quadrature point along a physical boundary the numerical flux $\tilde{\mathbf{F}}^*$ satisfies the bound [5]

$$\mathbf{U}^T \left\{ \tilde{\mathbf{F}}^* - \frac{1}{2} \overset{\leftrightarrow}{\tilde{\mathbf{F}}} \cdot \hat{n} \right\} \geq 0, \tag{26}$$

where $\overset{\leftrightarrow}{\tilde{\mathbf{F}}}$ is the polynomial interpolation of the contravariant flux from the interior, $\hat{n}$ is the reference space outward normal direction, and $\mathbf{U}$ is the approximation of the state vector. Since the contravariant fluxes are proportional to the normal fluxes [6], we can change the condition (26) to

$$B_L \equiv \mathbf{U}^T \left\{ \mathbf{F}^* - \frac{1}{2} \overset{\leftrightarrow}{\mathbf{F}} \cdot \vec{n} \right\} \geq 0, \tag{27}$$

For entropy stability of the nonlinear equations, the boundary stability condition shown in [5] is proportional to

$$B_{NL} \equiv \mathbf{W}^T \left(\mathbf{F}^* - \left(\overleftrightarrow{\mathbf{F}} \cdot \vec{n} \right) \right) + \left(\vec{F}^{\varsigma} \cdot \vec{n} \right) \geq 0, \tag{28}$$

where $\vec{F}^{\varsigma}$ is the polynomial interpolation of the entropy flux, $\vec{f}^{\varsigma}$, and $\mathbf{W}$ is the interpolation of the entropy variables.

Linear Stability of Wall Boundary Condition Approximations

To find linearly stable implementations of the wall condition $v_n = 0$, one needs only find a numerical flux that satisfies it and the condition (27). For the linear equations, the approximation of the state vector is $\mathbf{U} = [\rho' \;\; \vec{V} \;\; P']^T$ and the normal contravariant flux is proportional to

$$\overleftrightarrow{\mathbf{F}} \cdot \vec{n} = \underline{\vec{\mathrm{A}}} \cdot \vec{n}\, \mathbf{U} = \left[bV_n \;\; n_1 Q \;\; n_2 Q \;\; n_3 Q \;\; aV_n \right]^T, \tag{29}$$

where V_n is the approximation of the normal velocity at the wall computed from the interior, $Q = b\rho' + aP'$, and (n_1, n_1, n_3) are the three components of the physical space normal vector, $\vec{n}$. The numerical flux can be expressed as

$$\mathbf{F}^* = \underline{\vec{\mathrm{A}}} \cdot \vec{n}\, \mathbf{U}^* = \left[bV_n^* \;\; n_1 Q^* \;\; n_2 Q^* \;\; n_3 Q^* \;\; aV_n^* \right]^T. \tag{30}$$

It then remains only to find Q^* so that (27) is satisfied when the normal wall condition $V_n^* = 0$ is applied. When we substitute the fluxes (29) and (30) into (27),

$$B_L = \frac{1}{2} \left\{ Q \left(2V_n^* - V_n \right) + V_n \left(2Q^* - Q \right) \right\} = \frac{1}{2} \left\{ 2QV_n^* + 2V_n \left(Q^* - Q \right) \right\} \tag{31}$$

Substituting the wall boundary condition $V_n^* = 0$ yields the condition on Q^* for stability

$$V_n \left(Q^* - Q \right) \geq 0. \tag{32}$$

Neutral stability is thus ensured if ρ^* and P^* are computed from the interior, i.e. $\rho^* = \rho'$, $P^* = P'$ so that $Q^* = Q$.

In practice, the boundary condition is also implemented through the use of a Riemann solver and external state designed to imply the physical boundary condition to construct the numerical boundary flux. The exact upwind ($\varepsilon = 1$) normal Riemann flux and the central flux ($\varepsilon = 0$) for the linear system of equations is

$$\mathbf{F}^*\left(\mathbf{U}, \mathbf{U}^{\text{ext}}\right) = \frac{1}{2}\left\{\overleftrightarrow{\mathbf{F}}\left(\mathbf{U}\right)\cdot\vec{n} + \overleftrightarrow{\mathbf{F}}\left(\mathbf{U}^{\text{ext}}\right)\cdot\vec{n}\right\} - \frac{\varepsilon}{2}\left|\underline{\mathrm{A}}_n\right|\left(\mathbf{U}^{\text{ext}} - \mathbf{U}\right), \tag{33}$$

where $\underline{\mathrm{A}}_n \equiv \vec{\underline{\mathrm{A}}}\cdot\vec{n}$ is the normal coefficient matrix. The external state is set by using the interior values of the density and pressure and the negative of the value of the normal velocity,

$$\mathbf{U}^{\text{ext}} = \begin{bmatrix} \rho' & \left(\vec{V} - 2V_n\vec{n}\right) & P' \end{bmatrix}^T. \tag{34}$$

For $\varepsilon = 0$, using the central (averaged) numerical flux, the interior flux contribution cancels and condition (27) reduces to

$$\begin{aligned} B_{L,0} = \frac{1}{2}\mathbf{U}^T\underline{\mathrm{A}}_n\mathbf{U}^{\text{ext}} &= \begin{bmatrix} \rho' & \vec{V} & P' \end{bmatrix}\begin{bmatrix} 0 & n_1b & n_2b & n_3b & 0 \\ n_1b & 0 & 0 & 0 & n_1a \\ n_2b & 0 & 0 & 0 & n_2a \\ n_3b & 0 & 0 & 0 & n_3a \\ 0 & n_1a & n_2a & n_3a & 0 \end{bmatrix}\begin{bmatrix} \rho' \\ \\ \vec{V} - 2V_n\vec{n} \\ \\ P' \end{bmatrix} \\ &= Q\left(-V\cdot\vec{n}\right) + \left(V\cdot\vec{n}\right)Q = 0, \end{aligned} \tag{35}$$

which is neutrally stable, having no additional stabilizing dissipation. We note again, that the mean state for the linearization is chosen such that the normal mean velocity components are zero, resulting in the zeros on the diagonal of $\underline{\mathrm{A}}_n$.

Substituting the exact upwind flux where $\varepsilon = 1$ into (27) and rearranging,

$$B_{L,1} = -\mathbf{U}^T\left|\underline{\mathrm{A}}_n^-\right|\mathbf{U}^{\text{ext}} + \frac{1}{2}\mathbf{U}^T\left|\underline{\mathrm{A}}_n\right|\mathbf{U}, \tag{36}$$

where $\underline{\mathrm{A}}_n^- = \frac{1}{2}\left(\underline{\mathrm{A}}_n - \left|\underline{\mathrm{A}}_n\right|\right)$ is negative semidefinite. The second term is non-negative, depends only on the interior state, and adds stabilizing dissipation. From the matrix absolute value, the dissipation term is

$$\mathbf{U}^T\left|\underline{\mathrm{A}}_n\right|\mathbf{U} = \frac{1}{\bar{c}}Q^2 + \bar{c}^3\mathrm{Ma}_n^2, \tag{37}$$

where $\mathrm{Ma}_n = V_n/\bar{c}$ is the normal Mach number. Stability depends, then, on the value of the first term, which is where the boundary conditions are incorporated through the external state $\mathbf{U}^{\text{ext}}$ written in (34). Then

$$\mathbf{U}^T \left|\underline{\mathbf{A}}_n^-\right| \mathbf{U}^{\text{ext}} = \frac{1}{2\bar{c}} Q^2 - \frac{\bar{c}^3}{2} \text{Ma}^2. \tag{38}$$

Therefore, using the upwind numerical flux, (36) becomes

$$B_{L,1} = \bar{c}^3 \text{Ma}_n^2 \geq 0, \tag{39}$$

as required. The amount of dissipation depends on how far the interior computed normal velocity deviates from zero.

The combination of the reflective state and local Lax-Friedrichs flux is also linearly stable. In that case the exact matrix absolute value is replaced by a diagonal matrix, $\left|\underline{\mathbf{A}}_n\right| \approx |\lambda|_{\max} \underline{\mathbf{I}}$. The jump term is added to the central (averaged) flux so

$$B_{L,LF} = -\frac{|\lambda|_{\max}}{2} \mathbf{U}^T \left(\mathbf{U}^{\text{ext}} - \mathbf{U}\right) = \bar{c}^2 |\lambda|_{\max} \text{Ma}_n^2 \geq 0 \tag{40}$$

Finally, a dissipative version of the direct numerical flux (30) can be formed by looking at the reflective state approach. For instance, the equivalent to using the Lax-Friedrichs flux is to choose $\rho^* = \rho'$ and

$$P^* = P' + \frac{\bar{c}^3}{a} |\lambda|_{\max} \text{Ma}_n. \tag{41}$$

Then $Q^* = Q + \bar{c}^3 |\lambda|_{\max} \text{Ma}_n$ and

$$V_n \left(Q^* - Q\right) = \bar{c}^2 |\lambda|_{\max} \text{Ma}_n^2 \geq 0. \tag{42}$$

A similar, though more complicated, modified P^* can be made to be equivalent to the exact upwind flux.

Entropy Stability of Wall Boundary Condition Approximations

As in the linear approximation, the wall boundary condition can be imposed for the nonlinear equations either by directly specifying the numerical flux or by computing it through a Riemann solver using a reflection external state that enforces the normal wall condition implicitly. Note that in this section, the discrete variables $(\rho, \vec{V}, P)$ describe the full nonlinear state.

For the nonlinear equations, we construct the numerical flux for a slip-wall as

$$\left(\overleftrightarrow{\mathbf{F}} \cdot \vec{n}\right)^* = \begin{bmatrix} 0 \\ P^* \vec{n} \\ 0 \end{bmatrix} \tag{43}$$

where we imposed $V_n = 0$ leading to a flux with no mass or energy transfer, and we introduce a wall pressure P^*, whose value will be chosen to ensure consistency and stability.

After some manipulations, the discrete entropy stability condition (28) becomes

$$\begin{aligned} &-\rho V_n \left(\frac{\gamma - \varsigma}{(\gamma - 1)} - \beta ||\vec{V}||^2\right) \\ +2\beta V_n \left(P^* - P - \rho ||\vec{V}||^2\right) &+ 2\beta V_n \left(\rho E + P\right) - \frac{\rho \varsigma V_n}{(\gamma - 1)} = \\ -\rho V_n \left(\frac{\gamma}{(\gamma - 1)} - \beta ||\vec{V}||^2\right) &+ 2\beta V_n \left(P^* + \rho E - \rho ||\vec{V}||^2\right) = \\ \frac{\rho V_n}{P} \left(-\frac{\gamma}{(\gamma - 1)} P + \frac{1}{2}\rho ||\vec{V}||^2 + P^* \right. &\left. + \frac{P}{(\gamma - 1)} - \frac{1}{2}\rho ||\vec{V}||^2\right) = \\ &\rho V_n \left(\frac{P^*}{P} - 1\right) \geq 0 \end{aligned} \tag{44}$$

Therefore if we choose $P^* = P$, to be the internal pressure, the boundary flux does not contribute to the total entropy, independent of the inner normal velocity V_n. A value of P^* that leads to a dissipative boundary condition can be found either through exact solution of the Riemann problem at the boundary, or through the use of an external state and an approximate Riemann solver.

Exact Solution of the Riemann Problem

In [14] a symmetric 1D Riemann problem is exactly solved following Toro [13], to get the wall pressure P^*, accounting for the fact that V_n never vanishes discretely and therefore the wall pressure should be different from the interior pressure. The exact solution of the 1D Riemann problem reads as

$$\left(\frac{P^*}{P}\right)_{\mathrm{RP}} = \begin{cases} 1 + \gamma \mathrm{Ma}_n \left(\frac{(\gamma+1)}{4}\mathrm{Ma}_n + \sqrt{\left(\frac{(\gamma+1)}{4}\mathrm{Ma}_n\right)^2 + 1}\right) > 1 & \text{for} \quad V_n > 0 \\ \left(1 + \frac{1}{2}(\gamma - 1)\mathrm{Ma}_n\right)^{\frac{2\gamma}{(\gamma-1)}} \leq 1 & \text{for} \quad V_n \leq 0 \end{cases} \tag{45}$$

with the normal Mach number, $\mathrm{Ma}_n = \frac{V_n}{c}$, and the sound speed $c = \sqrt{\gamma \frac{P}{\rho}}$.

As shown by Toro [13], the solution for the rarefaction has a limiting vacuum solution for $\mathrm{Ma}_n \leq -2(\gamma - 1)^{-1}$. We will restrict our analysis to normal Mach numbers yielding strictly positive pressure solutions only ($\mathrm{Ma}_n > -5$ for $\gamma = \frac{7}{5}$).

It is easy to see that using P^* from (45), the entropy inequality (44) is still satisfied for $|V_n| \neq 0$, and the added entropy scales with the discrete value of V_n at the boundary. Hence, for $h \rightarrow 0$, the discrete boundary condition converges to its physical counterpart, since $V_n \rightarrow 0$. The choice of P^* from (45) appears to stabilize under-resolved simulations, which can be now explained by the fact that the boundary flux always adds entropy for $|V_n| \neq 0$.

Using Approximate Riemann Solvers for the Boundary Flux

A well known strategy in finite volume methods is to mirror only the velocity of the internal state and solve an approximate Riemann problem to get the boundary flux, mostly just because of a simpler implementation, since an approximate Riemann solver is already available and used for the fluxes between the elements. For DG methods, see also, for example, [2] and [12] where reflection conditions are proved to be entropy stable.

The mirror state is set so that the mass and energy flux are zero. Let the inner state be labeled L and the outer R. then the inner and outer states that satisfy the mirror condition are

$$\mathbf{U}^L = \begin{bmatrix} \rho & \rho \vec{V}_n & E \end{bmatrix}^T, \quad \mathbf{U}^R = \begin{bmatrix} \rho & \rho(\vec{V} - 2V_n\vec{n}) & E \end{bmatrix}^T \tag{46}$$

We show below under what conditions on the normal velocity V_n that the reflection condition is entropy stable for the Lax-Friedrichs, HLL and HLLC, Roe and EC-ES fluxes.

Lax-Friedrichs Flux

We start with the simplest approximate Riemann solver, the Lax-Friedrichs or Rusanov flux, which reads as

$$\left(\overleftrightarrow{\mathbf{F}} \cdot \vec{n}\right)^*_{\mathrm{LF}} = \frac{1}{2}\vec{n} \cdot \left(\overleftrightarrow{\mathbf{F}}(\mathbf{U}^L) + \overleftrightarrow{\mathbf{F}}(\mathbf{U}^R)\right) - \frac{|\lambda|_{\max}}{2}(\mathbf{U}^R - \mathbf{U}^L). \tag{47}$$

Inserting the states from (46), we get

$$\left(\overleftrightarrow{\mathbf{F}} \cdot \vec{n}\right)^*_{\mathrm{LF}} = \begin{bmatrix} 0 \\ (\rho V_n^2 + P)\,\vec{n} \\ 0 \end{bmatrix} - \frac{\lambda_{\max}}{2}\begin{bmatrix} 0 \\ -2\rho V_n \vec{n} \\ 0 \end{bmatrix} = \begin{bmatrix} 0 \\ (\rho V_n^2 + \rho V_n \lambda_{\max} + P)\,\vec{n} \\ 0 \end{bmatrix}. \tag{48}$$

The maximum wave speed is normally approximated from the largest leftgoing and rightgoing wave speed,

$$\lambda_{\max} = \max(|V_n^L| + c^L, |V_n^R| + c^R) = |V_n| + c\,, \quad \text{since} \quad c^L = c^R = c\,,$$
$$V_n = V_n^L = -V_n^R \tag{49}$$

and thus gives a definition of P^*

$$\left(\frac{P^*}{P}\right)_{\mathrm{LF}} = 1 + \gamma \mathrm{Ma}_n \left(\mathrm{Ma}_n + |\mathrm{Ma}_n| + 1\right)$$
$$= \begin{cases} 1 + \gamma \mathrm{Ma}_n (2\mathrm{Ma}_n + 1) > 1 & \text{for} \quad V_n > 0 \\ 1 + \gamma \mathrm{Ma}_n \qquad\qquad\qquad \leq 1 & \text{for} \quad V_n \leq 0 \end{cases}, \tag{50}$$

which shows that the Lax-Friedrichs flux satisfies the entropy inequality (44).

HLL and HLLC Flux

The HLL flux [13] is written as

$$\left(\overset{\leftrightarrow}{\mathbf{F}} \cdot \vec{n}\right)^*_{\mathrm{HLL}} = \frac{1}{S^R - S^L} \left(\vec{n} \cdot \left(S^R \overset{\leftrightarrow}{\mathbf{F}}(\mathbf{U}^L) - S^L \overset{\leftrightarrow}{\mathbf{F}}(\mathbf{U}^R)\right) + S^L S^R \left(\mathbf{U}^R - \mathbf{U}^L\right)\right). \tag{51}$$

The leftgoing and rightgoing wave speeds are $S^L = V_n^L - c^L = -V_n^R - c^R = -S^R$ and the HLL flux reduces to

$$\left(\overset{\leftrightarrow}{\mathbf{F}} \cdot \vec{n}\right)^*_{\mathrm{HLL}} = \frac{1}{2} \vec{n} \cdot \left(\overset{\leftrightarrow}{\mathbf{F}}(\mathbf{U}^L) + \overset{\leftrightarrow}{\mathbf{F}}(\mathbf{U}^R)\right) - \frac{S^R}{2} \left(\mathbf{U}^R - \mathbf{U}^L\right). \tag{52}$$

If we would choose S^R to be the maximum wave speed, the HLL flux would reduce to the Lax-Friedrichs flux. However, with $S^R = V_n^R + c^R = -V_n + c$, an even simpler relation for P^* is found, which also satisfies the entropy inequality

$$\left(\frac{P^*}{P}\right)_{\mathrm{HLL}} = 1 + \gamma \mathrm{Ma}_n \begin{cases} > 1 & \text{for} \quad V_n > 0 \\ \leq 1 & \text{for} \quad V_n \leq 0 \end{cases} \tag{53}$$

For the HLLC flux [13], one can show that since the Riemann problem is symmetric, the approximate wave speed of the contact discontinuity is $\lambda^* = 0$ and, choosing $S^R = -V_n + c$, HLLC reduces to the HLL flux.

$$\left(\frac{P^*}{P}\right)_{\mathrm{HLLC}} = 1 + \gamma \mathrm{Ma}_n = \left(\frac{P^*}{P}\right)_{\mathrm{HLL}} \tag{54}$$

Roe Flux

For the original Roe method without entropy fix [13], the mean values are

$$\tilde{V}_n = \frac{\sqrt{\rho^L} V_n^L + \sqrt{\rho^R} V_n^R}{\sqrt{\rho^L} + \sqrt{\rho^R}} = 0\,, \quad \tilde{V}_{t_1} = V_{t_1}\,, \quad \tilde{V}_{t_2} = V_{t_2}\,,$$

$$\tilde{c} = c\sqrt{1 + \frac{(\gamma - 1)}{2}\mathrm{Ma}_n^2}. \tag{55}$$

After some manipulations,

$$\left(\overset{\leftrightarrow}{\mathbf{F}} \cdot \vec{n}\right)^*_{\text{Roe}} = \left(\overset{\leftrightarrow}{\mathbf{F}} \cdot \vec{n}\right) + \tilde{\lambda}_1 \tilde{\alpha}_1 \tilde{\mathbf{K}}^1 = \left(\overset{\leftrightarrow}{\mathbf{F}} \cdot \vec{n}\right) + (-\tilde{c})\frac{\rho V_n}{\tilde{c}} \begin{bmatrix} 1 \\ -\tilde{c} \\ V_{t_1} \\ V_{t_2} \\ \frac{1}{\rho}(\rho E + P) \end{bmatrix}$$

$$= \begin{bmatrix} 0 \\ (\rho V_n^2 + \rho V_n \tilde{c} + P)\,\vec{n} \\ 0 \end{bmatrix}. \tag{56}$$

with $\tilde{\lambda_1} = \tilde{V}_n - \tilde{c} = -\tilde{c}$, $\alpha_1 = \rho V_n / \tilde{c}$ and $\tilde{\mathbf{K}}^1$ from [13]. This leads again to a definition of P^*

$$\left(\frac{P^*}{P}\right)_{\text{Roe}} = 1 + \gamma \mathrm{Ma}_n \left(\mathrm{Ma}_n + \sqrt{1 + \frac{(\gamma - 1)}{2}\mathrm{Ma}_n^2}\right), \tag{57}$$

which fulfills the entropy inequality as long as

$$\mathrm{Ma}_n \geq -\sqrt{\frac{2}{3 - \gamma}}\,, \quad \text{for}\, \gamma = \frac{7}{5} \quad \mathrm{Ma}_n > -1.12\,. \tag{58}$$

Thus, the Roe flux is entropy stable for shocks, but not for supersonic rarefactions.

EC-ES Fluxes

We can also apply an entropy conservative (EC) flux that is used for interior element interfaces and add an entropy stable dissipation term (ES) to compute the boundary flux via the mirrored states (46). This is exactly the strategy proposed in Parsani et al. [12] to get the boundary flux. Such an EC-ES flux is presented in Winters et al. [15]

$$\left(\overleftrightarrow{\mathbf{F}} \cdot \vec{n}\right)^*_{\text{ES}} = \overleftrightarrow{\mathbf{F}}_{\text{EC}}\left(\mathbf{U}_L, \mathbf{U}_R\right) \cdot \vec{n} - \frac{1}{2}\underline{\mathrm{D}}\,\underline{\mathrm{H}}\left(\mathbf{W}^R - \mathbf{W}^L\right) \tag{59}$$

where $\underline{\mathrm{D}}$ is a dissipation matrix and the matrix $\underline{\mathrm{H}}\,[\![\mathbf{w}]\!] \simeq [\![\mathbf{u}]\!]$ is carefully derived from the left and right states. Details are given in [15], where two approaches for the dissipation are distinguished. One is a Lax-Friedrichs-type dissipation, scaling with the maximum eigenvalue $\lambda_{\max} = |V_n| + c$ (referred to as 'EC-LF'). The other is a Roe-type dissipation computed via the eigenstructure of the matrix ($\underline{\mathrm{D}}\,\underline{\mathrm{H}}$) (referred to as 'EC-Roe').

If we carefully insert the two mirrored boundary states into (59), we again get an equation for the modified pressure

$$\left(\frac{P^*}{P}\right)_{\text{EC-LF}} = 1 + \gamma \mathrm{Ma}_n \left(|\mathrm{Ma}_n| + 1\right) \tag{60}$$

for the Lax-Friedrichs-type dissipation and

$$\left(\frac{P^*}{P}\right)_{\text{EC-Roe}} = 1 + \gamma \mathrm{Ma}_n \tag{61}$$

for the Roe-type dissipation. Both approaches lead to an entropy stable boundary flux when using a mirrored state. Note that the modified pressure of the EC-Roe flux (61) exactly matches the one of the HLL flux (53).

Discussion

In the previous section we have shown conditions under which a specified wall flux is stable. In the linear analysis, the central numerical flux adds no dissipation and is neutrally stable. In the nonlinear analysis, entropy is not generated if the numerical wall pressure is equal to the internal pressure, $P^* = P'$. For upwinded approximations, the amount of energy or entropy dissipation depends on the normal Mach number. Since the boundary condition is only imposed weakly through the numerical flux, the normal Mach number will not be exactly zero except in the convergence limit. In fact, flow computations (especially steady state ones) are usually initiated with an impulsive start, where the initial state is a uniform flow, and the normal Mach number is not zero. This has proved over time to be very robust in practice. The analysis above gives an explanation why.

In the linear analysis the dissipation due to imposing the boundary condition is proportional to the square of the normal Mach number. With an impulsive start initialization, this dissipation will be large. As the flow develops and the

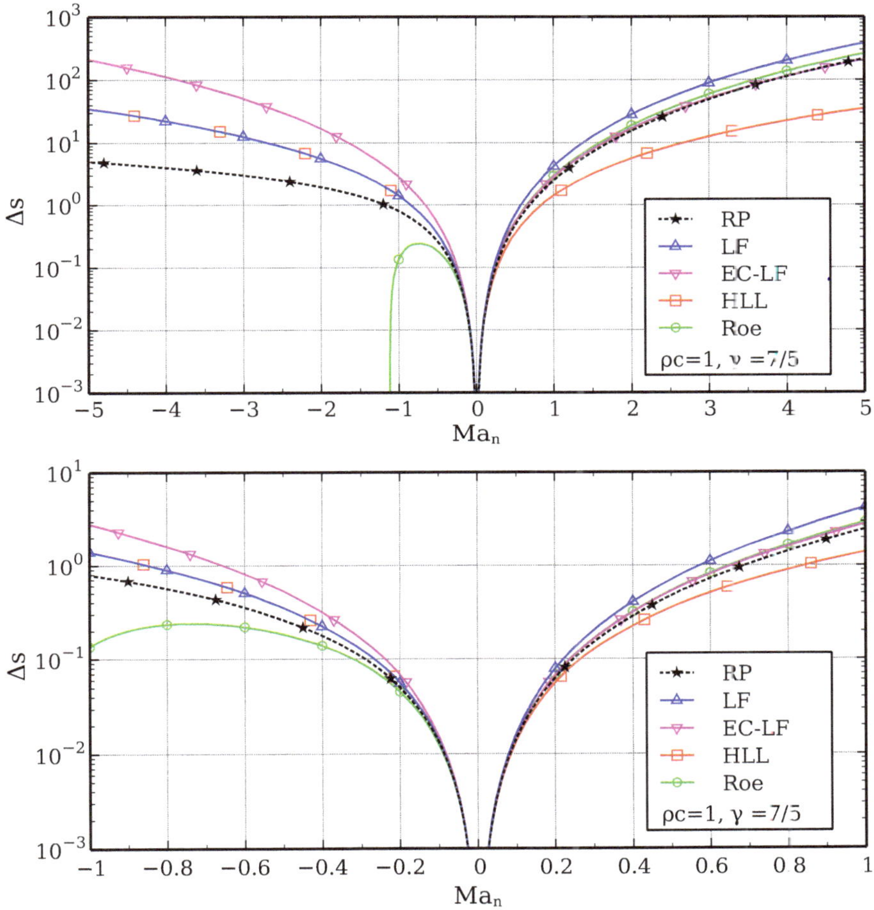

Fig. Entropy contribution Δs (62) produced by the wall boundary flux. RP refers to the exact Riemann problem (45), LF to (50), EC-LF to (60), HLL to (53) and Roe to (57). Plotted over the normal Mach number ranges $|\mathrm{Ma}_n| \leq 5$ on the top and restricted to $|\mathrm{Ma}_n| \leq 1$ on the bottom

boundary condition is better enforced, the dissipation reduces, going away only as the approximate solution converges.

A similar effect is observed for the use of the different approximate Riemann solvers in the nonlinear analysis. In Fig. , we compare the entropy contribution

$$\Delta s = (\rho c)\mathrm{Ma}_n \left(\frac{P^*}{P} - 1 \right) \tag{62}$$

for the different wall boundary fluxes, over a range of normal Mach numbers for $(\rho c) = 1$ and $\gamma = 7/5$. When the boundary condition is exactly fulfilled ($\mathrm{Ma}_n = 0$), the entropy contribution is zero. For low normal Mach numbers, all fluxes have the same behavior. Compared to the exact Riemann problem (RP), the Lax-Friedrichs flux and the EC-LF flux always produce more entropy whereas the HLL flux produces less entropy for impinging velocities $\mathrm{Ma}_n > 0$. The results of HLLC and EC-Roe fluxes are not plotted, as they coincide with the HLL flux. As shown in the analysis, the Roe flux produces a negative entropy change for supersonic rarefactions, implying that it is not suitable for all flow configurations.

Acknowledgements This work was supported by a grant from the Simons Foundation (#426393, David Kopriva). Gregor J. Gassner has been supported by the European Research Council (ERC) under the European Union's Eights Framework Program Horizon 2020 with the research project *Extreme*, ERC grant agreement no. 714487. Florian Hindenlang thanks Eric Sonnendrücker and the Max-Planck Institute for Plasma Physics in Garching for their constant support.

References

1. Black, K.: A conservative spectral element method for the approximation of compressible fluid flow. Kybernetika **35**(1), 133–146 (1999)
2. Chen, T., Shu, C.-W.: Entropy stable high order discontinuous Galerkin methods with suitable quadrature rules for hyperbolic conservation laws. J. Comput. Phys. **345**, 427–461 (2017)
3. Fisher, T., Carpenter, M.H., Nordström, J., Yamaleev, N.K., Swanson, C.: Discretely conservative finite-difference formulations for nonlinear conservation laws in split form: theory and boundary conditions. J. Comput. Phys. **234**, 353–375 (2013)
4. Gassner, G.J., Winters, A.R., Kopriva, D.A.: Split form nodal discontinuous Galerkin schemes with summation-by-parts property for the compressible Euler equations. J. Comput. Phys. **327**, 39–66 (2016)
5. Gassner, G.J., Winters, A.R., Hindenlang, F.J., Kopriva, D.A.: The BR1 scheme is stable for the compressible Navier–Stokes equations. J. Sci. Comput. (2018)
6. Kopriva, D.A.: Implementing Spectral Methods for Partial Differential Equations. Scientific Computation. Springer, Berlin (2009)
7. Kopriva, D.A., Gassner, G.: On the quadrature and weak form choices in collocation type discontinuous Galerkin spectral element methods. J. Sci. Comput. **44**(2), 136–155 (2010)
8. Kopriva, D.A., Gassner, G.J.: An energy stable discontinuous Galerkin spectral element discretization for variable coefficient advection problems. SIAM J. Sci. Comput. **36**(4), A2076–A2099 (2014)
9. Nordström, J.: Conservative finite difference formulations, variable coefficients, energy estimates and artificial dissipation. J. Sci. Comput. **29**(3), 375–404 (2006)
10. Nordström, J.: A roadmap to well posed and stable problems in computational physics. J. Sci. Comput. (2016). https://doi.org/10.1007/s10915-016-0303-9
11. Nordström, J., Svard, M.: Well-posed boundary conditions for the Navier–Stokes equations. SIAM J. Numer. Anal. **43**(3), 1231–1255 (2005)
12. Parsani, M., Carpenter, M.H., Nielsen, E.J.: Entropy stable wall boundary conditions for the three-dimensional compressible Navier–Stokes equations. J. Comput. Phys. **292**, 88–113 (2015)
13. Toro, E.F.: Riemann Solvers and Numerical Methods for Fluid Dynamics. Springer, Berlin (2009)

14. van der Vegt, J.J.W., van der Ven, H.: Slip flow boundary conditions in discontinuous Galerkin discretizations of the Euler equations of gas dynamics. In Mang, H.A., Rammerstorfer, F.G. (eds.) Proceedings of the 5th World Congress on Computational Mechanics (WCCM V), number NLR-TP in Technical Publications, pp. 1–16. National Aerospace Laboratory, NLR (2002)
15. Winters, A.R., Derigs, D., Gassner, G.J., Walch, S.: A uniquely defined entropy stable matrix dissipation operator for high Mach number ideal MHD and compressible Euler simulations. J. Comput. Phys. **332**, 274–289 (2017)

16

Greedy Kernel Methods for Center Manifold Approximation

Bernard Haasdonk, Boumediene Hamzi, Gabriele Santin, and Dominik Wittwar

Introduction

Center manifold theory plays an important role in the study of the stability of dynamical systems when the equilibrium point is not hyperbolic. It isolates the complicated asymptotic behavior by locating the center manifold which is an invariant manifold tangent to the subspace spanned by the eigenspace of eigenvalues on the imaginary axis. Then, the dynamics of the original system will be essentially determined by the restriction of this dynamics on the center manifold since the local dynamic behavior "transverse" to this invariant manifold is relatively simple as it corresponds to the flows in the local stable (and unstable) manifolds. In practice, one does not compute the center manifold and its dynamics exactly since this requires the resolution of a quasilinear partial differential equation which is not easily solvable. In most cases of interest, an approximation of degree two or three of the solution is sufficient. Then, the reduced dynamics on the center manifold can be determined, its stability can be studied and then conclusions about the stability of the original system can be obtained [1, 3, 4, 6, 8].

In this article, we use greedy kernel methods to construct a data-based approximation of the center manifold. The present work is a preliminary study that is intended to introduce our concept and algorithm, and to test it on some examples.

B. Haasdonk · G. Santin · D. Wittwar
Inst. of Applied Analysis and Numerical Simulation, University of Stuttgart, Stuttgart, Germany
e-mail: bernard.haasdonk@mathematik.uni-stuttgart.de;
gabriele.santin@mathematik.uni-stuttgart.de; dominik.wittwar@mathematik.uni-stuttgart.de

B. Hamzi (✉)
Department of Mathematics, Imperial College London, London, UK

Background

We consider a large dimensional dynamical system

$$\dot{x} = f(x), \quad x \in D, \tag{1}$$

where $f : D \to \mathbb{R}^n$ is a continuously differentiable function over the domain $D \subset \mathbb{R}^n$ such that $0 \in D$. We are interested in the study of the behavior of the system around an equilibrium point $x \in D$, i.e., $f(x) = 0$, possibly analyzing a smaller dimensional system.

Without loss of generality, we may assume that the equilibrium is $x = 0$, and, letting $L = \frac{\partial f}{\partial x}(x)|_{x=0}$, we can rewrite (1) as

$$\dot{x} = f(x) = Lx + N(x),$$

with a suitable nonlinear component N, and denote as $\sigma_{\mathbb{R}}(L)$ the set of real parts of the eigenvalues of L. A classical result relates the stability of the equilibrium with the spectrum of L, and in particular it is known that if L has all its eigenvalues with negative real parts, i.e., $\sigma_{\mathbb{R}}(L) \subset \mathbb{R}_{<0}$, then the origin is asymptotically stable, and if L has some eigenvalues with positive real parts, then the origin is unstable. If instead $\sigma_{\mathbb{R}}(L) \subset \mathbb{R}_{\leq 0}$, the linearization fails to determine the stability properties of the origin, and thus the analysis of this situation requires to employ additional tools.

In this case, we can first use a linear change of coordinates to separate the zero and the negative eigenvalues, i.e., we can rewrite (1) as

$$\begin{aligned} \dot{x} &= L_1 x + N_1(x, y) \\ \dot{y} &= L_2 y + N_2(x, y) \end{aligned} \tag{2}$$

where $L_1 \in \mathbb{R}^{d \times d}$ is such that $\sigma_{\mathbb{R}}(L_1) = \{0\}$ and $L_2 \in \mathbb{R}^{m \times m}$ with $m := n - d$ is such that $\sigma_{\mathbb{R}}(L_2) \subset \mathbb{R}_{<0}$. The nonlinear functions $N_1 : \mathbb{R}^d \times \mathbb{R}^m \to \mathbb{R}^d$ and $N_2 : \mathbb{R}^d \times \mathbb{R}^m \to \mathbb{R}^m$ are continuously differentiable. Intuitively, we expect the stability of the equilibrium to only depend on the nonlinear term $N_1(x, y)$. This intuition turns out to be correct, and indeed it can be properly formalized by means of the center manifold theorem.

We start by recalling a sufficient condition for the existence of a center manifold.

Theorem 1 ([1]) *If N_1 and N_2 are twice continuously differentiable and are such that*

$$N_i(0, 0) = 0, \quad \frac{\partial N_i}{\partial x}(0, 0) = 0, \quad \frac{N_i}{\partial y}(0, 0) = 0, \quad i = 1, 2,$$

and if the eigenvalues of L_1 have zero real parts, and all the eigenvalues of L_2 have negative real parts, then there exists a neighbourhood $\Omega \subset \mathbb{R}^d$ of the origin $0 \in \mathbb{R}^d$

and a center manifold $h : \Omega \to \mathbb{R}^m$ *for* (2), *i.e.,* $y = h(x)$ *is an invariant manifold for* (2),[1] h *is smooth, and*

$$h(0) = 0, \quad Dh(0) = 0. \tag{3}$$

Under the assumptions of this theorem, using (2) we deduce that h satisfies the PDE

$$L_2 h(x) + N_2(x, h(x)) = Dh(x)\left(L_1 x + N_1(x, h(x))\right), \tag{4}$$

and the following center manifold theorem ensures that there are smooth solutions to this PDE. Moreover, it also allows to deduce the stability of the origin of the full order system (2) from the stability of the origin of a reduced order system called the *center dynamics*.

Theorem 2 (Center Manifold Theorem [1]) *The equilibria* $x = 0$, $y = 0$ *of the original dynamics is locally asymptotically stable (resp. unstable) if and only if the equilibrium* $x = 0$ *of the center dynamics (dynamics on the center manifold)*

$$\dot{x} = L_1 x + N_1(x, h(x)), \tag{5}$$

is locally asymptotically stable (resp. unstable).

In particular, this result guarantees that, after solving the PDE (4), the problem of analyzing the stability properties of the system (2) reduces to analyzing the nonlinear stability of the lower dimensional system (5). This second problem is of smaller dimension and thus, provided the knowledge of h, the approach is attractive to obtain information on the system (1) via a reduced model.

Moreover, we remark that an exact knowledge of h is not required for this purpose, i.e., it is sufficient to have an approximate solution of the PDE (4). Indeed, it is frequently sufficient to compute only the low degree terms of the Taylor series expansion of h around $x = 0$, i.e., if $(\cdot)^{[k]}$ is the degree k part of the Taylor series of h, the approximation

$$h(x) \approx h^{[1]}x + h^{[2]}(x) + h^{[3]}(x) + \ldots + h^{[d-1]}(x) \tag{6}$$

is sufficient to obtain an approximation of the dynamics of order ε^d as $\|x\| \leq \varepsilon$. The approximation (6) can be obtained by coefficient comparison, thus rewriting

[1] A differentiable manifold $\mathcal{M}$ is said to be invariant under the flow of a vector field X if for $x \in \mathcal{M}$, $F_t(x) \in \mathcal{M}$ for small $t > 0$, where $F_t(x)$ is the flow of X.

the PDE (4) as a set of algebraic equations as

$$
\begin{aligned}
&L_2 h^{[1]} = h^{[1]} L_1 \\
&L_2 h^{[2]}(x) + N_2^{[2]}(x, h^{[1]}(x)) = \frac{\partial h^{[2]}}{\partial x}(x) \left(L_1 x_1 + N_1^{[2]}(x, h^{[1]}(x)) \right) \\
&L_2 h^{[3]}(x) + \left(N_2(x, h^{[2]}(x)) \right)^{[2]} = \frac{\partial h^{[2]}}{\partial x}(x) \left(L_1 x + \left(N_1(x, h^{[2]}(x)) \right)^{[2]} \right) \\
&\dots .
\end{aligned}
$$

We remark that this methodology is valid for parameterized dynamical systems and is used to study the stability of dynamical systems with bifurcations.

Nevertheless, even this approximated knowledge of h can be difficult to obtain in practice for a general ODE. To overcome this limitation, and since an approximated knowledge of the manifold is sufficient, our goal in this paper is to find a data-based approximation of the center manifold. This approximation is based solely on the knowledge of the splitting (1) and on the numerical computation of a set of trajectories of the system, and it provides an approximation of h which can be used to study the system stability.

Kernel Approximation

We want to build a surrogate model $s_h : \Omega \to \mathbb{R}^m$ which approximates the center manifold h on a suitable set $\Omega \subset \mathbb{R}^d$, in the sense that $s_h(x) \approx h(x)$ for all $x \in \Omega$. This model is constructed in a data-based way, i.e., we assume some knowledge of the map h on a finite set of input parameters, or training data. In practice, such values are computed from high-fidelity numerical approximations, which will be discussed in detail in the following.

The surrogate is based on kernel approximation, which allows the use of scattered data, i.e., we do not require any grid structure on the set of training data. Moreover, since the unknown function h is vector-valued, we employ here matrix-valued kernels. Details on kernel-based approximation can be found e.g. in [9], and the extension to the vectorial case is detailed e.g. in [5, 10]. We recall here only that a positive definite matrix-valued kernel on Ω is a function $K : \Omega \times \Omega \to \mathbb{R}^{m \times m}$ such that $K(x, y) = K(y, x)^T$ for all $x, y \in \Omega$ and $[K(x_i, x_j)]_{i,j=1}^N \in \mathbb{R}^{mN \times mN}$ is positive semidefinite for any set $\{x_1, \dots, x_N\} \subset \Omega$ of pairwise distinct points, for all $N \in \mathbb{N}$. Associated to a positive definite kernel there is a unique Hilbert space $\mathscr{H}$ of functions $\Omega \to \mathbb{R}^m$, named native space, where the kernel is reproducing, meaning that $K(\cdot, x)\alpha$ is the Riesz representer of the directional point evaluation $\delta_x^\alpha(f) := \alpha^T f(x)$, for all $\alpha \in \mathbb{R}^m$, $x \in \Omega$.

We consider here a twice continuously differentiable matrix-valued kernel k on Ω, and we use a specific functional formulation for our approximation and a specific cost function, in order to construct a surrogate that is well suited for the particular approximation task.

In detail, the approximant takes the form

$$s_h(x) = \sum_{i=1}^{n_1} K(x, x_i^{(1)})\alpha_i + \sum_{j=1}^{n_2}\sum_{i=1}^{m} \partial_i^{(2)} K(x, x_j^{(2)})\beta_{i,j},$$

with centers $x_i^{(1)} \in X^{(1)} = \left\{x_1^{(1)}, \ldots, x_{n_1}^{(1)}\right\}$, $x_j^{(2)} \in X^{(2)} = \left\{x_1^{(2)}, \ldots, x_{n_2}^{(2)}\right\}$ and coefficient vectors $\alpha_i, \beta_{i,j} \in \mathbb{R}^m$. Here the superscript $\partial^{(2)}$ denotes that the derivative with regards to the second kernel component is taken.

Subsequently, we assume to have a sufficient amount of data $X_{N^*} = \{x_1, \ldots, x_{N^*}\}$ and $Y_{N^*} = \{y_1, \ldots, y_{N^*}\}$ which, for example, is generated by running a numerical scheme to compute discrete trajectories for different initial values (x_0, y_0). For this step, we need to assume that the variable splitting (2) is known in advance. Note that this is not a severe restriction, as for a general ODE (1) the required state transformation can be determined by eigenvalue decomposition of L.

Observe that we do not know if a data pair (x_i, y_i) lies on the center manifold, i.e. if $y_i = h(x_i)$ holds. We only know that the data converges asymptotically to the center manifold as $x_i \to 0$. Thus, an interpolation-based surrogate which merely interpolates the data on a given subset $X \subset X_{N^*}$ seems ill-suited for our purposes. Instead we consider another set of conditions to define the approximant. First, we still require the conditions in (3) to be satisfied by our approximation. Moreover, for the given subsets $X = \{x_1, \ldots, x_N\}$ and $Y = \{y_1, \ldots, y_N\}$, we compute our approximant by minimizing the following functional $J : \mathscr{H} \to \mathbb{R}$ under the constraint $s(0) = 0$, $Ds(0) = 0$:

$$J(s) := \|s\|_{\mathscr{H}}^2 + \sum_{i=1}^{N} (s(x_i) - y_i)^T \omega_i (s(x_i) - y_i). \tag{7}$$

Here $\omega_i \in \mathbb{R}^{m\times m}$ is a positive definite weight matrix. It can be shown that (7) has a unique minimizer s_h (see [11]). In particular s_h and its derivative Ds_h have the form

$$s_h(x) = \sum_{i=1}^{N+1} K(x, x_i)\alpha_i + \sum_{i=1}^{m} \partial_i^{(2)} K(x, 0)\beta_i, \tag{8}$$

$$Ds_h(x) = \sum_{i=1}^{N+1} D^{(1)} K(x, x_i)\alpha_i + \sum_{i=1}^{m} D^{(1)} \partial_i^{(2)} K(x, 0)\beta_i,$$

where we set $x_{N+1} := 0$. The coefficient vectors α_i, β_i can be computed by solving the system

$$\begin{pmatrix} A+W & B \\ B^T & C \end{pmatrix} \begin{pmatrix} \boldsymbol{\alpha} \\ \boldsymbol{\beta} \end{pmatrix} = \begin{pmatrix} \boldsymbol{Y} \\ \boldsymbol{Z} \end{pmatrix}, \tag{9}$$

with

$$\begin{aligned}
A &:= \left(K(x_i, x_j)\right)_{i,j} \in \mathbb{R}^{m(N+1)\times m(N+1)}, \\
W &:= \operatorname{diag}\left(\omega_1^{-1}, \ldots, \omega_N^{-1}, 0\right) \in \mathbb{R}^{m(N+1)\times m(N+1)}, \\
B &:= \left(\partial_j^{(2)} K(x_i, 0)\right)_{i,j} \in \mathbb{R}^{m(N+1)\times m^2}, \\
C &:= \left(\partial_i^{(1)} \partial_j^{(2)} k(0,0)\right)_{i,j} \in \mathbb{R}^{m^2\times m^2}, \\
\boldsymbol{Y} &:= (y_1^T, \ldots, y_n^T, 0)^T \in \mathbb{R}^{m(N+1)}, \\
\boldsymbol{Z} &:= 0 \in \mathbb{R}^{m^2\times m}.
\end{aligned}$$

The weight matrices ω_i can either be chosen manually, or a regularizing function $r : \Omega \to \mathbb{R}^{m\times m}$ can be prescribed such that $\omega_i = r(x_i)$ is symmetric and positive definite. In our numerical examples in Sect. we chose a constant regularization function, i.e.

$$\omega_i = r(x_i) = \lambda I_m$$

for some $\lambda > 0$. However, one might consider a more general approach, where the weight increases as the data tends to the origin, i.e. $\omega_i \succeq \omega_j$ if $\|x_i\| \leq \|x_j\|$.

Greedy Approximation

If the technique of the previous section is used as it is, the surrogate (8) is given by an expansion with N^* terms, where N^* is the number of points in the training set. Therefore, the model evaluation might not be efficient enough if the model is built using a too large dataset. Furthermore, the computation of the coefficients in (8) requires the solution of the linear system (9), whose size again scales with the size of the training set, and which can be severely ill-conditioned for non well-placed points.

To mitigate both problems, we employ an algorithm that aims at selecting small subsets X_N, Y_N of points such that the surrogate computed with these sets is a sufficiently good approximation of the one which uses the full sets. The algorithm selects the points in a greedy way, i.e., one point at a time is selected and added to the current training set. In this way, it is possible to identify a good set without the need to solve a nearly infeasible combinatorial problem.

The selection is performed using the P-greedy method of [2] applied to the kernel K, such that the set of points is selected before the computation of the surrogate. The number of points, and therefore the expansion size and evaluation speed, is depending on a prescribed target accuracy $\varepsilon_{tol} > 0$. For details on the method implementation and its convergence properties we refer to [7].

Numerical Examples

We test now our method on three different examples. In each of them, we specify the setting and the parameters used to build the surrogate and visualize our approximation to the center manifold. Additionally, we compute the pointwise residual

$$r(x) = Ds_h(x)\left(L_1 x + N_1(x, s_h(x))\right) - \left(L_2 s_h(x) + N_2(x, s_h(x))\right),$$

which measures how well the surrogate s_h satisfies the ODE (4).

In all the three examples, the greedy algorithm is used to select a suitable subset of the points, and in all cases the procedure is stopped with a prescribed ε_{tol}. In the first two examples we set $\varepsilon_{tol} := 10^{-15}$, while $\varepsilon_{tol} := 10^{-10}$ is used in the last one.

Example 1

We consider the 2-dimensional system

$$\begin{aligned} \dot{x} &= L_1 x + N_1(x, y) = 0 + xy \\ \dot{y} &= L_2 y + N_2(x, y) = -y + x^2. \end{aligned} \tag{10}$$

We generate the training data by solving (10) with an implicit Euler scheme for initial time $t_0 = 0$, final time $T = 1000$ and with the time step $\Delta t = 0.1$. We initiate the numerical procedure with initial values $(x_0, y_0) \in \{\pm 0.8\} \times \{\pm 0.8\}$ and store the resulting data pairs in X and Y after discarding all data whose x-values are not contained in the neighborhood $[-0.1, 0.1]$ which results in $N^* = 38{,}248$ data pairs.

We run the greedy algorithm for the kernels $k_1(x, y) := \left(1 + xy/2\right)^4$ and $k_2(x, y) = e^{-(x-y)^2/2}$. This results in the sets X_1 and X_2 which contain 14 and 6 points, respectively. The corresponding approximations s_1 and s_2 for the constant

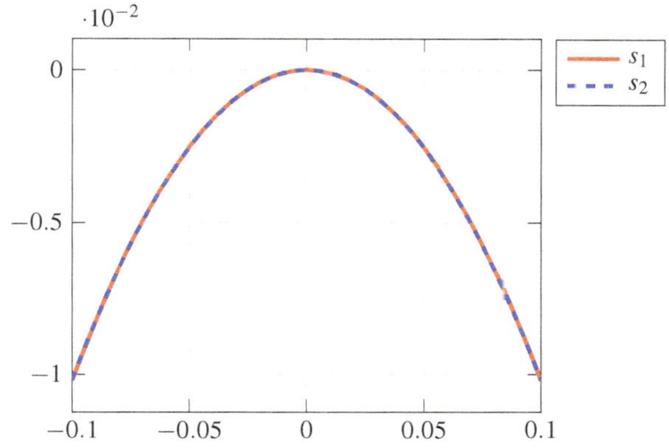

Fig. Approximations s_1 and s_2 of the center manifold

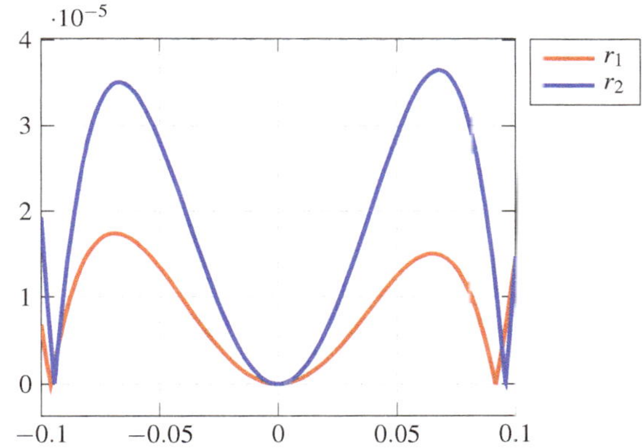

Fig. Residuals r_1 and r_2 of the center manifold

regularization function $r \equiv 10^{-10}$ are plotted in Fig. over the domain $[-0\ 1, 0.1]$. The pointwise residual is depicted in Fig.

Example 2

We consider the 2-dimensional system

$$\begin{aligned} \dot{x} &= L_1 x + N_1(x, y) = 0 - xy \\ \dot{y} &= L_2 y + N_2(x, y) = -y + x^2 - 2y^2. \end{aligned} \tag{11}$$

The training data is generated the same way as in Example 1. We again use the kernels k_1 and k_2. The greedy algorithm gives sets X_1 and X_2 of size 12 and 6, respectively. The evaluation of the approximations s_1 and s_2 over the neighborhood

Fig. Approximations s_1 and s_2 of the center manifold

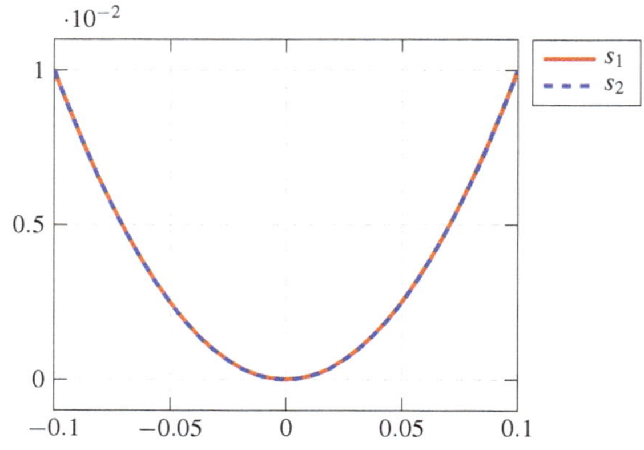

Fig. Residuals r_1 and r_2 of the center manifold

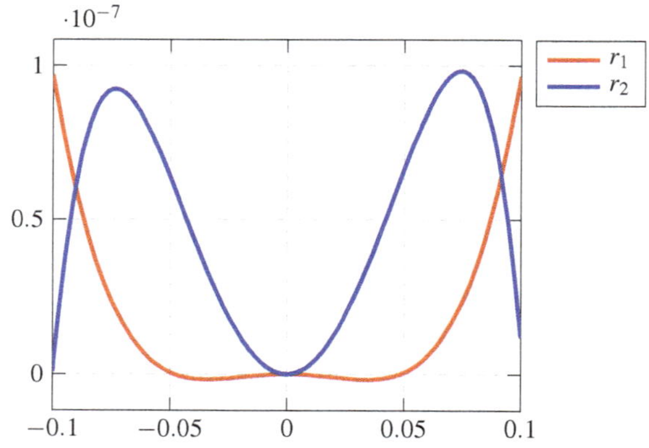

$[-0.1, 0.1]$ can be seen in Fig. , while the respective pointwise residuals are plotted in Fig.

Example 3

We consider the $(2+1)$-dimensional system

$$\begin{aligned} \dot{x} &= L_1 x + N_1(x, y) = \begin{pmatrix} 0 & -1 \\ 1 & 0 \end{pmatrix} \begin{pmatrix} x_1 \\ x_2 \end{pmatrix} + \begin{pmatrix} x_1 y \\ x_2 y \end{pmatrix} \\ \dot{y} &= L_2 y + N_2(x, y) = -y - x_1^2 - x_2^2 + y^2. \end{aligned} \tag{12}$$

We generate the training data in a similar fashion as before. We again use the implicit Euler scheme with start time $t = 0$, final time $T = 1000$ and with time step

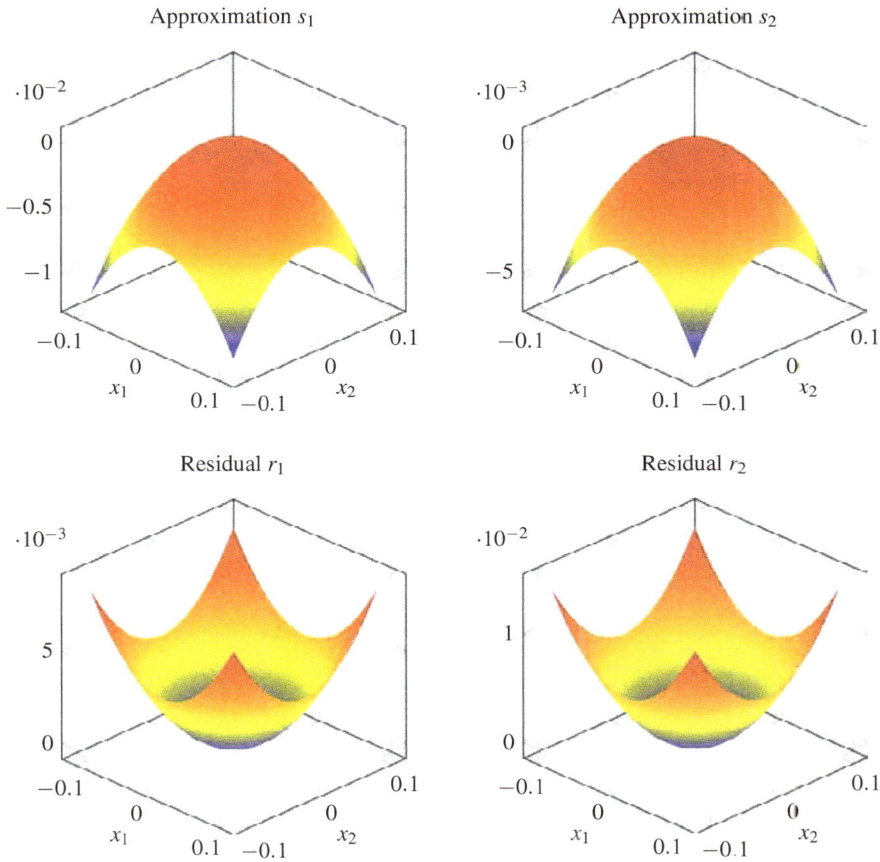

Fig. Approximations s_1 and s_2 of the center manifold and corresponding residuals r_1 and r_2

$\Delta t = 0.1$. The Euler method is performed for initial data $(x_0, y_0) \in \{\pm 0.8\}^3$ and the resulting trajectories are stored in X and Y, where only data with $x \in [-0.1, 0.1]^2$ was considered; this leads to $N^* = 78{,}796$ data pairs. We use the kernels $k_1(x, y) = (1 + x^T y/2)^4$ and $k_2(x, y) = e^{-\|x-y\|_2^2/2}$, and the greedy-selected sets have the size 21 (for k_1) and 25 (for k_2), respectively. The approximations s_1, s_2 and their corresponding residuals r_1 and r_2 computed over the domain $[-0.1, 0.1]^2$. The results can be seen in Fig.

We remark that in all the three experiments both kernels give comparable results in terms of error magnitude, and they both provide a good approximation of the manifold.

Conclusions

In this paper we introduced a novel algorithm to approximate the center manifold of a given ODE using a data-based surrogate.

This algorithm computes an approximation of the manifold from a set of numerical trajectories with different initial data. It is based on kernel methods, which allow the use of the scattered data generated by these simulations as training points. Moreover, an application-specific ansatz and cost function have been employed in order to enforce suitable properties on the surrogate.

Several numerical experiments suggested that the present method can reach a significant accuracy, and that it has the potential to be used as an effective model reduction technique. It seems promising to apply this approach to high dimensional systems as the approximation technique straightforwardly can be extended and is less prone to the curse of dimensionality than grid-based approximation techniques. An interesting extension would consist of determining the decomposition (2) in a data-based fashion by suitable processing of the trajectory data.

Acknowledgements The first, third, and fourth authors would like to thank the German Research Foundation (DFG) for support within the Cluster of Excellence in Simulation Technology (EXC 310/2) at the University of Stuttgart. The second author thanks the European Commission for financial support received through Marie Curie Fellowships.

References

1. Carr, J.: Applications of Centre Manifold Theory. Applied Mathematical Sciences, vol. 35. Springer, Berlin (1981)
2. De Marchi, S., Schaback, R., Wendland, H.: Near-optimal data-independent point locations for radial basis function interpolation. Adv. Comput. Math. **23**(3), 317–330 (2005)
3. Henry, D.: Geometric Theory of Semilinear Parabolic Equations. Lecture Notes in Mathematics, vol. 840. Springer, Berlin (1981)
4. Kelley, A.: The stable, center-stable, center, center-unstable, unstable manifolds. J. Differ. Equ. **3**, 546–570 (1967)
5. Micchelli, C.A., Pontil, M.: On learning vector-valued functions. Neural Comput. **17**(1), 177–204 (2005)
6. Pliss, V.A.: A reduction principle in the theory of stability of motion. Izv. Akad. Nauk SSSR Ser. Mat. **28**, 1297–1324 (1964)
7. Santin, G., Haasdonk, B.: Convergence rate of the data-independent P-greedy algorithm in kernel-based approximation. Dolomites Res. Notes Approx. **10**, 68–78 (2017)
8. Shoshitaishvili, A.N.: Bifurcations of topological type of singular points of vector fields that depend on parameters. Funkcional. Anal. i Priložen. **6**(2), 97–98 (1972)
9. Wendland, H.: Scattered Data Approximation. Cambridge Monographs on Applied and Computational Mathematics, vol. 17. Cambridge University Press, Cambridge (2005)
10. Wittwar, D., Santin, G., Haasdonk, B.: Interpolation with uncoupled separable matrix-valued kernels. Dolomites Res. Notes Approx. **11**, 23–29 (2018)
11. Wittwar, D., Santin, G., Haasdonk, B.: Weighted regularized interpolation with matrix valued kernels. Technical Report, University of Stuttgart, (2019, in preparation)

17

Model Reduction by Separation of Variables: A Comparison between Hierarchical Model Reduction and Proper Generalized Decomposition

Simona Perotto, Michele Giuliano Carlino, and Francesco Ballarin

Introduction

This paper is meant as a first attempt to compare two procedures which share the idea of exploiting separation of variables to perform model reduction, albeit with different purposes. Proper Generalized Decomposition (PGD) is essentially employed as a powerful tool to deal with parametric problems in several fields of application [3, 14, 23]. Parametrized models characterize multi-query contexts, such as parameter optimization, statistical analysis or inverse problems. Here, the computation of the solution for many different parameters demands, in general, a huge computational effort, and this justifies the development of model reduction techniques.

For this purpose, projection-based techniques, such as Proper Orthogonal Decomposition (POD) or Reduced Basis methods, are widely used in the literature [11]. The idea is to project the discrete operators onto a reduced space so that the problem can be solved rapidly in the lower dimensional space. PGD adopts a completely different way to deal with parameters. Here, parameters are considered

S. Perotto
MOX - Modeling and Scientific Computing, Dipartimento di Matematica, Politecnico di Milano, Milano, Italy
e-mail: simona.perotto@polimi.it

M. G. Carlino
Inria Bordeaux Sud-Ouest and Institut de Mathématiques de Bordeaux, University of Bordeaux, Talence, France
e-mail: michele-giuliano.carlino@inria.fr

F. Ballarin (✉)
MathLab, Mathematics Area, SISSA, Trieste, Italy
e-mail: francesco.ballarin@sissa.it

as new independent variables of the problem, together with the standard space-time ones [5]. Although the dimensionality of the problem is inevitably increased, PGD transforms the computation of the solution for new values of the parameters into a plain evaluation of the reduced solution, with striking computational advantages.

Hierarchical-Model (HiMod) reduction has been proposed to improve one-dimensional (1D) partial differential equation (PDE) solvers for problems defined in domains with a geometrically dominant direction, like slabs or pipes [6, 20]. The main applicative field of interest is hemodynamics, in particular the modeling of blood flow in patient-specific geometries. Purely 1D hemodynamic models completely drop the transverse dynamics, which, however may be locally important (e.g., in the presence of a stenosis or an aneurism). HiMod aims at providing a numerical tool to incorporate the transverse components of the 3D solution into a conceptually 1D solver. To do this, the driving idea is to discretize main and transverse dynamics in a different way. The latter are generally of secondary importance and can be described by few degrees of freedom using a spectral approximation, in combination, for instance, with a finite element (FE) discretization of the mainstream.

The parametric version of HiMod (namely, HiPOD) is a more recent proposal [4, 13]. On the other hand, PGD is not so widely employed in a non-parametric setting, despite its original formulation [12]. Nevertheless, for the sake of comparison, in this paper we consider the non-parametric as well as the parametric versions of both the HiMod and PGD approaches. The goal is to begin a preliminary comparative analysis between the two methodologies, to highlight the respective weaknesses and strengths. The main limit of PGD remains its inability to deal with non-Cartesian geometries without losing the computational benefits arising from the separability of the spatial coordinates. HiMod turns out to be more flexible from a geometric viewpoint. On the other hand, PGD turns out to be extremely effective for parametric problems thanks to the explicit expression of the PGD solution in terms of the parameters, while HiPOD can be classified as a projection-based method with all the associated drawbacks. In perspective, the ultimate goal is to merge HiMod with PGD to emphasize the good features and mitigate the intrinsic limits of the two methods taken alone.

The HiMod Approach

Hierarchical Model reduction proved to be an efficient and reliable method to deal with phenomena characterized by dominant dynamics [10]. In general, the computational domain itself exhibits an intrinsic directionality. We assume $\Omega \subset \mathbb{R}^d$ ($d = 2, 3$) to coincide with a d-dimensional fiber bundle, $\Omega = \bigcup_{x \in \Omega_{1D}} \{x\} \times \gamma_x$, where $\Omega_{1D} \subset \mathbb{R}$ denotes the supporting fiber aligned with the main stream, while $\gamma_x \subset \mathbb{R}^{d-1}$ is the transverse fiber at $x \in \Omega_{1D}$, parallel to the transverse dynamics. For the sake of simplicity, we identify Ω_{1D} with a straight segment, (x_0, x_1). We refer to [15, 21] for the case where Ω_{1D} is curvilinear. From a computational viewpoint, the idea is to exploit a map, $\Psi : \Omega \to \hat{\Omega}$, transforming the physical domain, Ω, into a reference domain, $\hat{\Omega}$, and to make explicit computations in $\hat{\Omega}$ only. Typically, $\hat{\Omega}$ coincides with a rectangle in 2D, with a cylinder with circular section in 3D. To define Ψ, for each $x \in \Omega_{1D}$, we introduce the map, $\psi_x : \gamma_x \to \hat{\gamma}_{d-1}$, from fiber γ_x to the reference transverse fiber, $\hat{\gamma}_{d-1}$, so that the reference domain coincides with $\hat{\Omega} = \bigcup_{x\in\Omega_{1D}}\{x\} \times \hat{\gamma}_{d-1}$. The supporting fiber is preserved by map Ψ, which modifies the lateral boundaries only.

We consider now the (full) problem to be reduced. Due to the comparative purposes of the paper, we focus on a scalar elliptic equation, and, in particular, on the associated weak formulation,

$$\text{find } u \in V : a(u, v) = F(v) \quad \forall v \in V, \tag{1}$$

where $V \subseteq H^1(\Omega)$, $a(\cdot,\cdot) : V \times V \to \mathbb{R}$ is a continuous and coercive bilinear form and $F(\cdot) : V \to \mathbb{R}$ is a continuous linear functional. To provide the HiMod formulation for problem (1), we introduce the hierarchical reduced space

$$V_m = \left\{ v_m(x, \mathbf{y}) = \sum_{k=1}^{m} \tilde{v}_k(x)\varphi_k(\psi_x(\mathbf{y})), \text{ with } \tilde{v}_k \in V_{1D}^h,\ x \in \Omega_{1D},\ \mathbf{y} \in \gamma_x \right\} \tag{2}$$

for a modal index $m \in \mathbb{N}^+$, where $V_{1D}^h \subseteq H^1(\Omega_{1D})$ is a discrete space of dimension N_h associated with a partition $\mathcal{T}_h$ of Ω_{1D}, while $\{\varphi_k\}_{k=1}^m$ denotes a modal basis of functions orthogonal with respect to the $L^2(\hat{\gamma}_{d-1})$-scalar product. Index m sets the hierarchical level of the HiMod space, being $V_m \subset V_{m+1}$, for any m. Concerning V_{1D}^h, we adopt here a standard FE space, although any discrete space can be employed (see, e.g., [21], where an isogeometric discretization is used). Functions in V_{1D}^h have to include the boundary conditions on $\{x_0\} \times \gamma_{x_0}$ and $\{x_1\} \times \gamma_{x_1}$; analogously, the modal functions have to take into account the boundary data along the horizontal sides. In Sect. further comments are provided about the selection of the modal basis and of the modal index m. The HiMod formulation for problem (1) thus reads

$$\text{find } u_m^{\text{HiMod}} \in V_m : a(u_m^{\text{HiMod}}, v_m) = F(v_m) \quad \forall v_m \in V_m. \tag{3}$$

To ensure the well-posedness of formulation (3) and the convergence of the HiMod approximation, u_m^{HiMod}, to the full solution, u, we endow the HiMod space with a conformity and a spectral approximability hypothesis, and we introduce a standard density assumption on the discrete space V_{1D}^h (see [20] for all the details).

The HiMod solution can be fully characterized by introducing a basis, $\{\theta_l\}_{l=1}^{N_h}$, for the space V_{1D}^h. Actually, each modal coefficient, $\tilde{u}_k$, of u_m^{HiMod} can be expanded in terms of such a basis, so that, we obtain the modal representation

$$u_m^{\text{HiMod}}(x,\mathbf{y}) = \sum_{k=1}^{m}\sum_{l=1}^{N_h} \tilde{u}_{k,l}\theta_l(x)\varphi_k(\psi_x(\mathbf{y})). \tag{4}$$

The actual unknowns of problem (3) become the mN_h coefficients $\{\tilde{u}_{k,l}\}_{k=1,l=1}^{m,N_h}$. With reference to the Poisson problem, $-\Delta u = f$, completed with full homogeneous Dirichlet boundary data, the corresponding HiMod formulation, after exploiting (4) in (3) and picking $v_m(x,\mathbf{y}) = \theta_i(x)\varphi_j(\psi_x(\mathbf{y}))$ with $i = 1,\ldots,N_h$ and $j = 1,\ldots,m$, reduces to the system of mN_h 1D equations in the mN_h unknowns $\{\tilde{u}_{k,l}\}_{k=1,l=1}^{m,N_h}$,

$$\sum_{k=1}^{m}\sum_{l=1}^{N_h}\tilde{u}_{k,l}\Bigg[\int_{\Omega_{1D}}\Bigg(\hat{r}_{jk}^{1,1}(x)\frac{d\theta_l}{dx}(x)\frac{d\theta_i}{dx}(x) + \hat{r}_{jk}^{1,0}(x)\frac{d\theta_l}{dx}(x)\theta_i(x) + \\ + \hat{r}_{jk}^{0,1}(x)\theta_l(x)\frac{d\theta_i}{dx}(x) + \hat{r}_{jk}^{0,0}(x)\theta_l(x)\theta_i(x)\Bigg)\Bigg]dx = \int_{\Omega_{1D}}\hat{f}_j(x)\theta_i(x)\,dx,$$

where $\hat{r}_{jk}^{a,b}(x) = \int_{\hat{\gamma}_{d-1}} r_{jk}^{a,b}(x,\hat{\mathbf{y}})|J|\,d\hat{\mathbf{y}}$ with $a,b = 0{,}1$, $J = \det\big(\mathcal{D}_2^{-1}(x,\psi_x^{-1}(\hat{\mathbf{y}}))\big)$ with $\mathcal{D}_2 = \mathcal{D}_2(x,\psi_x^{-1}(\hat{\mathbf{y}})) = \nabla_{\mathbf{y}}\psi_x$,

$$\begin{aligned} r_{jk}^{0,0}(x,\hat{\mathbf{y}}) &= \varphi_k'(\hat{\mathbf{y}})\varphi_j'(\hat{\mathbf{y}})\big(\mathcal{D}_1^2+\mathcal{D}_2^2\big), & r_{jk}^{0,1}(x,\hat{\mathbf{y}}) &= \varphi_k'(\hat{\mathbf{y}})\varphi_j(\hat{\mathbf{y}})\mathcal{D}_1, \\ r_{jk}^{1,0}(x,\hat{\mathbf{y}}) &= \varphi_k(\hat{\mathbf{y}})\varphi_j'(\hat{\mathbf{y}})\mathcal{D}_1, & r_{jk}^{1,1}(x,\hat{\mathbf{y}}) &= \varphi_k(\hat{\mathbf{y}})\varphi_j(\hat{\mathbf{y}}), \end{aligned}$$

with $\mathcal{D}_1 = \mathcal{D}_1(x,\psi_x^{-1}(\hat{\mathbf{y}})) = \partial\psi_x/\partial x$, and $\hat{f}_j(x) = \int_{\hat{\gamma}_{d-1}} f(x,\psi_x^{-1}(\hat{\mathbf{y}}))\varphi_j(\hat{\mathbf{y}})$ $|J|\,d\hat{\mathbf{y}}$. Information associated with the transverse dynamics are lumped in the coefficients $\{\hat{r}_{jk}^{a,b}\}$, so that the HiMod system is solved on the supporting fiber, Ω_{1D}. Collecting the HiMod unknowns, by mode, in the vector $\mathbf{u}_m^{\text{HiMod}} \in \mathbb{R}^{mN_h}$, such that

$$\mathbf{u}_m^{\text{HiMod}} = [\tilde{u}_{1,1},\tilde{u}_{1,2},\ldots,\tilde{u}_{1,N_h},\tilde{u}_{2,1},\ldots,\tilde{u}_{m,1},\ldots,\tilde{u}_{m,N_h}]^T, \tag{5}$$

we can rewrite the HiMod system in the compact form

$$A_m^{\text{HiMod}}\mathbf{u}_m^{\text{HiMod}} = \mathbf{f}_m^{\text{HiMod}}, \tag{6}$$

where $A_m^{\text{HiMod}} \in \mathbb{R}^{mN_h\times mN_h}$ and $\mathbf{f}_m^{\text{HiMod}} \in \mathbb{R}^{mN_h}$ are the HiMod stiffness matrix and right-hand side, respectively, with $[\mathbf{f}_m^{\text{HiMod}}]_{ji} = \int_{\Omega_{1D}}\hat{f}_j(x)\theta_i(x)dx$, and $[A_m^{\text{HiMod}}]_{ji,kl} = \sum_{a,b=0}^{1}\int_{\Omega_{1D}}\hat{r}_{jk}^{a,b}(x)\frac{d^a\theta_l}{dx}(x)\frac{d^b\theta_i}{dx}(x)dx$. According to (5), for each modal index j, between 1 and m, the nodal index, i, takes the values $1,\ldots,N_h$. Thus, HiMod reduction leads to solve a system of order mN_h, independently of the dimension of the full problem (1).

The PGD Approach

To perform PGD, we have to introduce on problem (1) a separability hypothesis with respect to both the spatial variables and the data [5, 22]. Thus, domain $\Omega \subset \mathbb{R}^d$ coincides with the rectangle $\Omega_x \times \Omega_y$ if $d = 2$, with the parallelepiped $\Omega_x \times \Omega_y \times \Omega_z$ (total separability) or with the cylinder $\Omega_x \times \Omega_{\mathbf{y}}$ (partial separability) if $d = 3$, for $\Omega_x, \Omega_y, \Omega_z \subset \mathbb{R}$ and $\Omega_{\mathbf{y}} \subset \mathbb{R}^2$, being $\mathbf{y} = (y, z)$. In the following, we focus on partial separability, since it is more suited to match HiMod reduction with PGD. Analogously, we assume that the generic problem data, $d = d(x, y, z)$, can be written as $d = d^x(x)d^{\mathbf{y}}(\mathbf{y})$. The separability is inherited by the PGD space

$$W_m = \left\{ w_m(x, \mathbf{y}) = \sum_{k=1}^{m} w_k^x(x) w_k^{\mathbf{y}}(\mathbf{y}), \text{ with } w_k^x \in W_h^x, \; w_k^{\mathbf{y}} \in W_h^{\mathbf{y}}, \; x \in \Omega_x, \; \mathbf{y} \in \Omega_{\mathbf{y}} \right\}, \tag{7}$$

where $W_h^x \subseteq H^1(\Omega_x)$ and $W_h^{\mathbf{y}} \subseteq H^1(\Omega_{\mathbf{y}}; \mathbb{R}^{d-1})$ are discrete spaces, with $\dim(W_h^x) = N_h^x$ and $\dim(W_h^{\mathbf{y}}) = N_h^{\mathbf{y}}$, associated with partitions, $\mathcal{T}_h^x$ and $\mathcal{T}_h^{\mathbf{y}}$, of Ω_x and $\Omega_{\mathbf{y}}$, respectively. In general, W_h^x and $W_h^{\mathbf{y}}$ are FE spaces, although, a priori, any discretization can be adopted. It turns out that W_m is a tensor function space, being $W_m = W_h^x \otimes W_h^{\mathbf{y}} \subseteq H^1(\Omega_x) \otimes H^1(\Omega_{\mathbf{y}}; \mathbb{R}^{d-1})$.

Index m plays the same role as in the HiMod reduction, setting the level of detail for the reduced solution (see Sect. for possible criteria to choose m). PGD exploits the hierarchical structure in W_m to build the generic function $w_m \in W_m$. In particular, w_m is computed as

$$w_m(x, \mathbf{y}) = w_m^x(x) w_m^{\mathbf{y}}(\mathbf{y}) + \sum_{k=1}^{m-1} w_k^x(x) w_k^{\mathbf{y}}(\mathbf{y}), \tag{8}$$

where w_k^x and $w_k^{\mathbf{y}}$ are assumed known for $k = 1, \ldots, m-1$, so that the enrichment functions, w_m^x and $w_m^{\mathbf{y}}$, become the actual unknowns. To provide the PGD formulation for the Poisson problem considered in Sect , we exploit representation (8) for the PGD approximation, u_m^{PGD}, and we pick the test function as $X(x)Y(\mathbf{y})$, with $X \in W_h^x$ and $Y \in W_h^{\mathbf{y}}$. The coupling between the unknowns, u_m^x and $u_m^{\mathbf{y}}$, leads to a nonlinear problem, which is tackled by means of the *Alternating Direction Strategy* (ADS) [5]. The idea is to look for u_m^x and $u_m^{\mathbf{y}}$, separately via a fixed point procedure. We introduce an auxiliary index to keep trace of the ADS iterations, so that, at the p-th ADS iteration we compute $u_m^{x,p}$ and $u_m^{\mathbf{y},p}$ starting from the previous approximations, $u_m^{x,g}$ and $u_m^{\mathbf{y},g}$ for $g = 1, \ldots, p-1$, following a two-step procedure. First, we compute $u_m^{x,p}$ by identifying $u_m^{\mathbf{y}}$ with $u_m^{\mathbf{y},p-1}$, and by selecting $Y(\mathbf{y}) = u_m^{\mathbf{y},p-1}$ in the test function. This yields, for any $X \in W_h^x$,

$$\begin{aligned}
&\int_{\Omega_x} \left(u_m^{x,p}\right)' X' dx \int_{\Omega_\mathbf{y}} \left[u_m^{\mathbf{y},p-1}\right]^2 d\mathbf{y} + \int_{\Omega_x} u_m^{x,p} X dx \int_{\Omega_\mathbf{y}} \left[\left(u_m^{\mathbf{y},p-1}\right)'\right]^2 d\mathbf{y} \\
&= \int_{\Omega_x} f^x X dx \int_{\Omega_\mathbf{y}} f^\mathbf{y} u_m^{\mathbf{y},p-1} d\mathbf{y} - \sum_{k=1}^{m-1} \int_{\Omega_x} \left(u_k^x\right)' X' dx \int_{\Omega_\mathbf{y}} u_k^\mathbf{y} u_m^{\mathbf{y},p-1} d\mathbf{y} \\
&- \sum_{k=1}^{m-1} \int_{\Omega_x} u_k^x X dx \int_{\Omega_\mathbf{y}} \left(u_k^\mathbf{y}\right)' \left(u_m^{\mathbf{y},p-1}\right)' d\mathbf{y},
\end{aligned} \tag{9}$$

where the separability of f is exploited (the dependence on the independent variables, x and $\mathbf{y}$, is omitted to simplify notation). Successively, we compute $u_m^{\mathbf{y},p}$, after setting u_m^x to $u_m^{x,p}$ and choosing function X as to $u_m^{x,p}$ in the test function, so that we obtain, for any $Y \in W_h^\mathbf{y}$,

$$\begin{aligned}
&\int_{\Omega_x} \left[\left(u_m^{x,p}\right)'\right]^2 dx \int_{\Omega_\mathbf{y}} u_m^{\mathbf{y},p} Y d\mathbf{y} + \int_{\Omega_x} \left[u_m^{x,p}\right]^2 dx \int_{\Omega_\mathbf{y}} \left(u_m^{\mathbf{y},p}\right)' Y' d\mathbf{y} \\
&= \int_{\Omega_x} f^x u_m^{x,p} dx \int_{\Omega_\mathbf{y}} f^\mathbf{y} Y d\mathbf{y} - \sum_{k=1}^{m-1} \int_{\Omega_x} \left(u_k^x\right)' \left(u_m^{x,p}\right)' dx \int_{\Omega_\mathbf{y}} u_k^\mathbf{y} Y d\mathbf{y} \\
&- \sum_{k=1}^{m-1} \int_{\Omega_x} u_k^x u_m^{x,p} dx \int_{\Omega_\mathbf{y}} \left(u_k^\mathbf{y}\right)' Y' d\mathbf{y}.
\end{aligned} \tag{10}$$

The algebraic counterpart of (9) and (10) is obtained by introducing a basis, $\mathcal{B}_x = \{\theta_\alpha^x\}_{\alpha=1}^{N_h^x}$ and $\mathcal{B}_\mathbf{y} = \{\theta_\beta^\mathbf{y}\}_{\beta=1}^{N_h^\mathbf{y}}$, for the space W_h^x and $W_h^\mathbf{y}$, respectively, so that $u_j^q(q) = \sum_{i=1}^{N_h^q} \tilde{u}_{ji}^q \theta_i^q(q)$, $u_m^{q,s}(q) = \sum_{i=1}^{N_h^q} \tilde{u}_{mi}^{q,s} \theta_i^q(q)$, with $q = x, \mathbf{y}$, $s = p, p-1$, $j = 1, \ldots, m-1$, and, likewise, $X(x) = \sum_{\alpha=1}^{N_h^x} \tilde{x}_\alpha \theta_\alpha^x(x)$ and $Y(\mathbf{y}) = \sum_{\beta=1}^{N_h^\mathbf{y}} \tilde{y}_\beta \theta_\beta^\mathbf{y}(\mathbf{y})$. Thanks to these expansions and to the arbitrariness of X and Y, we can rewrite (9) and (10) as

$$\begin{aligned}
&\left\{\left[\left(\mathbf{u}_m^{\mathbf{y},p-1}\right)^T M^\mathbf{y} \mathbf{u}_m^{\mathbf{y},p-1}\right] K^x + \left[\left(\mathbf{u}_m^{\mathbf{y},p-1}\right)^T K^\mathbf{y} \mathbf{u}_m^{\mathbf{y},p-1}\right] M^x\right\} \mathbf{u}_m^{x,p} = \left[\left(\mathbf{u}_m^{\mathbf{y},p-1}\right)^T \mathbf{f}^\mathbf{y}\right] \mathbf{f}^x \\
&- \sum_{k=1}^{m-1} \left\{\left[\left(\mathbf{u}_m^{\mathbf{y},p-1}\right)^T M^\mathbf{y} \mathbf{u}_k^\mathbf{y}\right] K^x + \left[\left(\mathbf{u}_m^{\mathbf{y},p-1}\right)^T K^\mathbf{y} \mathbf{u}_k^\mathbf{y}\right] M^x\right\} \mathbf{u}_k^x,
\end{aligned} \tag{11}$$

and

$$\begin{aligned}
&\left\{\left[\left(\mathbf{u}_m^{x,p}\right)^T K^x \mathbf{u}_m^{x,p}\right] M^\mathbf{y} + \left[\left(\mathbf{u}_m^{x,p}\right)^T M^x \mathbf{u}_m^{x,p}\right] K^\mathbf{y}\right\} \mathbf{u}_m^{\mathbf{y},p} = \left[\left(\mathbf{u}_m^{x,p}\right)^T \mathbf{f}^x\right] \mathbf{f}^\mathbf{y} \\
&- \sum_{k=1}^{m-1} \left\{\left[\left(\mathbf{u}_m^{x,p}\right)^T K^x \mathbf{u}_k^x\right] M^\mathbf{y} + \left[\left(\mathbf{u}_m^{x,p}\right)^T M^x \mathbf{u}_k^x\right] K^\mathbf{y}\right\} \mathbf{u}_k^\mathbf{y},
\end{aligned} \tag{12}$$

respectively, where vectors $\mathbf{u}_j^q$, $\mathbf{u}_m^{q,s} \in \mathbb{R}^{N_h^q}$ collect the PGD coefficients, being $\left[\mathbf{u}_j^q\right]_i = \tilde{u}_{ji}^q$, $\left[\mathbf{u}_m^{q,s}\right]_i = \tilde{u}_{mi}^{q,s}$ and $i = 1, \ldots, N_h^q$, K^x, $M^x \in \mathbb{R}^{N_h^x \times N_h^x}$ and $K^\mathbf{y}$, $M^\mathbf{y} \in \mathbb{R}^{N_h^\mathbf{y} \times N_h^\mathbf{y}}$ are the stiffness and mass matrices associated with x- and $\mathbf{y}$-variables, with $\left[K^x\right]_{\alpha l} = \int_{\Omega_x} \left(\theta_\alpha^x\right)' \left(\theta_l^x\right)' dx$, $\left[K^\mathbf{y}\right]_{\beta s} = \int_{\Omega_\mathbf{y}} \left(\theta_\beta^\mathbf{y}\right)' \left(\theta_s^\mathbf{y}\right)' d\mathbf{y}$, $\left[M^x\right]_{\alpha l} =$

$\int_{\Omega_x} \theta_\alpha^x \theta_l^x dx$, $\left[M^{\mathbf{y}}\right]_{\beta s} = \int_{\Omega_{\mathbf{y}}} \theta_\beta^{\mathbf{y}} \theta_s^{\mathbf{y}} d\mathbf{y}$, and where $\mathbf{f}^x \in \mathbb{R}^{N_h^x}$, $\mathbf{f}^{\mathbf{y}} \in \mathbb{R}^{N_h^{\mathbf{y}}}$, with $\left[\mathbf{f}^x\right]_l = \int_{\Omega_x} f^x \theta_l^x dx$, $\left[\mathbf{f}^{\mathbf{y}}\right]_s = \int_{\Omega_{\mathbf{y}}} f^{\mathbf{y}} \theta_s^{\mathbf{y}} d\mathbf{y}$, for $\alpha, l = 1, \ldots, N_h^x, \beta, s = 1, \ldots, N_h^{\mathbf{y}}$. Systems (11) and (12) are solved at each ADS iteration, so that the computational effort characterizing PGD is the one associated with the solution of two systems of order N_h^x and $N_h^{\mathbf{y}}$, respectively, for each ADS iteration. When a certain stopping criterion is met (see the next section for more details), ADS procedure yields vectors $\mathbf{u}_m^x$ and $\mathbf{u}_m^{\mathbf{y}}$ which identify the enrichment functions u_m^x and $u_m^{\mathbf{y}}$.

HiMod Reduction Versus PGD

Both HiMod reduction and PGD exploit the separation of variables and, according to [5], belong to the a priori approaches, since they do not rely on any solution to the problem at hand. Nevertheless, we can easily itemize features which distinguish the two techniques. The most relevant ones concern the geometry of Ω, the selection of the transverse basis and of the modal index, and the numerical implementation of the two procedures. Pros and cons of the two methods are then here highlighted.

Domain Geometry

HiMod reduction and PGD advance precise hypotheses on the geometry of the computational domain.

According to the HiMod approach, Ω is expected to coincide with a fiber bundle and to be mapped into the reference domain, $\hat{\Omega}$, by a sufficiently regular transformation. Actually, map Ψ is assumed differentiable, while map ψ_x is required to be a C^1-diffeomorphism, for all $x \in \Omega_{1D}$ [20]. These hypotheses introduce some constraints, in particular, on the lateral boundary of Ω which, e.g., cannot exhibit kinks. Additionally, geometries of interest in many applications, such as bifurcations or, more in general, networks are ruled out from the demands on ψ_x and Ψ. An approach based on the domain decomposition technique is currently under investigation as a viable way to deal with such geometries. The isogeometric version of HiMod (i.e., the HIgaMod approach) will play a crucial role in view of HiMod simulations for the blood flow modeling in patient-specific geometries [21].

The constraints introduced by PGD on the geometry of Ω are more restrictive. The separability hypothesis leads to consider essentially only Cartesian domains. This considerably reduces the applicability of PGD to practical contexts. Some techniques are available in the literature to overcome this issue. For instance, in [9] a generic domain is embedded into a Cartesian geometry, while in [7] the authors introduce a parametrization map for quadrilateral domains.

Overall, HiMod reduction exhibits a higher geometric flexibility with respect to PGD, in its straightforward formulation. As discussed in Sect. this limitation can be removed when considering a parametric setting.

Modeling of the Transverse Dynamics

In the HiMod expansion, **y**-components, $\varphi_k(\psi_x(\mathbf{y}))$, are selected before starting the model reduction. This choice, although coherent with an a priori approach, introduces a constraint on the dynamics that can be described, so that hints about the solution trend along the transverse direction can be helpful to select a representative modal basis. In the original proposal of the HiMod procedure, sinusoidal functions are employed according to a Fourier expansion [6, 20]. This turns out to be a reasonable choice when Dirichlet boundary conditions are assigned on the lateral surface, $\Gamma_{\text{lat}} = \{x\} \times \partial\gamma_x$, of Ω. Legendre polynomials, properly modified to include the homogeneous Dirichlet data and orthonormalized, are employed in [20] as an alternative to a trigonometric expansion. Nevertheless, Legendre polynomials require high-order quadrature rules to accurately compute coefficients $\{\hat{r}_{jk}^{a,b}\}$.

In [1], the concept of educated modal basis is introduced to impose generic boundary conditions on Γ_{lat}. The idea is to solve an auxiliary Sturm-Liouville eigenvalue problem on the transverse reference fiber $\hat{\gamma}_{d-1}$, to build a basis which automatically includes the boundary values on Γ_{lat}. The eigenfunctions of the Sturm-Liouville problem provide the modal basis. A first attempt to generalize the educated-HiMod reduction to three-dimensional (3D) cylindrical geometries is performed in [10], where the Navier-Stokes equations are hierarchically reduced to model the blood flow in pipes. This generalization is far from being straightforward due to the employment of polar coordinates. To overcome this issue, we are currently investigating the HIgaMod approach [21], which allows us to define the transverse basis as the Cartesian product of 1D modal functions, independently of the considered geometry.

Additionally, we remark that any modal basis can be precomputed on the transverse reference fiber before performing the HiMod reduction, thanks to the employment of map Ψ. This considerably simplifies computations.

When applying PGD, **y**-components are unknown as the ones associated with x. This leads to the nonlinear problems (9)–(10), thus loosing any advantage related to a precomputation of the HiMod modal basis. On the other hand, PGD does not constrain the transverse dynamic to follow a prescribed (e.g., sinusoidal) analytical shape as HiMod procedure does. The educated-Himod reduction clearly is out of this comparison, since the modal basis strictly depends on the problem at hand.

Finally, we observe that HiMod modes are orthonormal with respect to the $L^2(\hat{\gamma}_{d-1})$-norm. This property is not ensured by PGD.

Concerning the selection of the modal index m in (2) and (7), as a first attempt, both HiMod reduction and PGD resort to a trial-and-error approach, so that the modal index is gradually increased until a check on the accuracy of the reduced solution is satisfied. For instance, in [6, 20] a qualitative investigation of the contour plot of the HiMod approximation drives the choice of m. Concerning PGD, the check on the relative enrichment

$$\frac{\|u_m^x u_m^{\mathbf{y}}\|_{L^2(\Omega)}}{\|u_1^x u_1^{\mathbf{y}}\|_{L^2(\Omega)}} \leq \mathtt{TOL_E}, \tag{13}$$

is usually employed, with $\mathtt{TOL_E}$ a user-defined tolerance [5]. An automatic selection of index m can yield a significant improvement. In [17, 19], an adaptive procedure is proposed for HiMod, based on an a posteriori modeling error analysis. In particular, the estimator in [17] is derived in a goal-oriented setting to control a quantity of interest, and exploits the hierarchical structure (i.e., the inclusion $V_m \subset V_{m+d}$, $\forall m, d \in \mathbb{N}^+$) typical of a HiMod reduction. A similar modeling error analysis is performed in [2] for PGD, although no adaptive algorithm is here set to automatically pick the reduced model. Paper [19] generalizes the a posteriori analysis in [17] to an unsteady setting, providing the tool to automatically select m together with the partition $\mathcal{T}_h$ along Ω_{1D} and the time step.

Finally, HiMod allows to tune the modal index along the domain Ω, according to the local complexity of the transverse dynamics. In particular, m can be varied in different areas of Ω or, in the presence of very localized dynamics, in correspondence with specific nodes of the partition $\mathcal{T}_h$. We refer to these two variants as to *piecewise* and *pointwise* HiMod reduction, in contrast to a *uniform* approach, where the same number of modes is adopted everywhere [16, 18]. This flexibility in the choice of m is currently not available for PGD. Adaptive strategies to select the modal index are available for the three variants of the HiMod procedure [17, 19].

Computational Aspects

From a computational viewpoint, HiMod reduction and PGD lead to completely different procedures. Indeed, for a fixed value of m, we have to solve the only system (6) of order mN_h when applying HiMod, in contrast to PGD which demands a multiple solution of systems (11)–(12) of order N_h^x and $N_h^{\mathbf{y}}$, respectively because of the fixed point and the enrichment algorithms. Thus, the direct solution of a single system, in general of larger order, is replaced by an iterative solution of several and smaller systems. This heterogeneity makes a computational comparison between PGD and HiMod not so meaningful. We verify the reliability of the HiMod and PGD procedures on a common test case, by choosing in (1) $V = H_0^1(\Omega)$ with $\Omega = (0, 5) \times (0, 1)$, $a(u, v) = \int_\Omega \big[\mu \nabla u \cdot \nabla v + \mathbf{b} \cdot \nabla u\big] d\Omega$ for $\mu = 0.24$, $\mathbf{b} = [-5, 0]^T$, and $F(v) = \int_\Omega f v d\Omega$ with $f(x, y) = 50\big\{\exp\big[-\big((x-2.85)/0.075\big)^2 - \big((y-0.5)/0.075\big)^2\big] + \exp\big[-\big((x-3.75)/0.075\big)^2 - \big((y-0.5)/0.075\big)^2\big]\big\}$. For both the methods, we uniformly subdivide Ω_{1D} into 285 subintervals. We set the PGD discretization along y as well as the PGD and the HiMod index m in order

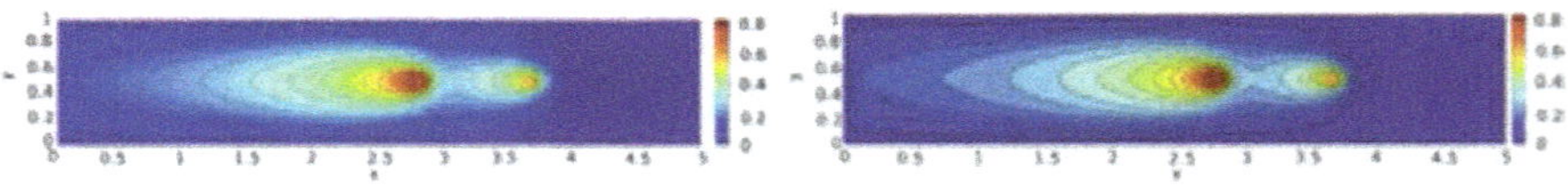

Fig. Qualitative comparison between a HiMod (left) and a PGD (right) approximations

to ensure the same accuracy, TOL, on the reduced approximations with respect to a reference FE solution, computed on a 2500×500 structured mesh. In particular, for $\texttt{TOL} = 8 \cdot 10^{-3}$, we have to subdivide interval $(0, 1)$ into 20 uniform subintervals, and to set m to 6 and to 9 in the PGD and the HiMod discretization, respectively. Sinusoidal functions are chosen for the HiMod modal basis. The ADS iterations are controlled in terms of the relative increment, as

$$\frac{\| u_m^{x,p} u_m^{y,p} - u_m^{x,p-1} u_m^{y,p-1} \|_{L^2(\Omega)}}{\| u_m^{x,p} u_m^{y,p} \|_{L^2(\Omega)}} \leq \texttt{TOL}_{\texttt{FP}}, \tag{14}$$

with $\texttt{TOL}_{\texttt{FP}} = 10^{-2}$. Figure shows the reduced approximations (which are fully comparable with the FE one, here omitted). The contourplots are very similar. The coarse PGD y-discretization justifies the slight roughness of the PGD contourlines.

Another distinguishing feature between HiMod and PGD is the domain discretization. Indeed, HiMod requires only the partition $\mathcal{T}_h$ along Ω_{1D}, independently of the dimension of Ω. No discretization is needed in the $\mathbf{y}$-direction, although we have to carefully select the quadrature nodes to compute coefficients $\{\hat{r}_{jk}^{a,b}\}$. This task becomes particularly challenging when dealing with polar coordinates [10]. With PGD to benefit of the computational advantages associated with a 1D discretization, we are obliged to assume the full separability of Ω; actually, a partial separability demands a 1D partition for Ω_x, and a two-dimensional partition of $\Omega_{\mathbf{y}}$. As explained in Sect. non-Cartesian domains require a 3D discretization of Ω.

Finally we analyze the interplay between the enrichment and the ADS iterations in the PGD reduction. We investigate the possible relationship between $\texttt{TOL}_{\texttt{FP}}$ in (14) and $\texttt{TOL}_{\texttt{E}}$ in (13), to verify if a small tolerance for the fixed point iteration improves the accuracy of the PGD approximation, thus reducing the number of enrichment steps. To do this, we adopt the same test case used above. Table gathers the number of ADS iterations, $\#\texttt{IT}_{\texttt{FP}}$, the number, m, of enrichment steps, and the CPU time[1] (in seconds) demanded by the PGD procedure, for two different values of $\texttt{TOL}_{\texttt{E}}$ and three different choices of $\texttt{TOL}_{\texttt{FP}}$. In particular, in column $\#\texttt{IT}_{\texttt{FP}}$ we specify the number of ADS iterations required by each enrichment step. As expected, there exists a link between the two tolerances, namely, when a higher accuracy constrains the fixed point iteration, a smaller number of enrichment steps is performed to ensure the accuracy $\texttt{TOL}_{\texttt{E}}$.

[1]The computations have been run on a Intel Core i5 Dual-Core CPU 2.7 GHz 8 GB RAM MacBook.

Table Quantitative analysis for PGD in terms of fixed point iterations and enrichment steps

	$\mathtt{TOL_E} = 2 \cdot 10^{-2}$			$\mathtt{TOL_E} = 8 \cdot 10^{-3}$		
	$\#\mathtt{IT_{FP}}$	m	CPU [s]	$\#\mathtt{IT_{FP}}$	m	CPU [s]
$\mathtt{TOL_{FP}} = 10^{-1}$	{2, 2, 2}	3	0.099640	{2, 2, 2, 2, 2}	5	0.337861
$\mathtt{TOL_{FP}} = 10^{-2}$	{4, 3}	2	0.046756	{4, 3, 2, 2, 4}	5	0.358555
$\mathtt{TOL_{FP}} = 10^{-3}$	{5, 5}	2	0.077958	{5, 5, 2, 7}	4	0.341748

HiMod Reduction and PGD for Parametrized Problems

The actual potential of PGD becomes more evident when considering a parametric setting, i.e., when problem (1) is replaced by the formulation

$$\text{find } u(\boldsymbol{\mu}) \in V : a(u(\boldsymbol{\mu}), v; \boldsymbol{\mu}) = F(v; \boldsymbol{\mu}) \quad \forall v \in V, \tag{15}$$

with $\boldsymbol{\mu}$ a parameter, which may represent any data of the problem, e.g., the coefficients of the considered PDE, the source term, a boundary value or the domain geometry.

The technique adopted by PGD to deal with the parametric dependence in (15) is very effective. Parameter $\boldsymbol{\mu}$ is considered as an additional independent variable which varies in a domain Ω_μ [5]. Thus, the PGD space (7) changes into the new one

$$W_m^{\boldsymbol{\mu}} = \Bigg\{ w_m(x, \mathbf{y}, \boldsymbol{\mu}) = \sum_{k=1}^{m} w_k^x(x) w_k^{\mathbf{y}}(\mathbf{y}) w_k^{\boldsymbol{\mu}}(\boldsymbol{\mu}), \text{ with} \\ w_k^x \in W_h^x,\ w_k^{\mathbf{y}} \in W_h^{\mathbf{y}},\ w_k^{\boldsymbol{\mu}} \in W_h^{\boldsymbol{\mu}}, x \in \Omega_x,\ \mathbf{y} \in \Omega_{\mathbf{y}},\ \boldsymbol{\mu} \in \Omega_\mu \Bigg\}, \tag{16}$$

with $W_h^{\boldsymbol{\mu}}$ a discretization of the space $L^2(\Omega_\mu; \mathbb{R}^Q)$, being Q the length of vector $\boldsymbol{\mu}$. Generalizing the enrichment paradigm in (8), at the m-th step of the PGD approach applied to problem (15) we have to compute three unknown functions, u_m^x, $u_m^{\mathbf{y}}$ and $u_m^{\boldsymbol{\mu}}$, by picking the test function as $X(x)Y(\mathbf{y})Z(\boldsymbol{\mu})$, with $X \in W_h^x$, $Y \in W_h^{\mathbf{y}}$, $Z \in W_h^{\boldsymbol{\mu}}$. Functions u_m^x, $u_m^{\mathbf{y}}$, $u_m^{\boldsymbol{\mu}}$ are computed by ADS, which now coincides with a three-step procedure. Thus, with reference to the Poisson problem, $-\nabla\cdot\big(\mu\nabla u\big) = f$ completed with full homogeneous Dirichlet boundary conditions and for $\boldsymbol{\mu} \equiv \mu$, we first compute $u_m^{x,p}$ by identifying $u_m^{\mathbf{y}}$ and $u_m^{\boldsymbol{\mu}}$ with the previous approximations, $u_m^{\mathbf{y},p-1}$ and $u_m^{\boldsymbol{\mu},p-1}$, respectively and by selecting $Y(\mathbf{y})Z(\boldsymbol{\mu}) = u_m^{\mathbf{y},p-1} u_m^{\boldsymbol{\mu},p-1}$ in the test function. This leads to a linear system which generalizes (11), namely

$$\begin{aligned}
&\left[\left(\mathbf{u}_m^{\boldsymbol{\mu},p-1}\right)^T M^{\boldsymbol{\mu}}\mathbf{u}_m^{\boldsymbol{\mu},p-1}\right]\\
&\left\{\left[\left(\mathbf{u}_m^{\mathbf{y},p-1}\right)^T M^{\mathbf{y}}\mathbf{u}_m^{\mathbf{y},p-1}\right]K^x+\left[\left(\mathbf{u}_m^{\mathbf{y},p-1}\right)^T K^{\mathbf{y}}\mathbf{u}_m^{\mathbf{y},p-1}\right]M^x\right\}\mathbf{u}_m^{x,p}\\
&=\left[\left(\mathbf{u}_m^{\mathbf{y},p-1}\right)^T\mathbf{f}^{\mathbf{y}}\right]\left[\left(\mathbf{u}_m^{\boldsymbol{\mu},p-1}\right)^T\mathbf{f}^{\boldsymbol{\mu}}\right]\mathbf{f}^x-\textstyle\sum_{k=1}^{m-1}\left[\left(\mathbf{u}_m^{\boldsymbol{\mu},p-1}\right)^T M^{\boldsymbol{\mu}}\mathbf{u}_k^{\boldsymbol{\mu}}\right]\\
&\left\{\left[\left(\mathbf{u}_m^{\mathbf{y},p-1}\right)^T M^{\mathbf{y}}\mathbf{u}_k^{\mathbf{y}}\right]K^x+\left[\left(\mathbf{u}_m^{\mathbf{y},p-1}\right)^T K^{\mathbf{y}}\mathbf{u}_k^{\mathbf{y}}\right]M^x\right\}\mathbf{u}_k^x,
\end{aligned} \tag{17}$$

where $M^{\boldsymbol{\mu}} \in \mathbb{R}^{N_h^{\boldsymbol{\mu}}\times N_h^{\boldsymbol{\mu}}}$ is the mass matrix associated with the parameter $\boldsymbol{\mu}$, with $\left[M^{\boldsymbol{\mu}}\right]_{ij} = \int_{\Omega_\mu}\boldsymbol{\mu}\theta_i^{\boldsymbol{\mu}}\theta_j^{\boldsymbol{\mu}}d\boldsymbol{\mu}$ for $i, j = 1, \dots, N_h^{\boldsymbol{\mu}}$ and $\mathcal{B}_{\boldsymbol{\mu}} = \{\theta_\gamma^{\boldsymbol{\mu}}\}_{\gamma=1}^{N_h^{\boldsymbol{\mu}}}$ a basis for the space $W_h^{\boldsymbol{\mu}}$, $\mathbf{f}^{\boldsymbol{\mu}} \in \mathbb{R}^{N_h^{\boldsymbol{\mu}}}$ with $\left[\mathbf{f}^{\boldsymbol{\mu}}\right]_l = \int_{\Omega_\mu} f^{\boldsymbol{\mu}}\theta_l^{\boldsymbol{\mu}}d\boldsymbol{\mu}$ for $l = 1, \dots, N_h^{\boldsymbol{\mu}}$ after assuming the separability $f = f^x f^{\mathbf{y}} f^{\boldsymbol{\mu}}$ for the source term f, and where we employ the same notation as in (11)–(12) to denote vectors $\mathbf{u}_w^{\boldsymbol{\mu}}$, $\mathbf{u}_m^{\boldsymbol{\mu},s}$, with $w = 1, \dots, m-1$, $s = p, p-1$, collecting the PGD coefficients associated with the basis $\mathcal{B}_{\boldsymbol{\mu}}$. Analogously, $u_m^{\mathbf{y},p}$ is computed by solving the generalization of the linear system (12) given by

$$\begin{aligned}
&\left[\left(\mathbf{u}_m^{\boldsymbol{\mu},p-1}\right)^T M^{\boldsymbol{\mu}}\mathbf{u}_m^{\boldsymbol{\mu},p-1}\right]\left\{\left[\left(\mathbf{u}_m^{x,p}\right)^T K^x\mathbf{u}_m^{x,p}\right]M^{\mathbf{y}}+\left[\left(\mathbf{u}_m^{x,p}\right)^T M^x\mathbf{u}_m^{x,p}\right]K^{\mathbf{y}}\right\}\mathbf{u}_m^{\mathbf{y},p}\\
&=\left[\left(\mathbf{u}_m^{x,p}\right)^T\mathbf{f}^x\right]\left[\left(\mathbf{u}_m^{\boldsymbol{\mu},p-1}\right)^T\mathbf{f}^{\boldsymbol{\mu}}\right]\mathbf{f}^{\mathbf{y}}-\textstyle\sum_{k=1}^{m-1}\left[\left(\mathbf{u}_m^{\boldsymbol{\mu},p-1}\right)^T M^{\boldsymbol{\mu}}\mathbf{u}_k^{\boldsymbol{\mu}}\right]\\
&\left\{\left[\left(\mathbf{u}_m^{x,p}\right)^T K^x\mathbf{u}_k^x\right]M^{\mathbf{y}}+\left[\left(\mathbf{u}_m^{x,p}\right)^T M^x\mathbf{u}_k^x\right]K^{\mathbf{y}}\right\}\mathbf{u}_k^{\mathbf{y}},
\end{aligned}$$

after setting $u_m^x = u_m^{x,p}$, $u_m^{\boldsymbol{\mu}} = u_m^{\boldsymbol{\mu},p-1}$ and $X(x)Z(\boldsymbol{\mu}) = u_m^{x,p}u_m^{\boldsymbol{\mu},p-1}$ for the PGD test function. Finally, we have the additional linear system used to compute $u_m^{\boldsymbol{\mu},p}$,

$$\begin{aligned}
&\left\{\left[\left(\mathbf{u}_m^{x,p}\right)^T K^x\mathbf{u}_m^{x,p}\right]\left[\left(\mathbf{u}_m^{\mathbf{y},p}\right)^T M^{\mathbf{y}}\mathbf{u}_m^{\mathbf{y},p}\right]+\left[\left(\mathbf{u}_m^{x,p}\right)^T M^x\mathbf{u}_m^{x,p}\right]\left[\left(\mathbf{u}_m^{\mathbf{y},p}\right)^T K^{\mathbf{y}}\mathbf{u}_m^{\mathbf{y},p}\right]\right\}\\
&M^{\boldsymbol{\mu}}\mathbf{u}_m^{\boldsymbol{\mu},p}=\left[\left(\mathbf{u}_m^{x,p}\right)^T\mathbf{f}^x\right]\left[\left(\mathbf{u}_m^{\mathbf{y},p}\right)^T\mathbf{f}^{\mathbf{y}}\right]\mathbf{f}^{\boldsymbol{\mu}}-\textstyle\sum_{k=1}^{m-1}\left\{\left[\left(\mathbf{u}_m^{x,p}\right)^T K^x\mathbf{u}_k^x\right]\left[\left(\mathbf{u}_m^{\mathbf{y},p}\right)^T M^{\mathbf{y}}\mathbf{u}_k^x\right]\right.\\
&\left.\left[\left(\mathbf{u}_m^{x,p}\right)^T M^x\mathbf{u}_k^x\right]\left[\left(\mathbf{u}_m^{\mathbf{y},p}\right)^T K^{\mathbf{y}}\mathbf{u}_k^{\mathbf{y}}\right]\right\}M^{\boldsymbol{\mu}}\mathbf{u}_k^{\boldsymbol{\mu}},
\end{aligned}$$

obtained for $u_m^x = u_m^{x,p}$, $u_m^{\mathbf{y}} = u_m^{\mathbf{y},p}$ and by selecting $X(x)Y(\mathbf{y}) = u_m^{x,p}u_m^{\mathbf{y},p}$ for the test function. From a computational viewpoint, at each ADS iteration, we have to solve now three linear systems of order N_h^x, $N_h^{\mathbf{y}}$, $N_h^{\boldsymbol{\mu}}$, respectively.

We investigate the reliability of PGD on problem (15), for $V = H^1_{\Gamma_{\text{in}}\cup\Gamma_{\text{up}}\cup\Gamma_{\text{down}}}(\Omega)$ with $\Omega = (0,3)\times(0,1)$, $\Gamma_{\text{in}} = \{0\}\times(0,1)$, $\Gamma_{\text{up}} = (0,3)\times\{1\}$, $\Gamma_{\text{down}} = (0,3)\times\{0\}$, $a(u,v) = \int_\Omega\left[\mu\nabla u\cdot\nabla v + \mathbf{b}\cdot\nabla u\right]d\Omega$ with $\mathbf{b} = [2.5, 0]^T$ and μ the parameter to be varied in $\Omega_\mu = [1, 5]$, $F(v) = \int_\Omega f v d\Omega$ with $f = 1$. The problem is completed with mixed boundary conditions, namely a homogeneous Dirichlet data on $\Gamma_{\text{up}}\cup\Gamma_{\text{down}}$, the non-homogeneous Dirichlet condition, $u = u_{\text{in}}$ with $u_{\text{in}} = y(1-y)$, on Γ_{in} and a homogeneous Neumann value on $\Gamma_{\text{out}} = \{3\}\times(0,1)$. We apply the PGD reduction for $m = 2$, and we uniformly subdivide Ω_x, Ω_y, Ω_μ, being $N_h^x =$

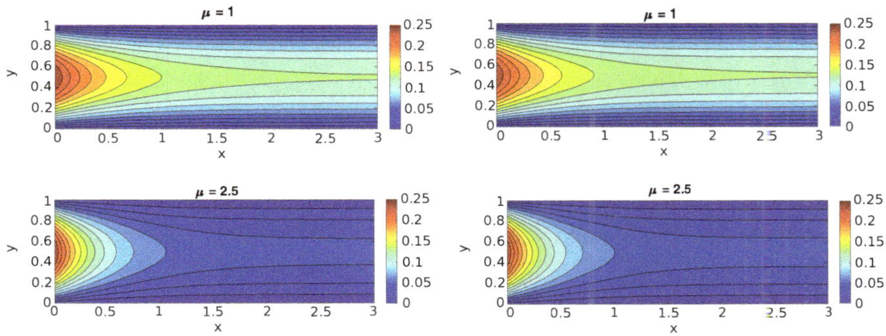

Fig. Qualitative comparison between the reference (left) and the PGD (right) solutions, for $\mu = 1$ (top) and $\mu = 2.5$ (bottom)

150, $N_h^y = 50$, $N_h^\mu = 500$. The tolerance in (14) is set to 10^{-2}. Figure compares the PGD approximation for $\mu = 1$ and $\mu = 2.5$ with a reference full solution coinciding with a linear FE approximation computed on a 300×100 structured mesh. The qualitative matching between the corresponding solutions is significant. From a quantitative viewpoint, the $L^2(\Omega)$-norm of the relative error associated with the PGD approximation does not vary significantly by increasing m, whereas a slight error reduction is detected by increasing μ.

The parametric counterpart of the HiMod reduction, known as HiPOD, merges HiMod with POD [4, 13]. HiPOD pursues a different goal with respect to PGD. Indeed, for a new value, $\boldsymbol{\mu}^*$, of the parameter, PGD provides an approximation for the full solution $u(\boldsymbol{\mu}^*)$, while HiPOD approximates the HiMod solution associated with $\boldsymbol{\mu}^*$. The offline/online paradigm of POD is followed also by HiPOD. The peculiarity is that the offline step is now performed in the HiMod setting to contain the computational burden typical of this stage and by relying on the good properties of HiMod in terms of reliability-versus-accuracy balance. Thus, we choose P different values, $\boldsymbol{\mu} = \boldsymbol{\mu}_i$ with $i = 1, \ldots, P$, for parameter $\boldsymbol{\mu}$, and we collect the HiMod approximation for the corresponding problem (15) into the response matrix, $\mathcal{S} = \left[\mathbf{u}_m^{\text{HiMod}}(\boldsymbol{\mu}_1), \mathbf{u}_m^{\text{HiMod}}(\boldsymbol{\mu}_2), \ldots, \mathbf{u}_m^{\text{HiMod}}(\boldsymbol{\mu}_P)\right] \in \mathbb{R}^{mN_h \times P}$, according to representation (5). Successively, we define the null-average matrix

$$\mathcal{V} = \mathcal{S} - \frac{1}{P}\sum_{i=1}^{P}\left[\mathbf{u}_m^{\text{HiMod}}(\boldsymbol{\mu}_i), \mathbf{u}_m^{\text{HiMod}}(\boldsymbol{\mu}_i), \ldots, \mathbf{u}_m^{\text{HiMod}}(\boldsymbol{\mu}_i)\right] \in \mathbb{R}^{mN_h \times P},$$

and we apply the Singular Value Decomposition (SVD) to $\mathcal{V}$, so that $\mathcal{V} = \Phi\Sigma\Psi^T$, where $\Phi \in \mathbb{R}^{(mN_h)\times(mN_h)}$ and $\Psi \in \mathbb{R}^{P\times P}$ are the unitary matrices of the left- and of the right-singular vectors of $\mathcal{V}$, respectively while $\Sigma = \text{diag}\,(\sigma_1, \ldots, \sigma_\rho) \in \mathbb{R}^{(mN_h)\times P}$ denotes the pseudo-diagonal matrix of the singular values of $\mathcal{V}$, being $\sigma_1 \geq \sigma_2 \geq \cdots \geq \sigma_\rho \geq 0$ and $\rho = \min(mN_h, P)$ [8]. The POD basis is identified by the first l left singular vectors, $\boldsymbol{\phi}_i$, of $\mathcal{V}$, so that the reduced POD space is $V^l_{\text{POD}} = \text{span}\{\boldsymbol{\phi}_1, \ldots, \boldsymbol{\phi}_l\}$, with $\dim(V^l_{\text{POD}}) = l$ and $l \ll mN_h$. In the numerical assessment below, value l coincides with the smallest integer such that $\sigma_l^2 < \varepsilon$, with ε a prescribed tolerance.

The online phase of HiPOD approximates the HiMod solution to problem (15) for a new value, $\boldsymbol{\mu}^*$, of the parameter by exploiting the POD basis instead of solving system (6). This is performed via a projection step. After assembling the HiMod stiffness matrix and right-hand side, $A_m^{\text{HiMod}}(\boldsymbol{\mu}^*)$ and $\mathbf{f}_m^{\text{HiMod}}(\boldsymbol{\mu}^*)$, associated with the new value of the parameter, we solve the POD system of order l

$$A_{\text{POD}}(\boldsymbol{\mu}^*)\mathbf{u}_{\text{POD}}(\boldsymbol{\mu}^*) = \mathbf{f}_{\text{POD}}(\boldsymbol{\mu}^*), \tag{18}$$

where $A_{\text{POD}}(\boldsymbol{\mu}^*) = (\Phi^l_{\text{POD}})^T A_m^{\text{HiMod}}(\boldsymbol{\mu}^*)\, \Phi^l_{\text{POD}}$ and $\mathbf{f}_{\text{POD}}(\boldsymbol{\mu}^*) = (\Phi^l_{\text{POD}})^T \mathbf{f}_m^{\text{HiMod}}(\boldsymbol{\mu}^*)$ denote the POD stiffness matrix and right-hand side, respectively with $\Phi^l_{\text{POD}} = [\boldsymbol{\phi}_1, \ldots, \boldsymbol{\phi}_l] \in \mathbb{R}^{(mN_h)\times l}$ the matrix collecting the POD basis vectors. The HiMod solution is thus approximated by vector $\Phi^l_{\text{POD}}\mathbf{u}_{\text{POD}}(\boldsymbol{\mu}^*) \in \mathbb{R}^{mN_h}$, i.e., after solving a system of order l instead of mN_h. Overall, HiPOD requires to solve P linear systems of order mN_h during the offline phase, additionally to a system of order l in the online phase.

To check the performances of HiPOD, we adopt the test case used above for PGD, for the same values of the parameters, $\mu^* = 1$ and $\mu^* = 2.5$. The reference solution is the corresponding HiMod approximation computed by using $m = 15$ sinusoidal functions in the y-direction, and a linear FE discretization along the mainstream based on a uniform subdivision of Ω_{1D} into 50 subintervals. The same HiMod discretization is adopted to build the response matrix. Concerning the HiPOD approximation, we pick $P = 100$ by uniformly sampling the interval $[1, 5]$, and we select $\varepsilon = 2.5 \cdot 10^{-15}$. This choice sets the dimension of the POD space to $l = 8$, so that we have to solve a system of order 8 instead of 750. The contour plots in Fig. qualitatively compare the HiMod solution with the HiPOD approximation for $l = 1$. The correspondence between the two approximations is good despite a single POD mode is employed (in such a case, system (18)

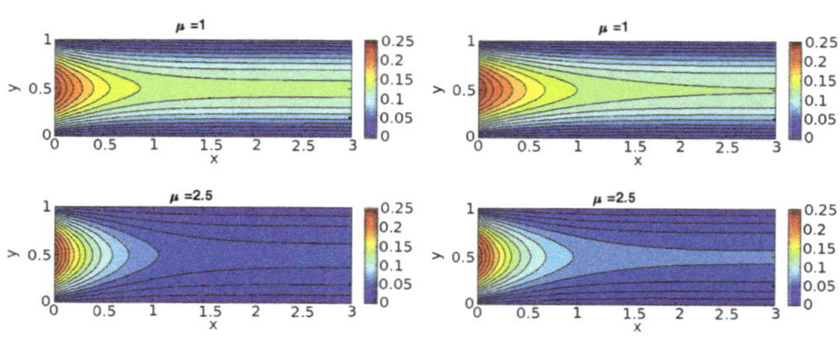

	$l = 1$	$l = 4$
$\mu^* = 1$	4.06e-02	2.53e-06
$\mu^* = 2.5$	2.74e-03	1.11e-07
random	7.19e-03	2.65e-07

	$l = 6$	$l = 8$
$\mu^* = 1$	1.79e-09	5.58e-12
$\mu^* = 2.5$	4.05e-10	1.21e-13
random	2.97e-10	3.66e-13

Fig. Contour plots: comparison between the reference HiMod solution (left) and the HiPOD approximation with $l = 1$ (right), for $\mu^* = 1$ (top) and $\mu* = 2.5$ (bottom). Table: relative error between HiMod and HiPOD solutions with respect to the $L^2(\Omega)$-norm reduces to a scalar equation). We do not provide the HiPOD approximations for $l = 8$ since they qualitatively coincide with the corresponding HiMod solution. The left panels can be additionally compared with the FE solutions in Fig. to verify the reliability of the HiMod procedure. Finally, the table in Fig. gathers the $L^2(\Omega)$-norm of the relative error between HiMod and HiPOD solutions, for four different POD bases and for three choices of the viscosity (1, 2.5 and the average over a sampling of 30 random values of μ). The error monotonically decreases for larger and larger values of l, independently of the choice for μ. If we compare the values for $\mu = 1$ and for $\mu = 2.5$ (one of the endpoints and the midpoint of the sampling interval, respectively), we notice a higher accuracy (of about one order of magnitude) for the latter choice. This is rather standard in projection-based reduced order modeling [11]. Concerning the computational saving in terms of CPU time, HiPOD method requires on average $O(10^{-3})$[s] to be compared with $O(10)$[s] demanded by HiMod, resulting in a speedup of 10^4.

Although PGD and HiPOD are not directly comparable due to the different purpose they pursue, we highlight the main pros and cons of the two methods. The explicit dependence of the approximation on the parameters makes PGD an ideal tool to efficiently deal with parametric problems. For any new parameter, a direct evaluation yields the corresponding PGD approximation. On the other hand, HiPOD suffers of the drawbacks typical of the projection-based methods. The main bottleneck is the assembling of the HiMod arrays involved in $A_{\text{POD}}(\boldsymbol{\mu}^*)$ and $\mathbf{f}_{\text{POD}}(\boldsymbol{\mu}^*)$.

When PGD is applied to parametric problems, we recover the possibility to deal with any geometric domain. In such a case, a partial separability is applied to the problem, so that the space independent variables are kept together whereas parameters are separated. This approach clearly looses the computational advantages due to space separability. On the contrary, HiPOD inherits the geometric flexibility of the HiMod reduction, without giving up the spatial dimensional reduction of the problem.

Acknowledgements The authors thank Yves Antonio Brandes Costa Barbosa for his support in the HiMod simulations. This work has been partially funded by GNCS-INdAM 2018 project on "Tecniche di Riduzione di Modello per le Applicazioni Mediche". F. Ballarin also acknowledges the support by European Union Funding for Research and Innovation, Horizon 2020 Program, in the framework of European Research Council Executive Agency: H2020 ERC Consolidator Grant 2015 AROMA-CFD project 681447 "Advanced Reduced Order Methods with Applications in Computational Fluid Dynamics" (P. I. G. Rozza).

References

1. Aletti, M.C., Perotto, S., Veneziani, A.: HiMod reduction of advection-diffusion-reaction problems with general boundary conditions. J. Sci. Comput. **76**(1), 89–119 (2018)
2. Ammar, A., Chinesta, F., Diez, P., Huerta, A.: An error estimator for separated representations of highly multidimensional models. Comput. Methods Appl. Mech. Eng. **199**(25–28), 1872–1880 (2010)

3. Ammar, A., Cueto, E., Chinesta, F.: Reduction of the chemical master equation for gene regulatory networks using proper generalized decompositions. Int. J. Numer. Methods Biomed. Eng. **28**(9), 960–973 (2012)
4. Baroli, D., Cova, C.M., Perotto, S., Sala, L., Veneziani, A.: Hi-POD solution of parametrized fluid dynamics problems: Preliminary results. In: Model Reduction of Parametrized Systems. MS&A: Modeling, Simulation and Applications, vol.17, pp. 235–254. Springer, Cham (2017)
5. Chinesta, F., Keunings, R., Leygue, A.: The Proper Generalized Decomposition for Advanced Numerical Simulations: A Primer. SpringerBriefs in Applied Sciences and Technology. Springer International Publishing, Berlin (2014)
6. Ern, A., Perotto, S., Veneziani, A.: Hierarchical model reduction for advection-diffusion-reaction problems. In: Numerical Mathematics and Advanced Applications, pp. 703–710. Springer, Berlin (2008)
7. Ghnatios, C., Ammar, A., Cimetiere, A., Hamdouni, A., Leygue, A., Chinesta, F.: First steps in the space separated representation of models defined in complex domains. In: 11th Biennial Conference on Engineering Systems Design and Analysis, pp. 37–42. Nantes (2012)
8. Golub, G.H., Van Loan, C.F.: Matrix Computations. Johns Hopkins Studies in the Mathematical Sciences, 4th edn. Johns Hopkins University Press, Baltimore (2013)
9. González, D., Ammar, A., Chinesta, F., Cueto, E.: Recent advances on the use of separated representations. Internat. J. Numer. Methods Eng. **81**(5), 637–659 (2010)
10. Guzzetti, S., Perotto, S., Veneziani, A.: Hierarchical model reduction for incompressible fluids in pipes. Internat. J. Numer. Methods Eng. **114**(5), 469–500 (2018)
11. Hesthaven, J.S., Rozza, G., Stamm, B.: Certified Reduced Basis Methods for Parametrized Partial Differential Equations. SpringerBriefs in Mathematics. Springer, Cham; BCAM Basque Center for Applied Mathematics, Bilbao (2016)
12. Ladevèze, P., Passieux, J.-C., Néron, D.: The LATIN multiscale computational method and the proper generalized decomposition. Comput. Methods Appl. Mech. Eng. **199**(21–22), 1287–1296 (2010)
13. Lupo Pasini, M., Perotto, S., Veneziani, A.: HiPOD: Hierarchical model reduction driven by a Proper Orthogonal Decomposition for parametrized advection-diffusion-reaction problems. In preparation
14. Niroomandi, S., González, D., Alfaro, I., Bordeu, F., Leygue, A., Cueto, E., Chinesta, F.: Real-time simulation of biological soft tissues: a PGD approach. Int. J. Numer. Methods Biomed. Eng. **29**(5), 586–600 (2013)
15. Perotto, S.: Hierarchical model (Hi-Mod) reduction in non-rectilinear domains. In: Domain Decomposition Methods in Science and Engineering XXI. Lecture Notes in Computational Science and Engineering, vol. 98, pp. 477–485. Springer, Cham (2014)
16. Perotto, S.: A survey of hierarchical model (Hi-Mod) reduction methods for elliptic problems. In: Numerical Simulations of Coupled Problems in Engineering. Computational Methods in Applied Sciences, vol. 33, pp. 217–241. Springer, Cham (2014)
17. Perotto, S., Veneziani, A.; Coupled model and grid adaptivity in hierarchical reduction of elliptic problems. J. Sci. Comput. **60**(3), 505–536 (2014)
18. Perotto, S., Zilio, A.: Hierarchical model reduction: three different approaches. In: Numerical Mathematics and Advanced Applications 2011, pp. 851–859. Springer, Heidelberg (2013)
19. Perotto, S., Zilio, A.: Space-time adaptive hierarchical model reduction for parabolic equations. Adv. Model. Simul. Eng. Sci. **2**, 25 (2015)
20. Perotto, S., Ern, A., Veneziani, A.: Hierarchical local model reduction for elliptic problems: a domain decomposition approach. Multiscale Model. Simul. **8**(4), 1102–1127 (2010)
21. Perotto, S., Reali, A., Rusconi, P., Veneziani, A.: HIGAMod: a hierarchical isogeometric approach for model reduction in curved pipes. Comput. Fluids **142**, 21–29 (2017)
22. Pruliere, E., Chinesta, F., Ammar, A.: On the deterministic solution of multidimensional parametric models using the proper generalized decomposition. Math. Comput. Simul. **81**(4), 791–810 (2010)
23. Signorini, M., Zlotnik, S., Díez, P.: Proper generalized decomposition solution of the parameterized Helmholtz problem: application to inverse geophysical problems. Int. J. Numer. Methods Eng. **109**(8), 1085–1102 (2017)

Permissions

All chapters in this book were first published in SHOMPDEI, by Springer Nature; hereby published with permission under the Creative Commons Attribution License or equivalent. Every chapter published in this book has been scrutinized by our experts. Their significance has been extensively debated. The topics covered herein carry significant findings which will fuel the growth of the discipline. They may even be implemented as practical applications or may be referred to as a beginning point for another development.

The contributors of this book come from diverse backgrounds, making this book a truly international effort. This book will bring forth new frontiers with its revolutionizing research information and detailed analysis of the nascent developments around the world.

We would like to thank all the contributing authors for lending their expertise to make the book truly unique. They have played a crucial role in the development of this book. Without their invaluable contributions this book wouldn't have been possible. They have made vital efforts to compile up to date information on the varied aspects of this subject to make this book a valuable addition to the collection of many professionals and students.

This book was conceptualized with the vision of imparting up-to-date information and advanced data in this field. To ensure the same, a matchless editorial board was set up. Every individual on the board went through rigorous rounds of assessment to prove their worth. After which they invested a large part of their time researching and compiling the most relevant data for our readers.

The editorial board has been involved in producing this book since its inception. They have spent rigorous hours researching and exploring the diverse topics which have resulted in the successful publishing of this book. They have passed on their knowledge of decades through this book. To expedite this challenging task, the publisher supported the team at every step. A small team of assistant editors was also appointed to further simplify the editing procedure and attain best results for the readers.

Apart from the editorial board, the designing team has also invested a significant amount of their time in understanding the subject and creating the most relevant covers. They scrutinized every image to scout for the most suitable representation of the subject and create an appropriate cover for the book.

The publishing team has been an ardent support to the editorial, designing and production team. Their endless efforts to recruit the best for this project, has resulted in the accomplishment of this book. They are a veteran in the field of academics and their pool of knowledge is as vast as their experience in printing. Their expertise and guidance has proved useful at every step. Their uncompromising quality standards have made this book an exceptional effort. Their encouragement from time to time has been an inspiration for everyone.

The publisher and the editorial board hope that this book will prove to be a valuable piece of knowledge for researchers, students, practitioners and scholars across the globe.

List of Contributors

Jiaxi Gu
Department of Mathematics, University at Buffalo, The State University of New York, Buffalo, NY, USA

Jae-Hun Jung
Department of Mathematics, University at Buffalo, The State University of New York, Buffalo, NY, USA
Department of Data Science, Ajou University, Suwon, South Korea

Ralf Hiptmair
Seminar for Applied Mathematics, ETH Zürich, Zürich, Switzerland

Clemens Pechstein
Dassault Systèmes, Darmstadt, Germany

Sheehan Olver
Department of Mathematics, Imperial College, London, UK

Alex Townsend
Department of Mathematics, Cornell University, Ithaca, NY, USA

Geoffrey M. Vasil
School of Mathematics and Statistics, University of Sydney, Sydney, NSW, Australia

Florian J. Hindenlang
Max Planck Institute for Plasma Physics, Garching, Germany

Gregor J. Gassner
Department for Mathematics and Computer Science, Center for Data and Simulation Science, University of Cologne, Cologne, Germany

Tracy Babb and Daniel Appelö
University of Colorado, Boulder, CO, USA

Per-Gunnar Martinsson
University of Texas, Austin, TX, USA

Gabriel R. Barrenechea
Department of Mathematics and Statistics, University of Strathclyde, Glasgow, UK

Michał Bosy
Department of Mathematics and Statistics, University of Strathclyde, Glasgow, UK
Dipartimento di Matematica "F. Casorati", Universitá degli Studi di Pavia, Pavia, Italy

Victorita Dolean
Department of Mathematics and Statistics, University of Strathclyde, Glasgow, UK University Côte d'Azur, CNRS, LJAD, Nice Cedex, France

Jan S. Hesthaven, Fabian Mönkeberg and Sara Zaninelli
École Polytechnique Fédérale de Lausanne (EPFL), Lausanne, Switzerland

Lorella Fatone
Dipartimento di Matematica, Università degli Studi di Camerino, Camerino, Italy

Daniele Funaro
Dipartimento di Scienze Chimiche e Geologiche, Università degli Studi di Modena e Reggio Emilia, Modena, Italy

Gianmarco Manzini
Group T-5, Applied Mathematics and Plasma Physics, Theoretical Division, Los Alamos National Laboratory, Los Alamos, NM, USA

Yi Zhang and Marc Gerritsma
Delft University of Technology, Delft, Netherlands

Artur Palha
Eindhoven University of Technology, Eindhoven, Netherlands

Yuanming Xiao
Department of Mathematics, Nanjing University, Nanjing, China

Fangman Zhai
Department of Applied Mathematics, Nanjing Forestry University, Nanjing, China

Linbo Zhang and Weiying Zheng
State Key Laboratory of Scientific and Engineering Computing, Academy of Mathematics and Systems Science, Chinese Academy of Sciences, Beijing, China
School of Mathematical Sciences, University of Chinese Academy of Sciences, Beijing, China

Rujie Gu and Thomas Hagstrom
Southern Methodist University, Dallas, TX, USA

Mike Stees
Department of Electrical Engineering and Computer Science, Information and Telecommunication Technology Center, University of Kansas, Lawrence, KS, USA

Suzanne M. Shontz
Department of Electrical Engineering and Computer Science, Bioengineering Program, Information and Telecommunication Technology Center, University of Kansas, Lawrence, KS, USA

Varun Jain, Joël Fisser, Artur Palha and Marc Gerritsma
Faculty of Aerospace Engineering, TU Delft, Delft, The Netherlands

Florian J. Hindenlang
Max Planck Institute for Plasma Physics, Garching, Germany

David A. Kopriva
Florida State University, Tallahassee, FL, USA
San Diego State University, San Diego, CA, USA

Bernard Haasdonk, Gabriele Santin and Dominik Wittwar
Inst. of Applied Analysis and Numerical Simulation, University of Stuttgart, Stuttgart, Germany

Boumediene Hamzi
Department of Mathematics, Imperial College London, London, UK

Simona Perotto
MOX - Modeling and Scientific Computing, Dipartimento di Matematica, Politecnico di Milano, Milano, Italy

Michele Giuliano Carlino
Inria Bordeaux Sud-Ouest and Institut de Mathématiques de Bordeaux, University of Bordeaux, Talence, France

Francesco Ballarin
MathLab, Mathematics Area, SISSA, Trieste, Italy

Index

CPSIA information can be obtained
at www.ICGtesting.com
Printed in the USA
LVHW082311100122
708212LV00001B/14

9 781639 894048